Phase Retrieval and Zero Crossings

Mathematics and Its Applications

Phase Retrieval and Zero Crossings

Mathematical Methods in Image Reconstruction

by

Norman E. Hurt

Zeta Associates, Fairfax, Virginia, U.S.A.

KLUWER ACADEMIC PUBLISHERS

DORDRECHT / BOSTON / LONDON

Library of Congress Cataloging in Publication Data

Hurt, Norman.
    Phase retrieval and zero crossings : mathematical methods in image
reconstruction / Norman E. Hurt.
        p.   cm. -- (Mathematics and its applications)
    Bibliography: p.
    Includes index.
    ISBN 0-7923-0210-9 (U.S.)
    1. Signal theory (Telecommunication)--Mathematics.   I. Title.
II. Series: Mathematics and its applications (Kluwer Academic
Publishers)
TK5102.5.H87   1989
621.38'043--dc19   89-2374

ISBN 0-7923-0210-9

Published by Kluwer Academic Publishers,
P.O. Box 17, 3300 AA Dordrecht, The Netherlands.

Kluwer Academic Publishers incorporates
the publishing programmes of
D. Reidel, Martinus Nijhoff, Dr W. Junk and MTP Press.

Sold and distributed in the U.S.A. and Canada
by Kluwer Academic Publishers,
101 Philip Drive, Norwell, MA 02061, U.S.A.

In all other countries, sold and distributed
by Kluwer Academic Publishers Group,
P.O. Box 322, 3300 AH Dordrecht, The Netherlands.

printed on acid free paper

To Susan, Michael and Jason

'Et moi, ..., si j'avait su comment en revenir,
je n'y serais point allé.'

Jules Verne

The series is divergent; therefore we may be
able to do something with it.

O. Heaviside

One service mathematics has rendered the
human race. It has put common sense back
where it belongs, on the topmost shelf next
to the dusty canister labelled 'discarded non-
sense'.

Eric T. Bell

Mathematics is a tool for thought. A highly necessary tool in a world where both feedback and non-
linearities abound. Similarly, all kinds of parts of mathematics serve as tools for other parts and for
other sciences.

Applying a simple rewriting rule to the quote on the right above one finds such statements as:
'One service topology has rendered mathematical physics ...'; 'One service logic has rendered computer science ...'; 'One service category theory has rendered mathematics ...'. All arguably true. And
all statements obtainable this way form part of the raison d'être of this series.

This series, *Mathematics and Its Applications*, started in 1977. Now that over one hundred
volumes have appeared it seems opportune to reexamine its scope. At the time I wrote

"Growing specialization and diversification have brought a host of monographs and
textbooks on increasingly specialized topics. However, the 'tree' of knowledge of
mathematics and related fields does not grow only by putting forth new branches. It
also happens, quite often in fact, that branches which were thought to be completely
disparate are suddenly seen to be related. Further, the kind and level of sophistication
of mathematics applied in various sciences has changed drastically in recent years:
measure theory is used (non-trivially) in regional and theoretical economics; algebraic
geometry interacts with physics; the Minkowsky lemma, coding theory and the structure
of water meet one another in packing and covering theory; quantum fields, crystal
defects and mathematical programming profit from homotopy theory; Lie algebras are
relevant to filtering; and prediction and electrical engineering can use Stein spaces. And
in addition to this there are such new emerging subdisciplines as 'experimental
mathematics', 'CFD', 'completely integrable systems', 'chaos, synergetics and large-scale
order', which are almost impossible to fit into the existing classification schemes. They
draw upon widely different sections of mathematics."

By and large, all this still applies today. It is still true that at first sight mathematics seems rather
fragmented and that to find, see, and exploit the deeper underlying interrelations more effort is
needed and so are books that can help mathematicians and scientists do so. Accordingly MIA will
continue to try to make such books available.

If anything, the description I gave in 1977 is now an understatement. To the examples of
interaction areas one should add string theory where Riemann surfaces, algebraic geometry, modu-
lar functions, knots, quantum field theory, Kac-Moody algebras, monstrous moonshine (and more)
all come together. And to the examples of things which can be usefully applied let me add the topic
'finite geometry'; a combination of words which sounds like it might not even exist, let alone be
applicable. And yet it is being applied: to statistics via designs, to radar/sonar detection arrays (via
finite projective planes), and to bus connections of VLSI chips (via difference sets). There seems to
be no part of (so-called pure) mathematics that is not in immediate danger of being applied. And,
accordingly, the applied mathematician needs to be aware of much more. Besides analysis and
numerics, the traditional workhorses, he may need all kinds of combinatorics, algebra, probability,
and so on.

In addition, the applied scientist needs to cope increasingly with the nonlinear world and the

vii

extra mathematical sophistication that this requires. For that is where the rewards are. Linear models are honest and a bit sad and depressing: proportional efforts and results. It is in the nonlinear world that infinitesimal inputs may result in macroscopic outputs (or vice versa). To appreciate what I am hinting at: if electronics were linear we would have no fun with transistors and computers; we would have no TV; in fact you would not be reading these lines.

There is also no safety in ignoring such outlandish things as nonstandard analysis, superspace and anticommuting integration, p-adic and ultrametric space. All three have applications in both electrical engineering and physics. Once, complex numbers were equally outlandish, but they frequently proved the shortest path between 'real' results. Similarly, the first two topics named have already provided a number of 'wormhole' paths. There is no telling where all this is leading - fortunately.

Thus the original scope of the series, which for various (sound) reasons now comprises five subseries: white (Japan), yellow (China), red (USSR), blue (Eastern Europe), and green (everything else), still applies. It has been enlarged a bit to include books treating of the tools from one subdiscipline which are used in others. Thus the series still aims at books dealing with:

- a central concept which plays an important role in several different mathematical and/or scientific specialization areas;
- new applications of the results and ideas from one area of scientific endeavour into another;
- influences which the results, problems and concepts of one field of enquiry have, and have had, on the development of another.

The topic of image and signal reconstruction from incomplete and noisy data is a vastly important one. The field also abounds in unsolved problems, many no doubt extremely difficult and also certainly many that still remain untreated simply because nobody has yet seriously looked at them. The field also abounds in algorithms that work (in certain cases) and it is not seldom that the reasons are largely open why they work well and what the limits of their applicability are. Finally, the mathematical ideas and techniques used are extremely varied and the open-minded researcher in the field is in some danger of getting in some stochastic walk-mode visiting known and unknown mathematical territories in random order.

Indeed, a brief casual glance at the table of contents already shows the presence of material on entire functions, algebraic topology, reproducing Kernel Hilbert spaces, the Riemann zeta function, ill-posed problems and (Tikhonov) regularization, irreducibility of polynomials criteria, all kinds of integral transforms, the complexity of algorithms, stochastic approximation, spin-glass models, simulated annealing, and the geometry of convex sets.

Working through this volume the attentive reader will find himself taken on an irregular grand tour through most of mathematics even though there is just one guiding problem area. All in all, an experience that I can rather recommend.

Perusing the present volume is not guaranteed to turn you into an instant expert, but it will help, though perhaps only in the sense of the last quote on the right below.

The shortest path between two truths in the real domain passes through the complex domain.	Never lend books, for no one ever returns them; the only books I have in my library are books that other folk have lent me.
J. Hadamard	Anatole France
La physique ne nous donne pas seulement l'occasion de résoudre des problèmes ... elle nous fait pressentir la solution.	The function of an expert is not to be more right than other people, but to be wrong for more sophisticated reasons.
H. Poincaré	David Butler

Bussum, March 1989   Michiel Hazewinkel

Preface

The goal of this volume is to try to capture the highlights of the study of signal reconstruction from incomplete, noisy data, especially in the areas of phase retrieval and signal recovery from zero crossings. Time, more than space, necessitates a restriction of topics to those of interest to the author. An effort was made to include a somewhat comprehensive bibliography to supplement the brief discussions in the text.

This volume would not have come about without tremendous support over the past ten years from many friends and colleagues. We can not list them all for fear of forgetting one. However, particular thanks go to Jim Fienup and Richard Barakat for all the help and conversations that they have provided me.

Finally, the author owes a special thanks to Wayne Lawton who saw more clearly the problems and solutions in phase retrieval. His work or conversations with the author touch on all the chapters in this volume.

The following figures are used with the permission of the Optical Society of America. Figure 10.1 is from A. Huiser and P. van Toorn, Optics Letters 5 (1980) 499-501; Figure 16.1 is from J. R. Fienup, Optics Letters 3 (1978) 27-29; and Figure 17.2 is from T. Sato et al., Applied Optics 20 (1981) 395-399.

Contents

Preface   ix

1 Introduction   1
1.1   Signal Recovery   1
1.2   Band-limited Extrapolation and Phase Retrieval   1
1.3   Spectrum Shaping and Kinoforms   3
1.4   Radar Signal Synthesis   4
1.5   Beam Shaping and Antenna Design   4
1.6   Optical and Microwave Beam Shaping   5
1.7   Wavefront Sensing   5
1.8   Spectroscopy   5
1.9   X-Ray Crystallography   5
1.10   Radio Astronomy   6
1.11   Tomography   6
1.12   Signal Reconstruction from Zero Crossings   6
1.13   Reconstruction and Resolution Limits   8
1.14   Ill-Posed Problems and Regularization   9

2 Polynomials: A Review   11
2.1   Introduction   11
2.2   Polynomials and Trigonometric Polynomials   15

3 Entire Functions and Signal Recovery   17
3.1   Introduction   17
3.2   Band-Limited Functions   20
3.3   Zeros of EFETs   21
3.4   Encoding of Information by Zeros   25
3.5   Far Field Distributions for Point Sources   25
3.6   Radar Waveform Construction a la Huffman   27
3.7   Hayes' Fundamental Theorem   28
3.8   Brightness Distribution (Bruck-Sodin)   29
3.9   Specified Autocorrelation Functions   30
3.10   Autocorrelation Function of Pulses   32
3.11   Autocorrelation Functions   33

3.12 Monochromatic Radiation   33
3.13 Coherence Theory   34
3.14 Blackbody Radiation   36
3.15 Holography and Zeros   36
3.16 Akutowicz Representation Theorem   37
3.17 Walther's Theorem and Zero Flipping   40
3.18 Summary of Phase Retrieval in One Dimension   45

4 Homometric Distributions   47
4.1 Introduction   47

5 Analytic Signals and Signal Recovery from Zero Crossings   55
5.1 Introduction   55
5.2 Amplitude Modulation   55
5.3 Single Sideband Modulation   56
5.4 Exponential Single Sideband Modulation   56
5.5 Distortionless Demodulation   57
5.6 Modulation Reconstruction from Zeros   59
5.7 Recovery from Zero Crossings   60
5.8 A Zero-Crossing Spectrum Analyzer   64

6 Signal Representation by Fourier Phase and Magnitude in
   One Dimension   67
6.1 Introduction   67
6.2 Arrival Time Determination   69
6.3 Fourier Representation by Fourier Magnitude   70
6.4 Representation by Fourier Signed Magnitude   71
6.5 Representation by One Bit of Fourier Phase   72

7 Recovery of Distorted Band-Limited Signals   73
7.1 Introduction   73
7.2 Landau's Convolution Equation   75
7.3 Time-Limited and Band-Limited Functions   77
7.4 Prolate Spheroidal Functions   79
7.5 Extrapolation by PSWFs   80
7.6 Riemann Zeta Function   82
7.7 Recovery of Stochastic Processes   83

8 Compact Operators, Singular Value Analysis and Reproducing Kernel Hilbert Spaces   87
8.1 Introduction   87
8.2 Singular Value Analysis   91
8.3 The Effect of Noise on Inversion   94
8.4 Laplace Transform Inversion   95

8.5 Resolution with Incoherent Illumination 98
8.6 Reproducing Kernel Hilbert Spaces 98

9 Kaczmarz Method, Landweber Iteration, Gerchberg-Papoulis
   and Regularization   103
   9.1 Introduction 103
   9.2 The Cimmino Method 104
   9.3 Cimmino for Fredholm 105
   9.4 Filtering and Landweber's Iteration 108
   9.5 Image Restoration by the Projections 109
   9.6 Van Cittert Deconvolution 109
   9.7 Gerchberg-Papoulis Algorithm 111
   9.8 Extrapolation by Gori's Algorithm 112
   9.9 Gerchberg's Method of Extrapolation 113
   9.10 Gerchberg-Papoulis Again 114
   9.11 Landweber's Iteration 116
   9.12 Discrete-Continuous Iterative Extrapolation 118
   9.13 Discrete-Discrete Extrapolation 118
   9.14 The Xu-Chamzas Algorithm 120
   9.15 The Discrete Xu-Chamzas Algorithm 122
   9.16 Adaptive Extrapolation 123
   9.17 Karhunen-Loeve Expansion 126
   9.18 Stopping Algorithms 126
   9.19 Regularization 127
   9.20 Tikhonov Regularization 128
   9.21 Landweber Regularization 128
   9.22 Regularization in Optics 129
   9.23 Moment Discretization 130
   9.24 Biraud's Algorithm 131

10 Two Dimensional Signal Recovery Problems   135
   10.1 Introduction 135
   10.2 Phase Retrieval from Magnitude 137
   10.3 Entire Functions in Two Dimensions 139
   10.4 Radial Geometry 143
   10.5 Factorization and Exponential Type 146
   10.6 Eisenstein Criterion for Irreducibility 148
   10.7 The Result of Sanz-Huang 149

11 Reconstruction Algorithms in Two Dimensions   151
   11.1 Canterakis Algorithm 151
   11.2 Berenyi-Deighton-Fiddy Algorithm 154
   11.3 Logarithmic Hilbert Transforms 155
   11.4 Algorithm of Nakajima and Asakura 157

11.5 Zero Crossings in 2-D 158
11.6 The Israelevitz-Lim Algorithm 162
11.7 The Rotem-Zeevi Algorithm 164

12 Nonexpansive Maps and Signal Recovery   165
12.1 Introduction 165
12.2 Time and Frequency Mappings 172
12.3 Time-Limited Extrapolation 173
12.4 Phase-Only Reconstruction 174
12.5 Minimal Phase Signals 175
12.6 Min and Max Constraints 176
12.7 Masks for Microlithography 176
12.8 Energy Reduction Algorithms 178

13 Projections on Convex Sets in Signal Recovery   179
13.1 Introduction 179
13.2 Alternating Projections 185
13.3 Reconstruction by Projections 187
13.4 Regularization 188
13.5 Convex Projection in Noise 189
13.6 Inconsistent Constraints 190
13.7 Algebraic Reconstruction Techniques 191
13.8 Nonnegative Spatial Constraints 193
13.9 Reconstruction from Zero Crossings 194
13.10 A Phase Retrieval Example 195
13.11 Lent-Tuy Algorithm 196
13.12 Example of Lent-Tuy 196
13.13 Divergence-Free Vector Fields 197

14 Method of Generalized Projections and Steepest Descent 199
14.1 Introduction 199
14.2 Phase Retrieval 200
14.3 Deconvolution by POCS 200
14.4 Fuzzy Sets 202
14.5 Method of Steepest Descent 203
14.6 Least Squares Solutions 203
14.7 Design of FIR Filters 207
14.8 Projection onto Nonconvex Sets 209

15 Closed Form Reconstruction of the Support and the Object 211
15.1 Introduction 211

16 Fienup's Input-Output Algorithms and Variations on this
   Theme   215
   16.1 Introduction   215

17 Topics and Applications of Signal Recovery   221
   17.1 Maximum Entropy Estimation   221
   17.2 Stellar Speckle Interferometry   222
   17.3 Radio Astronomy   223
   17.4 The CLEAN Method   223
   17.5 The Knox Method   225
   17.6 Filter Design 226
   17.7 Microwave Holographic Metrology   227
   17.8 Misell Algorithm   230
   17.9 Radio Holography   234
   17.10Pinhole Projections 235
   17.11Tomography 235
   17.12The Spin-Glass Model   237
   17.13Simulated Annealing 238
   17.14Robbins-Monro 238
   17.15Minimum Variance Algorithm   240
   17.16Newton-Direction Algorithms   241
   17.17Sandberg's Algorithm   242
   17.18FM Demodulation   243
   17.19Toeplitz Equations   244
   17.20Moses-Prosser Algorithm   245
   17.21The Baghlay Algorithm   246
   17.22Gonsalves' Algorithm   247
   17.23Neural Nets and POCS   247
   17.24Synthetic Aperture Radar   249
   17.25NMR Tomography   250
   17.26Multiple Threshold Crossings   251
   17.27Block-Iterative Methods 252

A The Geometry of Projections on Convex Sets   257

B Reference Summary   263

C References   267

   Index   300

Chapter 1

Introduction

1.1 Signal Recovery

Signal recovery problems which we deal with in this volume fall into two
general classes: (1) reconstruction problems and (2) synthesis problems.
In reconstruction problems the measured data, a priori knowledge or other
constraints often are not sufficient to permit reconstruction of the signal
directly. Iterative techniques have been developed to treat these cases.
When using iterative techniques, it is of central importance in reconstruction to know if the solution exists and is unique. Conditions for uniqueness
and existence are studied below for one dimensional and two dimensional
problems.

In synthesis problems uniqueness is usually not important; any solution which "best" satisfies the given measurements and constraints is an
admissible synthesis solution.

We outline now several of the general problems that will be treated in
detail in this volume.

1.2 Band-limited Extrapolation and Phase Retrieval

One classical signal recovery problem is the reconstruction of a Fourier
transform pair $f, F = \mathcal{F}(f)$ from partial data on either or both functions.
Examples of this problem, which we present as one dimensional cases, are:

(A) Extrapolation of band-limited signals. Given a noisy measurement
$\tilde{f}$ of f on the interval $A = [a_1, a_2]$ and knowledge that F vanishes outside
the bounded interval $B = [b_1, b_2]$, reconstruct f and F on the entire real
line.

1

Iterative techniques developed to solve this problem have been proposed by Gerchberg, Papoulis and others. The Gerchberg-Papoulis algorithm is very simple. Take $A = [-T, T]$ and $B = [-s, s]$; then

(1) let $w_1(t) = \begin{cases} \tilde{f}(t) & \text{if } |t| < T \\ 0 & \text{otherwise} \end{cases}$

(2) compute the Fourier transform $W_1(\omega)$ of $w_1(t)$

(3) set $F_1(\omega) = \begin{cases} W_1(\omega) & \text{for } |\omega| < s \\ 0 & \text{otherwise} \end{cases}$

(4) compute the inverse Fourier transform $f_1(t)$ of F_1

(5) let $w_2(t) = \begin{cases} \tilde{f}(t) & \text{for } |t| < T \\ f_1(t) & \text{for } |t| > T \end{cases}$

Thus, in general:

$$F_n(\omega) = \begin{cases} W_n(\omega) & \text{for } |\omega| < s \\ 0 & \text{otherwise} \end{cases}$$

and

$$w_{n+1}(t) = \begin{cases} \tilde{f}(t) & \text{for } |t| < T \\ f_n(t) & \text{for } |t| > T \end{cases}$$

The mean-square error at step n is

$$E_n = \int_{-\infty}^{\infty} [f(t) - f_n(t)]^2 dt$$

and it can be shown that E_n is a monotone decreasing function. For this reason the algorithm is sometimes called the error reduction algorithm.

(B) The two moduli phase retrieval problem. The data for this problem are noisy measurements m and n of $|f|$ on A and $|F|$ on B and the knowledge that F vanishes outside B; the problem is to reconstruct f and F over the entire real line.

Iterative algorithms here have been developed by Gerchberg, Saxton, Misell and others. The Gerchberg-Saxton algorithm has the form:

(1) $f_k(t) = m(t)e^{i\theta_k}$

(2) take the Fourier transform $F_k(\omega) = |F_k(\omega)|e^{i\phi_k(\omega)}$

(3) set $\tilde{F}(\omega) = \begin{cases} n(\omega)e^{i\phi_k(\omega)} & \text{for } |\omega| < s \\ 0 & \text{otherwise} \end{cases}$

(4) take the inverse Fourier transform to form $\tilde{f}_k(t) = |\tilde{f}_k(t)|e^{i\theta_{k+1}}$

(5) set $f_{k+1}(t) = m(t)e^{i\theta_{k+1}}$

(6) return to (2).

(C) Phase retrieval with nonnegativity constraints. Given a noisy measurement m of $|f|$ on A and the knowledge that F is nonnegative over B and vanishes outside B, reconstruct f and F over the real line.

There are many other examples of reconstruction of Fourier transform pairs in optics, electrical engineering, quantum physics, and astronomy.

The function $f(x)$ may be a wave function, an object function, a signal, an antenna array, a spectral density function, an electron density function or whatever. Clearly the variable x may represent time or a spatial coordinate or some other label. The function $f(x)$ may be real valued as for a sky-bright object or it may be complex valued as for an object in electron microscopy.

Two of the earliest examples of the reconstruction of Fourier transform pairs are those arising in astronomical interferometry and x-ray crystallography. Other examples occur in electron microscopy, spectroscopy, wave-front sensing, holography, kinoform generation, particle scattering, super resolution, radar signal synthesis, antenna synthesis and filter design. We briefly outline some of the problems in these fields.

1.3   Spectrum Shaping and Kinoforms

Spectrum shaping is a synthesis problem where, given $|f|$, the phase $\arg(f)$ is to be determined so that $\mathcal{F}(f(x))$ has a prescribed modulus $|F|$. An example of spectrum shaping arises in computer holography or kinoform generation where a version of the error reduction algorithm was proposed by Lesem, Jordan and Hersh and by Gallagher and Liu. In computer holography a transparency is to have the form $F = \mathcal{F}(f)$. In the kinoform problem the phase of F can be arbitrarily prescribed but the quantization levels of the modulus are limited. The object is to choose the phase $\arg(f)$ in such a way so as to reduce the variance, i.e. the dynamic range, of F.

In more detail, let $T(x, y)$ denote the transmittance of an image at (x, y). Let $\phi(x, y)$ be the phase function. The Fourier transform is written as

$$A(u, v)e^{i\psi(u,v)} = \int \int T(x, y)e^{i\phi(x,y)}e^{-i2\pi(xu+yv)/\lambda R}dxdy \qquad (1.1)$$

Suppose it is possible to find a function $\phi(x, y)$ such that

$$A(u, v) = \begin{cases} A & \text{for } |u|, |v| < W/2 \\ 0 & \text{otherwise;} \end{cases} \qquad (1.2)$$

then $e^{i\psi(u,v)}$ is called the kinoform of the image $T(x, y)$. The constant W represents a bandwidth constraint. The problem is to determine $\phi(x, y)$ so that the conditions (1.2) are most nearly satisfied.

Kinoforms are made by plotting $\psi(u, v)$ by a gray scale plotter which quantizes the values of $\psi(u, v)$ being plotted. The plot is photographed and etched so that the thickness of the resultant transparency is proportional to $\psi(u, v)$. The illuminated transparency then phase shifts the light passing through it by an amount $e^{i\psi(u,v)}$ thus reproducing the image $T(x, y)$.

The algorithm proposed by Gallagher and Liu is as follows:

(1) sample $T(x, y)$ to get $T(m, n)$ and apply random phases $\phi(m, n)$
(2) take the discrete Fourier transform (DFT) to form $A(p, q)e^{i\psi(p,q)}$
(3) replace $A(p, q)$ in (2) by

$$\tilde{A}(p, q) = \left\{ \begin{array}{ll} A & \text{for p,q on the object} \\ 0 & \text{otherwise} \end{array} \right.$$

(4) take the inverse DFT to form $\tilde{T}(m, n)e^{i\tilde{\phi}(m,n)}$
(5) replace $\tilde{T}(m, n)$ by $T(m, n)$ and return to (2).

1.4 Radar Signal Synthesis

In radar design one problem is to synthesize a radar signal $f(t)$ which is a pure phase function, i.e. $|f(t)| = 1$, over some interval of time and whose autocorrelation function approaches a delta function, i.e., its Fourier spectrum $|F(\omega)|^2$ is constant over the bandwidth of interest.

1.5 Beam Shaping and Antenna Design

A spectrum shaping problem arises in phasing the elements of an antenna array in order to achieve a specified far-field pattern – e.g. specified nulls. Consider a linear array of N elements located at $x_0, x_0 + d, ..., x_0 + (N+1)d$ along the x-axis. The space factor for this array is defined by

$$S = e^{ikx_0cos\alpha} \sum_{n=0}^{N-1} A(n)e^{ikndcos\alpha} \tag{1.3}$$

where α is the angle between the observer's direction and the axis of the array. Here $k = \frac{2\pi}{\lambda}$ for wavelength λ. Set $\psi = kdcos\alpha$. Then the magnitude $|S|$ represents the power flux in the far field zone. Here $|S|$ is periodic in ψ of period 2π. The complex elements $A(n)$ are proportional to the currents at the corresponding elements. The design problem is : for a specified $|S|$, determine the excitation coefficients $A(0), ..., A(N-1)$.

The function S can be rewritten as a polynomial in $z = e^{i\psi}$, viz.

$$S = z^{x_0/d} \sum_{n=0}^{N-1} A(n)z^n. \tag{1.4}$$

As a polynomial of degree N-1 over the complex field C, S can be factored as N-1 linear terms:

$$S = A(N-1)z^{x_0/d}(z - z_1)...(z - z_{N-1}). \tag{1.5}$$

We note that the zeros on the unit circle correspond to nulls in the radiation field. The second thing to note is that if any zero z_n is replaced by its image point $1/z_n^*$ then the relative pattern of the array is unchanged. However, the excitation coefficients will be modified.

The reader can check that if all the roots of $S(z)$ lie on the unit circle, then the solution to the array problem is unique and in this case $|S|$ goes to zero N-1 times in a 2π range of ψ.

1.6 Optical and Microwave Beam Shaping

A related beam shaping problem involves the transformation of a Gaussian laser beam into a beam having a more rectangular profile. Similar beam shaping problems have been considered for microwave beams by Gallagher.

1.7 Wavefront Sensing

The image $|f(x)|^2$ of a point source is recorded with an optical system. Assuming the aberration is a pure phase function, then $F(u)$ has modulus $|F(u)|$ equal to the aperture function of the optical system. The problem is to reconstruct the phase of $F(u)$.

1.8 Spectroscopy

Here the spectral density function is nonnegative and it is the Fourier transform of the complex degree of coherence, $\gamma(\tau)$, whose magnitude $|\gamma|$ is easily measured. The problem is to determine $\gamma(\tau)$.

1.9 X-Ray Crystallography

In this case the nonnegative electron density $\rho(x, y, z)$, which is periodic, is the Fourier transform of the structure factor $F(n, k, l)$, where $F(n, k, l)$ can be measured by a diffractometer. The problem is to determine ρ. In the language of crystallography the peaks in the Patterson function are related to the interatomic vectors of the structure. Patterson showed that the phase problem here may not always be unique. Patterson called two crystal structures which are distinct and yet have the same autocorrelation function "homometric". Later it was noted that the positivity condition here implies a set of inequalities involving the Fourier coefficients.

Consider the electron density function $\rho(x)$ which is periodic in 3 dimensions. The Fourier series of ρ is

$$\rho(x_1, x_2, x_3) = \sum_{h_1, h_2, h_3} c(h_1, h_2, h_3) e^{2\pi i \sum h_i b_i x_i} \qquad (1.6)$$

So $c(h_1, h_2, h_3) = \frac{1}{V_d} \int \int \int \rho(x_1, x_2, x_3) e^{-2\pi i \sum h_i b_i x_i} dx_1 dx_2 dx_3$ where $V_d = 1/b_1 b_2 b_3$ is the volume of the fundamental lattice. The Patterson function is given by

$$P(u_1, u_2, u_3) = \frac{1}{V_d} \int \int \int \rho(x_1+u_1, x_2+u_2, x_3+u_3)\rho(x_1, x_2, x_3) dx_1 dx_2 dx_3 \qquad (1.7)$$

whose Fourier coefficients are the intensities $|c(h_1, h_2, h_3)|^2 = c\bar{c}$ and

$$P(u_1, u_2, u_3) = \sum_{h_1, h_2, h_3} |c|^2 e^{2\pi i \sum h_i b_i u_i}. \qquad (1.8)$$

1.10   Radio Astronomy

In radio astronomy the radio brightness map is a two dimensional, real, non-negative function which is the Fourier transform of the complex visibility function. The visibility function is measured by radio interferometry. Due to missing data or gaps in long-baseline interferometry, the error reduction algorithm has been used to provide improved brightness maps.

1.11   Tomography

In tomographic imaging systems, many projections of the object are measured. The algebraic reconstruction technique in tomography, we will find, plays a fundamental role model in our analysis of signal recovery problems. Also, the error reduction algorithm has been used to improve tomographic imaging where there are gaps in the Fourier domain due to limited viewing angles.

1.12   Signal Reconstruction from Zero Crossings

A related set of signal reconstruction problems, which set we will show is intricately tied to the phase retrieval problem, arises in the problem of uniquely specifying signals by their zero crossings. As early as 1948 it was shown that hard-clipped speech (i.e., speech with only the zero crossing information preserved) retains much of the intelligibility of the original signal. For 2-D signals it has been suggested by Marr, Ullman and Poggio that image representation in vision may be accomplished through cortical extraction of zero crossings. The abstract problem is to recover the image from its zero crossing information.

Other applications of signal reconstruction from zero crossing data occur in filter design and antenna design where the zero crossing points are nulls of the filter response or antenna pattern and the object is to reconstruct the remainder of the response. Voelcker has studied zero crossings in modulation theory and has shown how modulation processes can be considered to be methods of manipulating or extracting zeros. For angle modulated signals, Voelcker showed that knowledge of the zero crossing locations of the modulated signal is equivalent to knowledge of the original signal before modulation; thus, if the signal is band limited, then the zero crossings of the modulated signal are sufficient to uniquely specify it.

Logan developed a new class of bandpass signals which are uniquely specified by their zero crossings. Viz., a signal with a bandwidth of less than one octave is uniquely specified by its zero crossings if it has no zeros in common with its Hilbert transform other than real simple zeros. However, this approach does not indicate how complex zeros can be deduced from real ones (i.e., the zero crossings) and thus, no algorithm is provided for signal reconstruction from zero crossing data.

Much of the early work in this area on 1-D signals is based on the fact that band limited signals are entire and are thus uniquely specified in their Hadamard expansion by their zeros (both real and complex) up to a constant and an exponential factor. So, an arbitrary band limited signal is uniquely specified by its zero crossings if all its zeros are guaranteed to be real. For example, if a 1-D complex signal has no energy for negative frequencies, then it is uniquely specified by the zero crossings of its real part if the complex signal has only zeros in the upper half plane. Another approach taken is to modify the signal so that all of its zeros become real. We develop these results in Chapter III.

Voelcker and Requicha have proposed iterative algorithms to reconstruct 1-D signals from their zero crossings. Related reconstruction algorithms have been advocated by Rotem and Zeevi for image or 2-D reconstruction from zero crossings. Curtis in her thesis has developed reconstruction algorithms for 2-D signals from zero crossing information. One of her algorithms is a Gerchberg-Saxton type algorithm which imposes the correct sign of the signal in the space domain and the correct bandwidth in the Fourier domain. Another algorithm is proposed which is based on reducing the reconstruction problem from zero crossings to a set of linear equations. This approach requires the exact zero crossing locations.

The intertwining of the zero crossing signal reconstruction problems and the phase retrieval reconstruction problems is provided in the Lim-Izraelevitz (LI) phase reconstruction algorithm. A basic step in the LI algorithm is to use the Curtis method to extract coefficients of an irreducible polynomial from its zeros.

1.13 Reconstruction and Resolution Limits

Light from a one-dimensional, space-limited object described by the function $O(x)$ for $|x| < X/2$ passes through a lens of finite aperture to form a band-limited object $I(y)$ which is given by

$$I(y) = \frac{1}{2\pi} \int_{-\Omega}^{\Omega} d\omega e^{-i\omega y} \int_{-X/2}^{X/2} e^{i\omega x} O(x) dx \qquad (1.9)$$

$$= \int_{-X/2}^{X/2} \frac{sin(\Omega(x-y))}{\pi(x-y)} O(x) dx$$

where Ω is the highest spatial frequency transmitted by the lens. The problem then is: given any image function $I(y)$ is it possible to reconstruct the exact object $O(x)$ from which it is obtained, i.e., invert the integral equation (1.9)? However, it is known that a lens with a finite spatial frequency bandwidth Ω has an associated resolution limit π/Ω and so the image formed by an object of spatial extent X can contain only a finite number $S = X\Omega/\pi$ of independent components or "degrees of freedom". Here S is called the Shannon number.

To understand the relationship of the resolution limit and the solution of (1.9) consider the eigenfunctions $\phi_n(y)$ which satisfy the integral equation

$$K\phi_n(y) = \int_{-X/2}^{X/2} \frac{sin(\Omega(x-y))}{\pi(x-y)} \phi_n(y) dy. \qquad (1.10)$$

As we will see below in Chapter IV these eigenfuctions of the compact integral operator K form a complete orthonormal basis over [-X/2,X/2]. If we expand the image function in terms of the eigenfunctions $\phi_n(y)$, viz. $I(y) = \sum b_n \phi_n(y)$ where $b_n = \int_{-X/2}^{X/2} I(y)\phi_n(y) dy$ and similarly $O(x) = \sum a_n \phi_n(x)$, then we find that $b_n = \lambda_n a_n$. Thus, we have the solution $O(x) = \sum b_n \phi_n(x)/\lambda_n$ which expresses the object function $O(x)$ in terms of the image function $I(y)$. However, in practice it is impossible to evaluate this expression due to the behavior of λ_n. These eigenvalues are essentially unity for n up to $S = X\Omega/\pi$ but fall off to zero extremely rapidly for higher values of n. This means that the terms b_n will be divided by extremely small numbers for these higher values of n and so noise or error will cause these terms to diverge rapidly. Thus, such components can not be determined and must be excluded to avoid corruption of the reconstruction. And in practice only S terms of the expansion can be calculated, so we are restricted to the above described resolution limit.

1.14 Ill-Posed Problems and Regularization

Hadamard defined a problem to be well-posed if (a) the solution exists, (b) is unique and (c) depends continuously on the data. For a mapping $K : X \rightarrow Y$, (a) means K is surjective and (b) means K is injective. So (a) and (b) imply K is invertible. And (c) states in this case that the inverse mapping is continuous. A problem which is not well-posed is called ill-posed. The band-limited extrapolation problem in 1-D satisfies (b) but it is easily seen to be ill-posed since the original function is analytic and any nonanalytic perturbation produces a function with no solution. We will see below that the phase retrieval problem in 1-D has multiple solutions even with nonnegativity constraints. Thus condition (b) fails and the problem is ill-posed.

The least squares solution of the equation $Kx = y$ is the solution to the variational problem, $min\|Kx - y\|_Y$, which is given by $K^*Kx = K^*y$. We will see below that if the range, $R(K)$, is closed, then a least squares solution exists but it is not unique when the null space, $N(K)$, is nontrivial. Thus the least squares problem is well-posed if K is injective. The minimal norm solution $x^\dagger$ to $K^*Kx = K^*y$ is given by the generalized inverse $x^\dagger = K^\dagger y$. And the minimum norm least squares solution is well-posed if $R(K)$ is closed.

Let $I = [-1, 1]$ and $cI = [-c, c]$ and define the projection operator

$$(P_c f)(v) = \begin{cases} f(v) & \text{if } v \in cI \\ 0 & \text{otherwise.} \end{cases}$$

The basic phase retrieval problems center on the study of the normal, compact operator $P_c \mathcal{F} P_c$. E.g., the phase retrieval problem, given magnitude m, can be formally stated as finding a solution pair (g, G), where G is the Fourier transform of g and $K(G) = |P_c \mathcal{F} P_c G| = (LP_c \mathcal{F} P_c)(G) = |P_c g| = m$. Here $L(g) = |g|$. The operator K is the composition of L and the compact operator $P_c \mathcal{F} P_c$. One can show that the Fréchet derivative of $L(g)$ is the bounded linear operator $D_1(g)h = Re(g^* h)/|g|$ and the Fréchet derivative $D(G)$ of K is $D_1(G)P_c \mathcal{F} P_c$. Since this is the composition of a bounded operator with a compact operator, it follows that $D(G)$ is compact. Thus, in general, the types of signal recovery problems represented by phase retrieval and band-limited extrapolation deal with the the solution of compact operator equations.

The discussion in the last section can be formalized for any compact operator $KG = g, K : H \rightarrow H$. K is described by its singular value system

$\{u_i, v_i, \sigma_i\}$, which we will develop later in this volume. If $g = \sum a_i v_i$ then $G = \sum \frac{a_i}{\sigma_i} u_i$. A small perturbation ϵ in a high frequency component a_i of g results in a perturbation ϵ/σ_i. Thus, for any given ϵ and δ and solution pair g, G there exists an infinite dimensional set of solutions g', G' such that $KG' = g', \|g - g'\| < \epsilon, \|G - G'\| < \delta$; and we see that the inversion of compact operators is ill-posed.

One approach to reduce an ill-posed problem to a well-posed problem is the regularization method of Tikhonov, which provides a one-parameter family of continuous operators R_λ such that $x_\lambda = R_\lambda y$ and $lim_{\lambda \to 0} R_\lambda y = K^\dagger y$. The regularized solution x_λ is found by minimization of $\|K(x) - y\| + \lambda \Omega(x)$ for some functional $\Omega(x)$. The basic results of Tikhonov state that under certain weak conditions on K, Ω, x, y, then x_λ minimizes $\|K(x) - y\| + \lambda \Omega(x)$ if and only if (1) there is a $\delta_\lambda > 0$ such that x_λ minimizes $\|K(x) - y\|$ with $\Omega(x) < \delta_\lambda$ and (2) there is an $\epsilon_\lambda > 0$ such that x_λ minimizes $\Omega(x)$ with $\|K(x) - y\| \le \epsilon_\lambda$.

Taking $\Omega(x) = \|x\|^2$ is equivalent to replacing $Kx = y$ by the equation of the second kind $(K^*K + \lambda I)x = K^*y$. The operator $K^*K + \lambda I$ is one to one and onto; hence its inverse is continuous and we set $x_\lambda = (K^*K + \lambda I)^{-1}K^*y$. And we see that as $\lambda \to 0, x_\lambda \to K^\dagger y$.

Another approach to regularization is to modify the spaces X, Y and their topologies. E.g., for the case $K : H \to H$ where H is $L^2(0,1)$, we will see later that we can replace H by a reproducing kernel Hilbert space H_Q where $K : H \to H_Q$ is now onto. So for each y in $H_Q, Kx = y$ has a unique least squares solution of minimum norm and the solution depends continuously on y in the topology of H_Q; i.e. $K^\dagger : H_Q \to H$ is bounded.

All of these fundamental ideas permeate the problems of signal recovery. We turn now to the development of the tools necessary to treat these problems in greater detail.

Chapter 2

Polynomials: A Review

2.1 Introduction

Given a commutative ring R and an indeterminate x over R then an expression of the form $P = a(d)x^d + a(d-1)x^{d-1} + ... + a(1)x + a(0)$ with coefficients $a(d), ..., a(0)$ in R is called a polynomial in x over R. If $a(d)$ is not equal to 0 in P, then P is said to have degree $d = degP$ and $a(d)$ is called the leading coefficient of P. Under the natural operations of addition and multiplication of polynomials the set $R[x]$ of polynomials in x forms a commutative ring with R as a subring.

DEFINITION 2.1. If R has a unity element 1, and the leading coefficient of P is 1, then P is called a monic polynomial .

DEFINITION 2.2. If R is a ring with unity, then an element r in R for which there is another element r^{-1} in R such that $rr^{-1} = 1$ is called an invertible or unit of R.

The reader can check that for a field R an element P of $R[x]$ is an invertible if and only if P is an invertible of R.

DEFINITION 2.3. If $ab = 0$, with a not equal to 0 and b not equal to 0 in a ring R, then a and b are said to be zero divisors in R. And a commutative ring R with unity 1 not equal to 0 and with no zero divisors is called an integral domain.

Since every nonzero element of a field is an invertible, a field has no zero divisors and so is an integral domain. We leave it to the reader to show that:

THEOREM 2.4. If R is an integral domain, then $R[x]$ is an integral domain.

11

DEFINITION 2.5. For a ring R a nonzero element P in $R[x]$ is said to be reducible if there exist S, T in $R[x]$ such that $P = ST$ and $degS < degP$ and $degT < degP$. A polynomial is irreducible if it has positive degree and is not reducible.

Clearly if P is in $R[x]$ and $degP = 1$, then P is irreducible.

If r, s are elements in R, then we write $s|r$ if there is an element t in R such that $r = st$. In other words s is a divisor of r or r is a multiple of s in R. If both $a|b$ and $b|a$ in R, then a and b are said to be associates in R. We leave it as an exercise to check that if R is an integral domain, the polynomials P, Q in $R[x]$ are associates if and only if $P = VQ$ where V is an invertible in R; so in this case $degP = degQ$.

DEFINITION 2.6. In a ring R with unity let P be an element that is not zero and not invertible. If every divisor of P is either invertible or an associate of P, then P is said to prime in R.

THEOREM 2.7. Let R be a field. Then a polynomial P is prime in $R[x]$ if and only if P is irreducible in $R[x]$.

Proof. Let P be an irreducible polynomial and let $P = ST$ in $R[x]$. Since P is irreducible either $degS \geq degP$ or $degT \geq degP$. Say $degS \geq degP$; then $degP = deg(ST) = degS + degT$, which implies $degT = 0$. So T is a nonzero element of R and is then an invertible in R and $R[x]$. This implies S is an associate of P; thus, every divisor of P is either an invertible or an associate of P; i.e., P is a prime. We leave the converse to the reader.

THEOREM 2.8. Let R be a ring with unity and let P, S belong to $R[x]$ with S monic. Then there is a unique ordered pair Q, r in $R[x]$ such that $P = QS + r$ and either $r = 0$ or $degr < degS$.

DEFINITION 2.9. Let R be a subring of a commutative ring R' and let P belong to $R[x]$. An element s or R' such the $P(s) = 0$ is called a zero in R' of P. In other words s is a root of the polynomial P.

THEOREM 2.10. Let R be a ring with unity and let P belong to $R[x]$. Then r in R is a zero of P if and only if $x - r$ is a divisor of P in $R[x]$.

COROLLARY 2.11. If R is an integral domain and if $r_1, ... r_n$ are distinct zeros of P in $R[x]$, then there is a polynomial S in $R[x]$ such that $P = (x - r_1)(x - r_2)...(x - r_n)S$.

COROLLARY 2.12. If P, not equal to 0, has n distinct zeros, then $degP \geq n$.

DEFINITION 2.13. If R is a ring with unity and m is a positive integer, then if $(x - r)^m$ is a divisor or P in $R[x]$ and $(x - r)^{m+1}$ is not a divisor of P in $R[x]$, then r is said to be a zero with multiplicity m.

THEOREM 2.14. Let R be a commutative ring. Then for P in $R[x]$ every P of positive degree is expressible in the form

$$P = S_1 S_2 ... S_n \qquad (2.1)$$

where S_j are irreducible.

We have noted above that if R is a field then irreducible polynomials are the same as primes.

THEOREM 2.15. If R is a field, every polynomial P of positive degree in $R[x]$ is uniquely expressible in the form

$$P = V S_1^{n_1} ... S_r^{n_r}, \qquad (2.2)$$

where V is a nonzero element of R, S_j are distinct monic primes in $R[x]$ and n_j are positive integers.

We will be interested mostly in the polynomial ring $C[x]$ over the complex numbers C. Since C is algebraically closed, we can use the fundamental theorem of algebra to provide a more explicit factorization result.

THEOREM (Gauss' Fundamental Theorem of Algebra) 2.16. If P is a polynomial of positive degree in $C[x]$ then $P = 0$ has a root in C.

Proof. This follows from Rouché's theorem which we state for completeness.

THEOREM (Rouché) 2.17. Let $f(z)$ and $g(z)$ be analytic in a simply connected domain D containing a Jordan contour CT. Let $|f(z)| > |g(z)|$ on CT. Then f and $f + g$ have the same number of zeros inside CT.

Gauss' theorem follows by taking $f(z) = a(n)z^n$ and

$$g(z) = a(n-1)z^{n-1} + ... + a(0)$$

and CT a circle of large radius.

As a corollary to Gauss' theorem we have

THEOREM 2.18. The primes or irreducible polynomials in $C[x]$ are just the polynomials of degree one.

Combining this result with the unique factorization theorem we have

THEOREM 2.19. Let P be a polynomial of positive degree d in $C[x]$. Then P has d zeros $r_1, ... r_d$ (not necessarily distinct) in C, and

$$P = a(x - r_1)...(x - r_d)$$

, with $a \neq 0$ in C.

We introduce one last bit of terminology from polynomial rings.

DEFINITION 2.20. If R is an integral domain, then R is called a unique factorization domain (UFD) if for each nonunity P in R, $P \neq 0$, there is a factorization $P = S_1...S_n$ where S_j are primes; and if $P = S_1...S_k = T_1...T_l$ are two factorizations, then $k = l$ and for some permutation of subscripts S_i and T_k are associates.

Clearly from the results above, for fields F, $F[x]$ is a UFD.

DEFINITION 2.21. Let R be a commutative ring. Then the ideal $A = \{ra|r \in R\}$ is called a principal ideal. We denote it by $A = (a)$. A ring (domain) R is called a principal ideal ring (domain) if every ideal of R is principal.

By the division algorithm it follows that:

THEOREM 2.22. If R is a field, then $R[x]$ is a principal ideal ring (in fact a principal ideal domain since $R[x]$ is an integral domain).

DEFINITION 2.23. An ideal P in a commutative ring R is called prime if whenever ab belongs to P with a, b in R then either a belongs to P or b belongs to P. An ideal M in R is called maximal if $M \subset R$ and whenever N is an ideal in R such that $M \subseteq N \subseteq R$, then either $N = M$ or $N = R$.

The principal ideal (p) for prime p in the integer ring Z is both a prime ideal and a maximal ideal (i.e., if $a \in (p)$, then $\{ma + np|m, n \in Z\} = Z$). Similarly for a field R, the principal ideal $(x - r)$ in $R[x]$ is both prime and maximal; and if $P(x)$ is irreducible, then $(P(x))$ is a maximal ideal in $R[x]$. For any commutative ring with unit, if M is a maximal ideal of R, then M is a prime ideal.

THEOREM 2.24. If R is a principal ideal domain (PID), then R is a unique factorization domain.

DEFINITION 2.25. Let R be an integral domain and let $d : R \to Z$. Then R is called an Euclidean domain if $d(P) = 0$ if and only if $P = 0$, $d(ST) = d(S)d(T)$ for all S, T in R and for any a, b in R there exist q, r in R such that $a = bq + r$ and $d(r) < d(b)$.

Examples of Euclidean domains are: (1) Z with $d(m) = m$, (2) $F[x]$ for field F with $d(f(x)) = 2^{deg f(x)}$ and (3) field F with $d(a) = 1$ for $a \neq 0$ and $d(0) = 0$.

One can show that:

THEOREM 2.26. Every Euclidean domain is a PID and every PID is a UFD.

THEOREM 2.27. If R is a UFD, then so is $R[x]$.

COROLLARY 2.28. If R is a UFD, then so is $R[x_1, ..., x_n]$.

However, the reader can check that for field F, the ring $F[x, y]$ of polynomials in two indeterminates is a UFD but not a PID; viz., the ideal $(x, y) = \{ax + by | a, b \in F\}$ is not principal.

2.2 Polynomials and Trigonometric Polynomials

Consider the polynomial $P(z)$ of degree n in $C[z]$,

$$P(z) = \sum_{k=0}^{n} a(k) z^k. \tag{2.3}$$

Let $P(N)$ denote the complex vector space of all polynomials of degree less than or equal to N. Then $P(N)$ forms a vector space of dimension $N + 1$ over C.

A polynomial of degree N possesses precisely N roots or zeros $\{z_1, ... z_N\}$ and may be factored as

$$P(z) = a(N) \prod_{k=1}^{N} (z - z_k). \tag{2.4}$$

We assume the zeros are ordered so that

$$|z_1| \leq |z_2| \leq \cdots \tag{2.5}$$

and if $|z_k| = |z_{k+1}|$, then $arg(z_k) \leq arg(z_{k+1})$. For each $P(z)$ in $P(N)$ we associate a sequence $(a(N), z_1, ... z_n)$ $0 \leq n \leq N$ where $a(N)$, not equal to 0, is the coefficient of the highest power of $P(z)$ and z_k are the n zeros of $P(z)$. This zero-based representation is unambiguous and unique.

DEFINITION 2.29. Trigonometric polynomials (tp's) are periodic functions of the form

$$f(z) = \sum_{k=n_1}^{n_2} C(k) e^{ik\Omega z} \tag{2.6}$$

where $n_1 \leq n_2$ and Ω is a positive real number called the fundamental frequency. Here $i = \sqrt{-1}$. The closed interval $[n_1\Omega, n_2\Omega]$ is called the spectral interval of $f(z)$ and $n\Omega$ where $n = n_2 - n_1$ is called the spectral width of $f(z)$.

Notice that $f(z)$ is periodic with period $X = 2\pi/\Omega$, i.e.,

$$f(z) = f(z \pm kX) \text{ k} = 0, 1, ... \tag{2.7}$$

We may associate a polynomial say $g(w)$ with $f(z)$ where $w = e^{i\Omega z}$. Viz.,

$$g(w) = C(n_2)w^{n_1} \sum_{k=n_1}^{n_2} \frac{C(k)w^{k-n_1}}{C(n_2)}. \tag{2.8}$$

If $w_1, ... w_n$ denotes the n zeros of $g(w)$ and $w_k = e^{i\Omega z_k}$, then

$$f(z) = C(n_2)e^{in_1\Omega z} \prod_{k=1}^{n} (e^{i\Omega z} - e^{i\Omega z_k}). \tag{2.9}$$

Let $T(N_1, N_2)$ denote the complex vector space of tps with spectral support $[N_1\Omega, N_2\Omega]$. This vector space has dimension N+1 where $N = N_2 - N_1$. Again there is an unambiguous and unique representation of tps of spectral width N given by the map

$$f(z) \rightarrow (C(n_2), n_1, z_1, ... z_n)\ 0 \leq n \leq N, \tag{2.10}$$

where $C(n_2)$ is the Fourier coefficient of the highest frequency component present in f, $n_1\Omega$ is the lowest frequency present and $z_1, ... z_n$ are the zeros of $f(z)$ where $0 \leq Re(z) < X$.

Chapter 3

Entire Functions and Signal Recovery

3.1 Introduction

A function $f(z)$ is entire if it is analytic for all finite z. An entire function (EF) $f(z)$ has a Taylor series representation

$$f(z) = \sum_{n=0}^{\infty} b(n)z^n \tag{3.1}$$

which converges for all finite z.

EXAMPLES 3.1. The polynomials $P(z) = \sum_{n=0}^{N} b(n)z^n$ are entire functions as are the transcendental functions e^z, $\sin(z)$ and $\cos(z)$. If $g(z)$ is an entire function, then $e^{g(z)}$ is an entire function.

Weierstrass has shown that every entire function is specified by its zeros in the following sense: let $\{a(n)\}$ be a set of zeros; then every entire function $f(z)$ with these and no other zeros can be written in the form

$$f(z) = z^m e^{g(z)} \prod_{n=1}^{\infty} (1 - \frac{z}{a(n)}) e^{\frac{z}{a(n)} + \frac{1}{2}\frac{z^2}{a(n)^2} + \cdots + \frac{1}{m_n}\frac{z^{m_n}}{a(n)^{m_n}}}, \tag{3.2}$$

where the product is taken over all $a(n) \neq 0$, m_n are integers and $g(z)$ is an entire function.

Let $M(r)$ denote the maximum modulus of an entire function $f(z)$ on a disk of radius r, i.e.,

$$M(r) = max_{|z|=r}|f(z)|. \tag{3.3}$$

17

The order of an entire function f is given by

$$\rho = lim_{r \to \infty} loglog M(r)/log(r) \qquad (3.4)$$

and the type b of f is given by

$$b = lim_{r \to \infty} log M(r)/r^\rho. \qquad (3.5)$$

An entire function of order ρ and type b with $\rho \neq 0$ and $b \neq 0$ grows asymptotically as fast as e^{br^ρ}, i.e., as $|z| \to \infty$ then

$$|f(z)| \leq e^{b|z|^\rho}. \qquad (3.6)$$

An entire function is said to be of zero, normal or infinite type as $b = 0, 0 < b < \infty$ or $b = \infty$ respectively.

If f is an entire function of order less than one or of order one and finite (i.e., zero or normal type), then f is called an entire function of exponential type (EFET). For such a function the maximum rate of growth occurs along the imaginary axis and the minimum rate of growth occurs along the real axis.

EXAMPLES 3.2. Clearly any algebraic polynomial is an entire function of order zero. The function $sin(\pi z)$ is an entire function of order one and type π.

Entire functions of exponential type are the simplest generalization of polynomials and like polynomials they can be represented by their zeros. This is given by the Hadamard factorization theorem:

THEOREM (Hadamard) 3.3. Let $f(z)$ be an entire function with zeros a_k and assume $f(z)$ is of finite order ρ; then $f(z)$ can be represented in the form

$$f(z) = z^m e^{Q(z)} \prod_k (1 - \frac{z}{a_k}) e^{\frac{z}{a_k} + \frac{1}{2}(\frac{z}{a_k})^2 + \cdots + \frac{1}{\rho}(\frac{z}{a_k})^\rho}, \qquad (3.7)$$

where $Q(z)$ is a polynomial of degree $q < \rho$ and m is the multiplicity of the zero at the origin.

COROLLARY (Hadamard Factorization Theorem) 3.4. An EFET $f(z)$ can be represented in the form

$$f(z) = z^m e^{az+b} \prod_{k=1}^{\infty} (1 - \frac{z}{a_k}) e^{z/a_k} \qquad (3.8)$$

where a_k are the nontrivial zeros of $f(z)$ and m is the order of the zero at the origin. This infinite product converges absolutely in the entire z-plane.

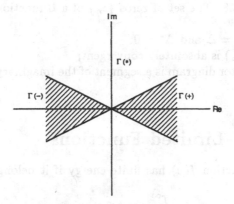

Figure 3.1: Angular Sectors

EXAMPLE 3.5. The entire function $sin(\pi z)$ has zeros at the integers and admits the Hadamard factorization

$$sin(\pi z) = \pi z \prod_{n \neq 0} (1 - \frac{z}{n})e^{z/n}. \qquad (3.9)$$

We note in this case that the zeros of $sin(\pi z)$ are all real and equidistant along the real axis.

The indicator function $h(\theta)$ for an entire function $f(z)$ of order one is given by

$$h(\theta) = \lim_{r \to \infty} \sup \frac{\log|f(re^{i\theta})|}{r}. \qquad (3.10)$$

We will associate with $h(\theta)$ a bounded, closed, convex set of points called the indicator diagram, to be described below.

DEFINITION 3.6. An EFET is said to have zeros close to the real axis when $\sum_{n=1}^{\infty} |Im(1/z_n)| < \infty$ where z_n are the set of zeros.

We let $\Gamma(*), \Gamma(+)$, and $\Gamma(-)$ denote the angular sectors shown in Figure 3.1. And if $n^*(r)$ is the number of zeros with $|z| < r$ in $\Gamma(*)$, then an EFET with zeros close to the real axis has zero angular density in $\Gamma(*)$ – i.e.,

$$\Delta^* = \lim_{r \to \infty} n^*(r)/r = 0. \qquad (3.11)$$

Let $\Delta^{\pm} = \lim_{r \to \infty} n^{\pm}(r)/r$. Finally, let $n(r)$ denote the number of zeros whose moduli are less than $r = |z|$ and set $\Delta = \lim_{r \to \infty} n(r)/r$.

DEFINITION 3.7. A B-function is an EFET with zeros close to the real axis and whose indicator function satisfies $h(0) = h(\pi) = 0$.

THEOREM 3.8. The set of zeros $\{z_n\}$ of a B-function has the properties:

(1) $\Delta^+ = \Delta^- = \Delta$ and $\Delta^* = 0$.

(2)$\sum Im(1/z_n)$ is absolutely convergent;

(3) the indicator diagram is a segment of the imaginary axis with length $d = 2\pi\Delta$.

3.2   Band-Limited Functions

A continuous function $f(x)$ has finite energy if it belongs to $L^2(-\infty, \infty)$ –i.e., if

$$\int_{-\infty}^{\infty} |f(x)|^2 dx < \infty. \tag{3.12}$$

Let $F(s)$ denote the Fourier transform of $f(x)$

$$F(s) = \frac{1}{2\pi} \int f(x)e^{-isx} dx. \tag{3.13}$$

When $F(s)$ is zero outside a finite interval [a,b], $f(x)$ is said to be band-limited. Thus a band-limited finite energy function may be written as

$$f(z) = \int_a^b F(s)e^{isz} dz. \tag{3.14}$$

The interval [a,b] is called the spectral interval of F and W = b-a is called the spectral width.

THEOREM (Paley-Wiener) 3.9. The Fourier transform of a finite energy function $F(s)$ is null outside [a,b] if and only if $F(z)$ is an EFET of type (b-a)/2. If [a,b] cannot be replaced by a smaller interval in (3.14), then the vertical segment [-ib,-ia] on the imaginary axis is the indicator diagram for $F(z)$.

The indicator function for $F(z)$ in the Paley-Wiener theorem is $h(\theta) = ((b-a)/2)|sin(\theta)|$. So $h(0) = h(\pi) = 0$, in this case.

By the Paley-Wiener theorem we see that any finite energy EFET is a B-function. And conversely finite energy B-functions are band-limited.

THEOREM 3.10. A finite energy function $f(t)$ is a B-function with indicator diagram [-ib,-ia] if and only if $f(t)$ is band-limited to [a,b].

The simplest example of the Paley-Wiener theorem is given by taking $f(t)$ to be the constant function $1/2\pi$ over $[-\pi, \pi]$. Then the Fourier transform is given by

$$F(z) = sin(z\pi)/(z\pi) \tag{3.15}$$

which can be written as

$$F(z) = \prod_{n \neq 0}(1 - \frac{z}{n})e^{z/n}. \qquad (3.16)$$

In this case the zeros are all real and equidistant. For the interval (-b,b) the zeros occur at $z_n = n\pi/b$ and the space between adjacent zeros is π/b.

Titchmarsh has shown that these facts extend in some sense to every EFET.

3.3 Zeros of EFETs

Let f be an integrable function which vanishes outside (a,b). Set $F(z) = \int_a^b e^{zt}f(t)dt$.

THEOREM (Polya) 3.11. The function $F(z)$ has an infinity of zeros $z_k = r_k e^{i\theta_k}$ such that the series $\sum_1^\infty 1/r_k$ is divergent.

Proof. We assume $F(0)$ is not zero. If $F(0) = 0$, then

$$F(z) = -z \int_a^b e^{zt}g(t)dt \qquad (3.17)$$

where $g(t) = \int_a^t f(u)du$. If the series were convergent, then the Weierstrass product gives

$$F(z) = F(0)e^{(\alpha+i\beta)z} \prod_1^\infty (1 - z/z_n) = e^{(\alpha+i\beta)z}G(z) \qquad (3.18)$$

where for every positive ε

$$|G(z)| = \mathcal{O}(e^{\varepsilon|z|}). \qquad (3.19)$$

However, one checks that when $r \to \infty$

$$|F(re^{i\theta})| \neq \mathcal{O}(e^{r(|cos\theta|-\delta)}) \qquad (3.20)$$

for

$$\delta > 0$$

for any value of θ. This gives rise to a contradiction. Hence, the series is divergent.

THEOREM (Titchmarsh) 3.12. The series $\sum cos(\theta_n)/r_n$ is absolutely convergent.

Proof. By the formula of Carleman we have

$$\sum_{l<r_n<R}(\frac{1}{r_n}-\frac{r_n}{R^2})cos\theta_n < K \qquad (3.21)$$

and the result follows since one checks that $e^{-z}F(z)$ is bounded for $Re(z) \geq 0$ and $e^z F(z)$ is bounded for $Re(z) \leq 0$. Viz.,

THEOREM 3.13. $|F(re^{i\theta})| < \delta e^{r|cos\theta|}$ as $r \to \infty$, uniformly with respect to θ.

Proof. If x is real and positive

$$|F(x)| =$$

$$|\int_{-1}^{1-\delta} e^{xt}f(t)dt + \int_{1-\delta}^{1} e^{xt}f(t)dt| \leq e^{(1-\delta)x}\int_{1}^{1-\delta}|f(t)|dt + e^x\int_{1-\delta}^{1}|f(t)|dt. \qquad (3.22)$$

We select δ so small that

$$\int_{1-\delta}^{1}|f(t)|dt < \varepsilon/2, \qquad (3.23)$$

and then select x_0 so large that

$$e^{-\delta x}\int_{-1}^{1-\delta}|f(t)|dt < \varepsilon/2 \qquad (3.24)$$

for $x > x_0$. So $|F(x)| < \varepsilon e^x$ for $x > x_0$. Again

$$F(iy) = \int_{-1}^{1} cos(yt)f(t)dt + i\int_{-1}^{1} sin(yt)f(t)dt \qquad (3.25)$$

which tends to zero as $y \to \infty$ by the Riemann-Lebesgue theorem. The result follows from the Phragman-Lindelof theorem applied to $G(z) = e^{-z}F(z)$.

THEOREM (Titchmarsh) 3.14. If $n(r)$ is the number of zeros of $F(z)$ for $|z| \leq r$, then $n(r) \sim \frac{b-a}{\pi}r$.

We will not complete the proof of this result. However, the first step is:

THEOREM 3.15. Let $N(r) = \int_0^r \frac{n(x)}{x}dx$. Then $N(r) < 2r/\pi$ for $r > r_0$.

Proof. By Jensen's formula we have

$$N(r) = \frac{1}{2\pi}\int log|F(re^{i\theta})|d\theta - log|F(0)| \qquad (3.26)$$

and by theorem 3.13

$$|F(re^{i\theta})| < \delta e^{r|cos\theta|} \qquad (3.27)$$

for $r > r_0$. Hence $log|F(re^{i\theta})| < log\delta + r|cos\theta|$ and

$$N(r) < log\delta + \frac{r}{2\pi} \int_{-\pi}^{\pi} |cos\theta|d\theta - log|F(0)| = \frac{2r}{\pi} - log(|F(0)|/\delta); \quad (3.28)$$

so the result holds if $\delta < |F(0)|$.

COROLLARY 3.16. $n(r) < Kr$.
Proof. $n(r)logk \leq \int_r^{kr} \frac{n(x)}{x}dx \leq N(kr) < Kr$.

THEOREM (Titchmarsh) 3.17. If x is real

$$|F(x)| = |F(0)| \prod |1 - \frac{xe^{-i\theta_n}}{r_n}|. \qquad (3.29)$$

Proof. By the Weierstrass product we have

$$F(z) = F(0)e^{(\alpha+i\beta)z} \prod (1 - z/z_n)e^{z/z_n}. \qquad (3.30)$$

So if x is real

$$|F(x)| = |F(0)|e^{\alpha x} \prod |1 - \frac{xe^{-i\theta_n}}{r_n}| = e^{x cos(\theta_n)/r_n}. \qquad (3.31)$$

By theorem 3.12, the product $\prod e^{x cos\theta_n/r_n}$ is convergent. Hence,

$$|F(z)| = |F(0)|e^{\gamma x} \prod |1 - \frac{xe^{-i\theta_n}}{r_n}| \qquad (3.32)$$

where $\gamma = \alpha + \sum_1^{\infty} cos(\theta_n)/r_n$. We leave it to the reader to show that $\gamma = 0$.

THEOREM (Titchmarsh) 3.18. The series $\sum sin(\theta_n)/r_n$ is conditionally convergent.

THEOREM (Titchmarsh) 3.19.

$$\sum_1^{\infty} \frac{cos\theta_n}{r_n} = (a+b)/2 - Re(F'(0)/F(0)) \qquad (3.33)$$

$$\sum_1^{\infty} \frac{sin\theta_n}{r_n} = Im(F'(0)/F(0)).$$

EXAMPLE 3.20. Take

$$f(t) = \begin{cases} 1 & -1 < t \le 0 \\ i & 0 < t < 1. \end{cases} \qquad (3.34)$$

Then

$$F(z) = (e^z - 1)(e^{-z} + i)/z.$$

The zeros of $F(z)$ are $\pm 2i\pi, \pm 4i\pi, \ldots$ and $i\pi/2, -3i\pi/2, 5i\pi/2, \ldots$ Hence, $\sum \frac{sin\theta_n}{r_n} = \frac{2}{\pi}(1 - \frac{1}{3} + \frac{1}{5} - \frac{1}{7}\ldots) = 1/2$.

THEOREM (Titchmarsh) 3.21. We have

$$F(z) = F(0)e^{(a+b)z/2} \prod_1^\infty (1 - z/z_n) \qquad (3.35)$$

where the product is conditionally convergent.

This follows from the results developed. And as a corollary to theorem 3.14 we have

THEOREM 3.22. If $\phi(t)$ and $\psi(t)$ are integrable functions such that

$$\int_0^x \phi(t)\psi(x - t)dt = 0 \qquad (3.36)$$

a.e. in the interval $0 < x < K$, then $\phi(t) = 0$ a.e. in $(0, \lambda)$ and $\psi(t) = 0$ a.e. on $(0, \mu)$ where $\lambda + \mu \ge K$.

The proof is left to the reader.

The fact that the series $\sum_1^\infty 1/r_j$ is divergent while $\sum cos(\theta_j)/r_j$ is absolutely convergent implies that the zeros tend to be located near the real axis for large n. Furthermore, from the result

$$n(t) \sim \frac{(b - a)r}{\pi} \qquad (3.37)$$

we have, in the case a = -b, the density of zeros $\Delta = 2b/\pi$. Thus, the spacing between two adjacent zeros must approach π/b for large n, as we saw above.

If the function $f(t)$ is real and nonnegative, we have the following theorem due to Polya:

THEOREM 3.23. If $f(t) \ge 0$ is a nonincreasing (nondecreasing) integrable function, all the zeros of $F(z)$ have positive (negative) imaginary parts.

THEOREM 3.24. If $f(t)$ is continuous, positive and differentiable except at a finite number of points and $c_1 \le f'(t)/f(t) \le c_2$ for $a < t < b$, then all the zeros of F(z) lie in the strip $c_1 \le Re(z) \le c_2$.

EXAMPLE 3.25. Let $f(t) = Ae^{bt}$; then the zeros of $F(z) = \int_{-a}^{a} f(t)e^{-izt}dt$ are given by $z = \frac{n\pi}{a} - ib$.

If we know the behavior of $f(t)$ near the end points (-a,a), there are results regarding the asymptotic distribution of zeros:

THEOREM (Cartwright) 3.26. If $f(t)$ is continuous on (-a,a) and $f(\pm a) \neq 0$ and if $f'(t)$ is integrable, then the zeros of $F(z) = \int_{-a}^{a} f(t)e^{-izt}dt$ are given by

$$z_n \sim \frac{n\pi}{a} + \frac{i}{2a}ln(f(-a)/f(a)) + \varepsilon_n \qquad (3.38)$$

where $\varepsilon_n \to 0$ as $n \to \infty$.

3.4   Encoding of Information by Zeros

Consider the simple case of a finite slit $(-\pi, \pi)$ with Fourier transform $F(x) = sin(\pi x)/x$. The zeros of $F(x)$ lie at $x = \pm j, j = 1, 2, 3...$ If N zeros are displaced to new positions, we write

$$G(x) = \frac{sin(\pi x)}{x} \prod_{1}^{N} \frac{1 - x/z_j}{1 - x/j}. \qquad (3.39)$$

The associated object is

$$g(t) = rect(t/\pi)(1 + \sum_{1}^{N} a_k(-1)^k e^{-ikt}) \qquad (3.40)$$

where $a_k = \prod_{j=1}^{N} \frac{1-k/z_j}{1-k/j}$. Clearly G is unchanged if one or more of the complex z_j are flipped to their conjugates z_j^*. Of course a new function g(t) will be generated.

3.5   Far Field Distributions for Point Sources

Consider an array of point sources of radiation along the real axis

$$c_{-1} \to c_0 \to c_1.$$

We can represent this array of radiating sources by the distribution

$$f(x) = c_{-1}\delta(x + a) + c_0\delta(x) + c_1\delta(x - a). \qquad (3.41)$$

The far field distribution of these sources is given by the Fourier transform $F(u) = \mathcal{F}(f(x))$. Recall that

$$\mathcal{F}(\delta(x - a)) = < \delta_{x-a}, e^{-2\pi ux} > = e^{-2\pi iau}. \qquad (3.42)$$

We note that the Fourier transform of a distribution with point supports is an entire function. More generally we have the Paley-Wiener-Schwartz theorem:

THEOREM 3.27. If f(x) is a distribution with bounded support, then its Fourier transform is an entire function.

Returning to the introductory example, the far field distribution is given by

$$F(z) = \sum_{n_1}^{n_2} c(k) e^{kbiz} \qquad (3.43)$$

which is a trigonometric polynomial with fundamental frequency $b =_\setminus \pi$. The closed interval $[n_1 b, n_2 b]$ is the spectral interval of $F(z)$ and Nb is the spectral width where $N = n_2 - n_1$. Note that $F(z)$ is periodic with period $X = 2\pi/b$; i.e.,

$$F(z) = F(z \pm kX) \qquad (3.44)$$

for k = 0,1,... This periodicity implies that $F(z)$ is completely determined by the values it takes on any z-plane strip of width X, parallel to the imaginary axis; e.g., the strip $0 \le Re(x) \le X$.

When the far-field image is recorded on film, then the image is described by the modulus $|F(z)|^2$. We note that $|F(z)|^2$ is also a trigonometric polynomial which is now nonnegative.

Can every nonnegative trigonometric polynomial be realized as the far field image of a series of point sources? This is answered affirmatively by the following theorem of Fejer and Riesz.

THEOREM 3.28. Every nonnegative trigonometric polynomial $t(u)$ can be represented as the square modulus of another trigonometric polynomial $F(u)$:

$$t(u) = |F(u)|^2. \qquad (3.45)$$

Proof. Consider the case $t(u) > 0$ and let $t(u) = \sum_{-n}^{n} c(k) e^{iku}$. Note that $c(k) = c(k)^*$. We assume $c(n) \ne 0$. Let $P(z)$ denote the polynomial

$$P(z) = c(-n) + c(-n+1)z + ... + c(n)z^{2n}. \qquad (3.46)$$

Then $P(z) = z^{2n} P(1/z^*)$. We can write $t(u) = e^{-inu} P(e^{iu})$. $P(z)$ can have no zeros on $|z| = 1$. Let $a_1, a_2, ...$ denote the zeros on the interior of the unit circle and let $b_1, b_2, ...$ denote the zeros on the exterior. Let $r_1, r_2, ...$ and $s_1, s_2, ...$ denote the multiplicities of these zeros. Thus, we have the factorization

$$P(z) = c \prod_k (z - a_k)^{r_k} z^s \prod_j \left(\frac{1}{z} - \frac{1}{b_j}\right)^{s_j} \qquad (3.47)$$

where $s = \sum s_j$. We see that if b is an exterior zero, then $a = 1/b^*$ is an interior zero of the same multiplicity. The converse is also true. Thus, we can label the roots so that $b_k = 1/a_k^*$ and $r_k = s_k$. Setting $s = \sum s_k = \sum r_k = \frac{1}{2}\sum(s_k + r_k) = n$ we have

$$t(u) = e^{-inu}P(e^{iu}) = c\prod_k(e^{iu} - a_k)^{r_k}\prod_k(e^{-iu} - a_k^*)^{r_k}. \qquad (3.48)$$

The constant c is positive and we have

$$F(u) = \sqrt{c}\prod_k(e^{iu} - a_k)^{r_k}. \qquad (3.49)$$

EXERCISE 3.29. Show that any periodic EFET is a trigonometric polynomial.

3.6 Radar Waveform Construction a la Huffman

The objective in radar waveform design is to select a signal waveform which when correlated with the reflected signal will produce a high correlation at zero time shift and falls rapidly to zero for other shifts. Let $P(z)$ denote the finite pulse train with amplitude $c(k)$

$$P(z) = \sum_{k=0}^{N} c(k)z^k. \qquad (3.50)$$

We assume both $c(0)$ and $c(N)$ are nonzero. Then

$$P(z) = c(0)\prod_{k=1}^{N}(z - z_k) \qquad (3.51)$$

where z_k are the roots of $P(z)$. Let $Q^*(z)$ denote the polynomial

$$Q^*(z) = c(0)^* + c(1)^*z + ... + c(N)^*z^N. \qquad (3.52)$$

Then Q^* can be represented as

$$Q^*(z) = c(0)^*(-1)^N\prod z_j^*\prod(z - 1/z_j^*). \qquad (3.53)$$

The coefficients of the polynomial PQ^* can be equated to the values of the autocorrelation function. For the autocorrelation polynomial PQ^* it is unimportant as to which roots belong to P and which belong to Q^*; for each root z_j of P there is a root $1/z_j^*$ of Q^*.

If z_j is on the unit circle, then $z_j = 1/z_j^*$. For the roots z_j not on the unit circle with multiplicity k_j there are $\prod_j (k_j + 1)$ distinct ways of assigning half the total number of roots of PQ^* to P. If there are N distinct roots, then there are 2^N distinct assignments – i.e., 2^N different pulse trains having the given autocorrelation function.

EXERCISE 3.30. Show that if the autocorrelation is to be realizable with a sequence of real numbers, then the complex roots of P must occur in conjugate pairs.

3.7 Hayes' Fundamental Theorem

Let $x(n)$ be a real sequence which is nonzero only for finitely many values of n. Let $F(n)$ denote the set of such sequences. The z-transform of $x(n)$ is given by

$$X(z) = \sum_n x(n) z^{-n}. \tag{3.54}$$

As a polynomial in z^{-1}, $X(z)$ admits a factorization

$$X(z) = \alpha z^{-n_0} \prod X_k(z) \tag{3.55}$$

where $X_k(z)$ are nontrivial irreducible polynomials in z^{-1}.

The Fourier transform of $x(n)$ is given by taking $z = e^{i\omega}$; viz.,

$$X(\omega) = \sum x(n) e^{-in\omega}. \tag{3.56}$$

The inverse Fourier transform of $|X(\omega)|^2$ is the autocorrelation $r_x(n)$ of $x(n)$:

$$r_x(n) = x(n) \star x(-n). \tag{3.57}$$

The z-transform of $r_x(n)$ is

$$R_x(z) = X(z) X(z^{-1}). \tag{3.58}$$

The general form of $R_x(z)$ is

$$R_x(z) = \alpha^2 \prod_{k=1}^{p} X_k(z) X_k(z^{-1}). \tag{3.59}$$

If $\prod_{k=1}^{p} X_k(z)$ is a polynomial of degree N is z^{-1}, then multiplying $R_x(z)$ by z^{-N} yields a polynomial in z^{-1} of degree 2N:

$$Q_x(z) = z^{-N} R_x(z) = \alpha^2 \prod X_k(z) \tilde{X}_k(z) \tag{3.60}$$

where $\tilde{X}(z) = z^{-N}Z(z^{-1})$. Note that $\tilde{X}(z)$ is the z-transform of the time reversed version of $x(n)$.

Thus, $Q_x(z)$ and $|X(\omega)|$ contain the same information about $x(n)$. So the problem to uniquely recover $x(n)$ form $|X(\omega)|$ is equivalent to the problem to uniquely recover $X(z)$ from $Q_x(z)$.

THEOREM (Hayes) 3.31. Let $x \in F(n)$ have a z-transform

$$X(z) = \alpha z^{-n_0} \prod_{k=1}^{p} X_k(z)$$

where $X_k(z)$ are nontrivial irreducible polynomials. If y belongs to F(n) and $|X(\omega)| = |Y(\omega)|$ for all ω, then $Y(z)$ must be of the form

$$Y(z) = \pm\alpha z^m \prod_{k \in I} X_k(z) \prod_{k \in I^c} X_k(z) \tag{3.61}$$

where I is a subset of the integers [1,p] with complement I^c.

3.8 Brightness Distribution (Bruck-Sodin)

Consider the case of a nonnegative object of finite extent. The spectrum of the object in the finite case is given by

$$F(u) = \sum_{j=0}^{n} f(j)e^{-i2\pi\Delta ju} \tag{3.62}$$

If we substitute $z = e^{-2\pi i \Delta u}$, then the spectrum is given by the polynomial $F(z) = \sum_{j=0}^{n} f(j)z^j$. The polynomial $F(z)$ and $AF(z)$ for constant A represent the same brightness distribution; and the polynomial $F(z)z^s$ is the same distribution, just shifted by s steps along the x-axis.

The reconstruction problem is to determine the function f from the modulus $|F|$. The modulus $|F|$ is known for real values of u, i.e., for $|z| = 1$. This implies knowledge of the product

$$F(z)F^*(z) = F(z)F(z^{-1}). \tag{3.63}$$

Equivalently, this is specified by the autocorrelation polynomial

$$Q(z) = F(z)F(z^{-1})z^n. \tag{3.64}$$

Consider the case $F(z) = \prod_{j=1}^{n}(z - z_j)$. Then $Q(z) = z^n \prod(z - z_j)(z - z_j^{-1})$, i.e., both z_j and $1/z_j$ are roots of $Q(z)$. The solution to the reconstruction problem for a given autocorrelation polynomial is the determination of all polynomials $F(z)$ with nonnegative coefficients where either z_j

or $1/z_j$ are roots for j = 1,...n. If all $|z_j| \neq 1$, then there are no more than 2^n solutions. If there are r roots with $|z_j| = 1$, then the problem has no more than 2^{n-r} solutions.

EXAMPLE (Binary Star) 3.32. Let $a, 1 + a^2$ and a be the readings of the autocorrelation function. Thus, $Q(z) = a + (1 + a^2)z + az^2$. There are two solutions, viz. $P_1(z) = 1 + az$ and $P_2(z) = z + a$:

$$1 \to az$$

$$a \to z.$$

EXAMPLE (Triple Star) 3.33. Let $Q(z) = (z+a)(z+b)(z+a^{-1})(z+b^{-1})$ where a and b are real and not equal. In this case there are four solutions:

$$P_1(z) = (z + a)(z + b) \qquad (3.65)$$

$$P_2(z) = (z + a^{-1})(z + b)$$

$$P_3(z) = (z + a)(z + b^{-1})$$

$$P_4(z) = (z + a^{-1})(z + b^{-1}).$$

3.9   Specified Autocorrelation Functions

Up to now we have dealt with discrete distributions of sources. In this section we extend the above analysis to nondiscrete functions.

DEFINITION 3.34. A pulse is defined to be any complex-valued time function $f(t)$ which satisfies

$$f(t) = 0 \text{ for } |t| > T/2 \qquad (3.66)$$

and

$$\int_{-T/2}^{T/2} |f(t)|^2 dt < \infty. \qquad (3.67)$$

Let $F(z)$ denote the Laplace transform

$$F(z) = \int_{-T/2}^{T/2} f(t)e^{-zt} dt. \qquad (3.68)$$

By the Paley-Wiener theorem $F(z)$ is an entire function.

DEFINITION 3.35. The autocorrelation function $R_f(t)$ of the pulse $f(t)$ is defined by

$$R_f(t) = \int_{-T/2}^{T/2} f(s) f^*(s+t) ds. \tag{3.69}$$

THEOREM (Hofstetter) 3.36. (1) $R_f(t) = 0$ for $|t| > T$;
(2) $\int_{-T}^{T} |R_f(t)|^2 dt < \infty$
(3) $E_f(z) = \int_{-T}^{T} R(t) e^{-zt} dt$ is the entire function $E_f(z) = F^*(z^*) F(-z)$.

Given a pulse $f(t)$ with autocorrelation function, the basic reconstruction problem is to find all other pulses of the same duration whose autocorrelation functions are equal to R_f. If f_1 and f_2 have the same autocorrelation function, then $F_1^*(z^*) F_1(-z) = F_2^*(z^*) F_2(-z)$. Upon setting $z = -iv$ we have

$$|F_1(iv)| = |F_2(iv)|. \tag{3.70}$$

Set

$$f_2(t) = \frac{1}{2\pi} \int_{-\infty}^{\infty} F_1(iv) e^{i\phi(v)} e^{ivt} dv \tag{3.71}$$

where $\phi(v)$ is a real function. Here, $f_2(t)$ is not necessarily a pulse.

Let z_o be a zero of $F_1(z)$ and let $f_2(t) = \int_{-T/2}^{T/2} f_1(t) h(t-s) ds$ where h is the impulse response of the transfer function

$$H(z) = \frac{z + z^*}{z - z_o} \tag{3.72}$$

and $H(s)$ is an all-pass filter (i.e., $|H(iv)| = 1$). Here $h(t) = (z_o + z_o^*) e^{z_o t} U_{-1}(t)$ where $U_{-1}(t)$ is the unit step function. Thus,

$$f_2(t) = \begin{cases} 0 & t < -T/2 \\ (z + z_o^*) \int_{-T/2}^{T/2} f_1(\tau) e^{z_o(t-\tau)} d\tau & -T/2 \le t \le T/2 \\ (z + z_o^*) \int_{-T/2}^{T/2} f_1(\tau) e^{z_o(t-\tau)} d\tau & t > T/2. \end{cases} \tag{3.73}$$

However, for $t > T/2$, $\int_{-T/2}^{T/2} f(\tau) e^{z_o(t-\tau)} d\tau = e^{z_o t} F_1(z_o) = 0$. Thus, f_2 is a pulse limited to [-T/2,T/2].

We return now to the problem of determining pulses with the same autocorrelation function.

THEOREM 3.37. If f_1 and f_2 are finite energy pulses with $R_{f_1} = R_{f_2}$; then

$$F_1(z) = A_1 z^{n_1} e^{a_1 z} \prod_{k=1}^{\infty} (1 - z/z_k) e^{z/z_k} \tag{3.74}$$

$$F_2(z) = A_2 z^{n_2} e^{a_2 z} \prod_{k=1}^{\infty} (1 - z/t_k) e^{z/t_k}$$

where z_k and t_k are the zeros of F_1 and F_2 respectively. Here $|A_1| = |A_2|, n_1 = n_2, Im(a_1) = Im(a_2)$ and the zeros can be ordered so that $t_k = z_k$ or $t_k = -z_k^*$ for each k = 1,2,...

The proof follows directly from the Paley-Wiener theorem and the factorization theorem of Hadamard.

THEOREM 3.38. If f_1 is a finite energy pulse and z_k are the zeros of $F_1(s)$, then for any choice of t_k, k = 1,2,.. where $t_k = z_k$ or $t_k = -z_k^*$, a real constant τ_o can be found such that

$$F_2(z) = e^{z\tau_o} F_1(z) \prod_{k=1}^{\infty} \frac{1 - z/t_k}{1 - z/z_k} e^{z(\frac{1}{t_k} - \frac{1}{z_k})} \tag{3.75}$$

has the property that the associated pulse f_2 satisfies $R_{f_1} = R_{f_2}$.

Thus, except for a scale factor and/or a delay, all pulses f_2 having the same autocorrelation as a given pulse f_1 can be obtained by passing f_1 through all-pass filters of the form

$$H(z) = e^{z\tau_o} \prod_{1}^{\infty} \frac{1 - z/t_k}{1 - z/z_k} e^{z(1/t_k - 1/z_k)} \tag{3.76}$$

where t_k and z_k are as above.

3.10 Autocorrelation Function of Pulses

The conditions on a function for it to be the autocorrelation function of a pulse are given by:

THEOREM 3.39. A function $R(t)$ is the autocorrelation of a finite energy pulse on [-T/2,T/2] if and only if
(1) $R(t) = 0$ for $|t| > T$;
(2) $R(t) = R^*(-t)$;
(3) $\int_{-T}^{T} |R(t)|^2 dt < \infty$;
(4) $E(iv) = \int_{-T}^{T} R(t) e^{-ivt} dt \geq 0$;
(5) $\int_{-\infty}^{\infty} E(iv) dv < \infty$.

EXERCISE 3.40. Let $f_1(t) = e^{at}$ for $|t| \leq 1$ and 0 for $|t| > 1$. Then show that $F_1(z) = \int_{-1}^{1} e^{t(a-z)} dt = 2 sinh(z-a)/(z-a)$. The zeros of $F_1(z)$ are $z_k = a + ik$ for k = ±1, ±2, ... Let

$$H(z) = \frac{(z+a)^2 + \pi^2}{(z-a)^2 + \pi^2} \tag{3.77}$$

and $F_2(z) = F_1(z)H(z)$. Show that F_2 corresponds to the pulse

$$f_2(t) = \frac{4a}{\pi}e^{at}(\frac{a}{\pi}cos(\pi t) - sin(\pi t) + \frac{a}{\pi}) + e^{at} \qquad (3.78)$$

for $|t| \leq 1$ and 0 for $|t| > 1$ and check that $R_{f_1} = R_{f_2}$.

3.11 Autocorrelation Functions

By the Riesz-Fejer theorem we have seen that every nonnegative trigonometric polynomial can be factorized in terms of the squared modulus of
another trigonometric polynomial. This result has been extended to entire
functions by Boas and Kac. They showed that an entire function of exponential type b, bounded and nonnegative on the real axis, can be expressed
as the square modulus of an entire function of exponential type b/2 with
its zeros in the closed upper half plane. More precisely, let $E_\Omega^1 = \{h|h$ in
$L_1(R)$ with $\hat{h}(\omega) = 0$ for $|\omega| \geq \Omega\}$. Let $E_\Omega^2 = \{h|h \in L_2(R), \hat{h}(\omega) = 0$ for
$|\omega| > \Omega\}$. Let $E_\Omega^{1,+}$ denote the subset of real-valued, nonegative functions.

THEOREM (Boas-Kac) 3.41. Any function $h(x)$ in $E_\Omega^{1,+}$ has the rep-
resentation $h(x) = |f(x)|^2$ where $f(x)$ belongs to $E_{\Omega/2}^2$ where $f(z)$ has its
zeros in $Im(z) \geq 0$ and is determined up to a linear phase factor. Moreover,
$h(z)$ and $f(z)$ are of class A – i.e., their zeros satisfy $\sum |Im(1/z_k)| < \infty$.

3.12 Monochromatic Radiation

Let $V(r,t)$ denote a complex scalar wavefunction. The mutual coherence
function is defined by

$$\Gamma(r_1, r_2, \tau) = < V(r_1, t + \tau)V^*(r_2, t) > \qquad (3.79)$$

where $<>$ represents the time average. The complex degree of coherence
is given by

$$\gamma(r_1, r_2, \tau) = \Gamma(r_1, r_2, \tau)/\sqrt{\Gamma(r_1, r_1, 0)\Gamma(r_2, r_2, 0)} \qquad (3.80)$$

THEOREM 3.42. If the complex degree of coherence is unimodular,
$|\gamma| = 1$, then γ is of the form

$$\gamma(r_1, r_2, \tau) = e^{i(\alpha(r_1) - \alpha(r_2) - 2\pi\nu_0\tau)} \qquad (3.81)$$

where $\alpha(r)$ is a real function of r and ν_o is a positive constant.

The proof is based on the fact that γ is nonnegative definite which implies that there are no zeros for $\gamma(r, r, \tau)$ in the lower half plane. In particular if $\gamma(r, r, \tau)$ is unimodular, then one shows that the self coherence function is of the form

$$\gamma(r, r, \tau) = e^{-2\pi i \nu_o \tau} \tag{3.82}$$

where ν_o is a positive constant.

3.13   Coherence Theory

For quasi-monochromatic light (i.e., light for which the effective spectral range $\Delta\nu$ is small compared to the mean frequency $\bar{\nu}$), the visibility $\mathcal{V}(t)$ of the interference fringes, obtained by dividing a beam into two partial beams and letting them interfere after a time delay t has been introduced between them, is relate to the degree of coherence $\gamma(t)$ by the formula

$$\mathcal{V}(t) = A|\gamma(t)|. \tag{3.83}$$

Here $A = 2\sqrt{I_1 I_2}/(I_1 + I_2)$ where I_i, i = 1,2, are the time averaged intensities of the two partial beams. The complex degree of coherence $\gamma(t)$ is given by

$$\gamma(t) = \int_0^\infty g(\nu) e^{-2\pi i \nu t} d\nu \tag{3.84}$$

where $g(\nu)$ is the spectral energy density. The spectral density $g(\nu)$ is real and nonnegative. We set $\gamma(t) = |\gamma(t)| e^{i\phi(t)}$.

The complete determination of the energy spectrum of light from measurement of the visibility curve, which was first explicity stated by Rayleigh in 1892, is a phase retrieval problem due to the relation 3.84.

When the spectrum is known to be symmetric about its mean frequency $\bar{\nu}$, it is possible to reconstruct $\gamma(t)$ from $|\gamma|$ and $\bar{\nu}$ – viz., the phase is essentially given by $e^{-i\bar{\nu}t}$.

Since the integral defining $\gamma(z)$ contains only positive frequencies, $\gamma(z)$ can be analytically continued as a regular function of z in the lower half complex plane. Viz.,

THEOREM (Titchmarsh) 3.43. If a function $F(z)$ fulfills one of the following conditions, it fulfills them all:

(1) the inverse Fourier transform $g(v)$ of $F(z)$ vanishes for $v < 0$;

(2) $F(z)$ is for almost all x the limit as $y \to 0^+$ of an analytic function $F(z)$ regular for $y > 0$ and is square integrable over any line in the upper half plane which is parallel to the real axis

$$\int_{-\infty}^\infty |F(z)|^2 dx < c \tag{3.85}$$

for $y > 0$;

(3) ReF and ImF are related by the Hilbert transform or dispersion relations

$$ReF(x') = \frac{1}{\pi}PV \int_{-\infty}^{\infty} \frac{ImF(x)dx}{x - x'}; \tag{3.86}$$

(4) similarly

$$ImF(x') = \frac{-1}{\pi}PV \int_{-\infty}^{\infty} \frac{ReF(x)dx}{x - x'}. \tag{3.87}$$

Finally, the fact that $g(\nu)$ is real then implies that $\gamma(-z) = \gamma^*(z^*)$. If we assume $\gamma(t)$ satisfies the Paley-Wiener condition,

$$\int_{-\infty}^{\infty} \frac{|ln|\gamma(t)||dt}{1 + t^2} < \infty \tag{3.88}$$

then it is possible to relate $ln|\gamma(t)|$ and $\phi(t)$ by a dispersion relation

$$\phi(z) = \phi_m(z) + \phi_B(z) - 2\pi\nu_o z. \tag{3.89}$$

Here

$$\phi_m = \frac{2zPV}{\pi} \int_0^{\infty} \frac{ln|\gamma(z')|dz'}{z'^2 - t^2},$$

where PV denotes the Cauchy principal value,

$$\phi_B(z) = \sum_n arg(\frac{z - z_n}{z - z_n^*}), \tag{3.90}$$

and ν_o is real and nonnegative. The sum extends over all zeros of $\gamma(z)$ in the upper half plane. This implies that $\gamma(t)$ has the representation

$$\gamma(t) = |\gamma(t)|e^{i\phi_m(t)} \prod_n \frac{z - z_n}{z - z_n^*} \tag{3.91}$$

where $B(z) = \prod_n \frac{z - z_n}{z - z_n^*}$ is called the Blaschke product. The function $\phi_m(t)$ represents the minimal phase or Hilbert phase.

The phase retrieval problem for coherency theory is reduced to the determination of the zeros z_n under these conditions. Note that since $g(\nu)$ is real, if z_n is a zero, then so is $-z_n^*$; so the distribution of zeros is symmetrical with respect to the imaginary axis. Since g is nonnegative, there cannot be any zeros on the imaginary axis.

3.14   Blackbody Radiation

The real coherence tensor $\mathcal{E}_{ij}^{(r)}(r,t)$ where $r = r_2 - r_1$ and $t = t_2 - t_1$ relating the correlations at times t_1, t_2 of the electric field vectors at two points r_1, r_2 in a volume filled with black body radiation has been evaluated by Bourret. In the case of temporal coherence, i.e., $r = 0$, because of symmetry the off-diagonal elements of the temporal coherence tensor vanish. The diagonal terms are all equal to each other. Viz.,

$$\mathcal{E}(t') = \mathcal{E}_{ii}(0,t) = \frac{8(kT)^4}{3\pi(\hbar c)^3} \int_0^\infty \frac{x^3 e^{-x(1+it')} dx}{1 - e^{-x}} \tag{3.92}$$

where $t' = kTt/\hbar$, k = Boltzmann's constant, $\hbar = h/2\pi$, h = Planck's constant, T = absolute temperature, and c = speed of light. Recall that the Riemann zeta function is given by

$$\zeta(s,a) = \sum_{n=0}^\infty \frac{1}{(a+n)^s}. \tag{3.93}$$

Thus, $\mathcal{E}(t') = \frac{16(kT)^4 \zeta(4, 1+it')}{\pi(\hbar c)^3}$. Normalizing by $\mathcal{E}(0)$ and using the fact that $\zeta(4,1) = \pi^4/90$ we have

$$\gamma(t') = 90\zeta(4, 1 + it')/\pi^4. \tag{3.94}$$

The coherence function of black body radiation thus has no zeros in the lower half plane, so knowledge of $|\gamma(t)|$ suffices to reconstruct the spectrum.

3.15   Holography and Zeros

Consider the addition of an offset reference beam to a function, viz., $f(x) + A\delta(x - c)$. The Fourier transform is then $\hat{f}(x) + Ae^{icz}$. If f has support $[0,b]$ and $c > b$ and if we select $|A| > |f(x)|$ for all x, then since e^{icz} has no zeros in the upper half plane, by application of Rouche's theorem to the complex extension $F_s(z) = \hat{f}(z) + Ae^{icz}$ we find that $F_s(z)$ has no zeros in the upper half plane. In this case the phase of the $F_s(z)$ is referred to as the Hilbert phase. And dispersion relations have been developed which relate $ln|F_s(z)|$ or $ln|F_s(z)|/z$ and the Hilbert phase. Viz., if $F_s(z)$ has all its zeros on the lower half plane or real axis, then

$$\phi_H(x') = \frac{-x' PV}{\pi} \int_{-\infty}^\infty \frac{ln|F_s(x)| dx}{x(x - x')} + \phi_H(0). \tag{3.95}$$

In the general phase retrieval problem we nominally record the magnitude of the Fourier transfrom without reference points. We turn in the next section to this more general problem of reconstruction from the magnitude of the Fourier transform.

3.16 Akutowicz Representation Theorem

THEOREM (Akutowicz) 3.44. Let $C(a)$ denote the class of all functions fulfilling the following conditions:

(1) ϕ delongs to $L^1 \cap L^2$

(2) $\phi(t)$ vanishes almost everywhere for $t < 0$

(3) $a(x)$ is a fixed function such that $a(x) = |\hat{\phi}|$, where $\hat{\phi}$ is the Fourier transform of ϕ. Then if ϕ_1, ϕ_2 belong to $C(a)$, we have

$$e^{ic_1 + ib_1 x} B_1(x) \hat{\phi}_1(x) = e^{ic_2 + ib_2 x} B_2(x) \hat{\phi}_2(x)$$

where c_1, c_2, b_1, b_2 are real, b_1, b_2 are nonnegative and B_1, B_2 are limits as $y \to 0+$ of certain Blaschke products in the upper half-plane. Here $B_1(x)$ and $B_2(x)$ are holomorphic functions of modulus one.

This result is based on the following facts.

THEOREM 3.45. If $F(z)$ is holomorphic for $y > 0$, $|F(z)| \leq 1$, and

$$lim_{y \to 0} \int \frac{log|F(x+iy)|dy}{1+x^2} = 0 \tag{3.96}$$

then $F(z)$ is of the form

$$F(z) = e^{ic + ibz} B(z) \tag{3.97}$$

where c,b are real, b is nonegative and B is the Blaschke product formed with the zeros of F. Conversely, if B(t) is any Blaschke product in the upper half-plane, then 3.96 holds with F replaced by B.

THEOREM (Paley-Wiener) 3.46. If ϕ belongs to $L^2(R)$ and vanishes on a half-line, then

$$\int_{-\infty}^{\infty} \frac{log|\hat{\phi}(x)|}{1+x^2} dx < \infty. \tag{3.98}$$

THEOREM (Kryloff) 3.47. Let $\Phi(z)$ be a holomorphic function for $y > 0$ for which $\int_{-\infty}^{\infty} |\Phi(x+iy)|^2 dx \leq M < \infty$ where M is independent of y, then

$$\Phi(z) = e^{ic + ibz} B(z) D(z) G(z) \tag{3.99}$$

where c is real, b is nonnegative real, B(z) is the convergent Blaschke product formed of zeros of Φ

$$D(z) = e^{\frac{1}{\pi i} \int_{-\infty}^{\infty} \frac{(1+ts)log|\Phi(t)|dt}{(t-s)(1+t^2)}} \tag{3.100}$$

where $\Phi(z) = \lim_{y \to 0} \Phi(x + iy)$ almost everywhere

$$\int_{-\infty}^{\infty} \frac{|\log|\Phi(x)||dx}{1 + x^2} < \infty \tag{3.101}$$

$$\int |\Phi(x)|^2 dx < \infty$$

$$G(z) = e^{i \int_{-\infty}^{\infty} \frac{1+tz}{t-z} dE(t)}$$

where E(t) is real bounded increasing function with $E'(t) = 0$ almost everywhere. Conversely, every function of the form 3.99 is holomorphic for $y > 0$ and satisfies $\int_{-\infty}^{\infty} |\Phi(x + iy)|^2 dx \leq M < \infty$.

To prove 3.44 we set

$$\Phi(z) = \frac{1}{\sqrt{2\pi}} \int_0^{\infty} e^{i(x+iy)} \phi(t) dt. \tag{3.102}$$

Then by the Parseval identity and Schwarz inequality

$$\int_{-\infty}^{\infty} |\Phi(x + iy)|^2 dy = \frac{1}{2\pi} \int_0^{\infty} e^{ixt - y(t)} \phi(t) dt = \tag{3.103}$$

$$\frac{1}{2\pi} \int_{-\infty}^{\infty} |\int (2/\pi)^{1/2} \frac{y}{y^2 + (x + \lambda} \bar{\hat{\phi}} d\lambda|^2 dx \leq$$

$$\frac{1}{2\pi} \int dx \int \frac{y}{y^2 + (x + \lambda)^2} d\lambda \int \frac{y}{y^2 + (x + \lambda)^2} |a(\lambda)|^2 d\lambda =$$

$$\frac{1}{\pi} \int |a(\lambda)|^2 d\lambda.$$

By Kryloff's theorem we have

$$\Phi(z) = e^{ic+ibz} B(z) D(z) G(z). \tag{3.104}$$

Here $|D(z)| = e^{\frac{1}{\pi} \int_{-\infty}^{\infty} \frac{y \log(a(t)) dt}{y^2 + (t-x)^2}}$ because $\Phi(x + iy) \to \hat{\phi}(x)$ as $y \to 0$. By the Poisson integral representation, if $\phi(t)/(1 + t^2)$ belongs to $L^1(R)$ and is continuous at x_o, then

$$u(x, y) = \frac{1}{\pi} \int_{-\infty}^{\infty} \frac{y \phi(t) dt}{y^2 + (x - t)^2} \tag{3.105}$$

has $\lim u(x_o, y) = \phi(x_o)$. Thus we see that $|D(z)| \to a(x)$ as $y \to 0$. We leave it to the reader to show that $G(z) = 1$, i.e., $E(t) = $ constant. Furthermore, $D(x) = a(x) e^{iH(x)}$ where

$$H(x) = \frac{1}{\pi} \int \frac{(1 + tx) \log a(t) dt}{(t - x)(1 + t^2)}. \tag{3.106}$$

The Blaschke product $B_k(z)$ has as zeros the set of zeros z_{nk} of the function

$$\Phi_k(z) = \frac{1}{\sqrt{2\pi}} \int_0^\infty e^{izt} \phi_k(t) dt \qquad (3.107)$$

and is given by the product

$$B_k(z) = [(z-i)/(z+i)]^{m_k} \prod \frac{|z_{nk}-i|}{z_{nk}-i} \frac{|z_{nk}+i|}{z_{nk}+i} \frac{z-z_{nk}}{z-\bar{z}_{nk}} \qquad (3.108)$$

where m_k is a nonnegative integer. This product converges if and only if $\sum Imz_{nk}/(1+|z_{nk}|^2) < \infty$.

The following example shows that if ϕ is nonnegative and ϕ is in $\mathcal{C}(a)$, then ϕ is not uniquely determined. Viz., set $b = \alpha + i\beta, \beta > 0, \alpha \neq 0, |4\beta/\alpha| < 1$. Let $\phi_1(t) = e^{-\beta t}, t \geq 0$, and $\phi_1(t) = 0$ for $t < 0$. Set $a(x) = |1/(ix - \beta)|$. Define ϕ_2 by

$$\hat{\phi}_2 = \frac{(x+\bar{b})(x-b)}{(x+b)(x-\bar{b})} \hat{\phi}_1(x) \qquad (3.109)$$

Clearly ϕ_1, ϕ_2 belong to $\mathcal{C}(a)$. We leave it to the reader to show that $\phi_2(t) = e^{-\beta t}((1 - 4\frac{\beta}{\alpha}sin\alpha t) + 4\frac{\beta^2}{\alpha^2}(1 - cos\alpha t))$ for $t \geq 0$ (so $\phi_2(t) \geq 0$).

Let $\mathcal{C}_o(a)$ denote the functions which satisfy (1),(3), (4) of theorem 3.44 and for which (2) is replaced by

(2') ϕ is zero outside a finite interval.

For ϕ in $\mathcal{C}_o(a)$ suppose (-T,T) is the smallest interval where ϕ is nonzero. Set

$$\Phi(z) = \frac{1}{\sqrt{2\pi}} \int_{-T}^{T} e^{-zt} \phi(t) dt. \qquad (3.110)$$

Then by theorem 3.47 we can write

$$\Phi(z) = e^{i\alpha + \beta z} B(z) D(z) \qquad (3.111)$$

where B(z) is the Blaschke product in the upper half-plane and D(z) is given by 3.100. Since $\Phi(z)$ is entire of exponential type we have

$$\Phi(z) = e^{c(z)} P(z) \qquad (3.112)$$

where $c(z) = c_o + c_1 z$ for complex constants c_o, c_1 and P(z) is the canonical product

$$P(z) = \prod_n (1 - z/z_n) e^{z/z_n} \qquad$$

A direct check then verifies:

COROLLARY 3.48. Any two functions ϕ_1, ϕ_2 belonging to $\mathcal{C}_o(a)$ are related by an equation of the form

$$\hat{\phi}_1(x) = e^{i\alpha + i\beta x} B_2^*(x) B_1(x) \hat{\phi}_1(x). \qquad (3.113)$$

EXAMPLE 3.49. Let

$$\phi_1(t) = \begin{cases} e^{-t} & -1 \le t \le 1 \\ 0 & \text{otherwise} \end{cases} \tag{3.114}$$

$$\phi_2(t) = \begin{cases} e^{t} & -1 \le t \le 1 \\ 0 & \text{otherwise} \end{cases}$$

$$\hat{\phi}_1(x) = (2/\pi)^{1/2} \frac{1}{ix-1} sinh(ix - 1)$$

$$\hat{\phi}_2(x) = (2/\pi)^{1/2} \frac{1}{ix+1} sinh(ix + 1).$$

So $|\hat{\phi}_1(x)| = |\hat{\phi}_2(x)|$ and the zeros associated with ϕ_1 are $z_n = -i + n\pi, n = \pm 1, \pm 2, ...$ and those with ϕ_2 are $z_n = i + n\pi, n = \pm 1, \pm 2, ..$

3.17 Walther's Theorem and Zero Flipping

Let f be a complex-valued measurable function on the real line R. The support of f is the smallest closed subset of R outside of which f is zero almost everywhere. The inteval of support $I(f) = [a_f, b_f]$ of f is the smallest closed interval containing $Supp(f) = S(f)$. We set $c_f = (a_f + b_f)/2$.

Let $L_o^2(R)$ denote the space of all complex-valued square integrable functions on R with compact support. For f, g in $L_o^2(R)$ the convolution of f and g is given by

$$f \star g(x) = \int_{-\infty}^{\infty} f(y)g(x-y)dy. \tag{3.115}$$

We note that $L_\varrho^2(R)$ is closed under convolution.

Set $\tilde{f}(x) = \bar{f}(-x)$. Then the autocorrelation of f is defined by

$$Auto(f) = f \star \tilde{f}. \tag{3.116}$$

The Fourier-Laplace transform of f is defined by

$$F(w) = \int_{-\infty}^{\infty} f(x)e^{-iwx}dx \tag{3.117}$$

for w in C. We define F^* by $F^*(x) = \bar{F}(\bar{w})$. The properties of the Fourier-Laplace transform are:

THEOREM 3.50. (1) If F is the transform of f, then F^* is the transform of $\tilde{f}$

(2) If G is the transform of g, then the transform of $f \star g$ is FG.

COROLLARY 3.51. The transform of Auto(f) is FF^*.

For a,c real numbers the functions $f(x)$, $f(x+a)e^{ic}$, and $\tilde{f}(a+c)e^{ic}$ all have the same Fourier transform modulus. If these are the only functions with that Fourier modulus we say $f(x)$ is unique and its Fourier transform is unambiguous. If $g = f(x+a)e^{ic}$ or $g(x) = \tilde{f}(x+a)e^{ic}$, we say f and g are equivalent and write $f \sim g$.

THEOREM (Paley-Wiener) 3.52. The entire function $F(w)$ is of exponential type b and its restriction to the real axis belongs to $L^2(R)$ if and only if it is given by

$$F(w) = \int_a^b f(x)e^{iwx}dx \qquad (3.118)$$

where f(x) belongs to $L^2(a,b)$ with $0 \le |a| \le b < \infty$.

For any entire function F we define η_F by:
(1) if w is not a zero of F, then $\eta_F(w) = 0$
(2) if w is a non-real zero of F, then $\eta_F(w)$ is its order
(3) if w is a real zero of F, then $\eta_F(w)$ is one-half its order.
We set $Z(F) = \{w|\eta_F(w) > 0$ for w not real$\}$ and $D(F) = \{w|Im(w) \ge 0, \eta_{FF^*}(w) > 0, w \ne 0\}$, $\sigma_F(w) = \eta_F(w) - \eta_F(\bar{w})$, and $W(F) = \{w|\sigma_F(w) > 0\}$.

Since FF^* is entire of exponential type $D(F)$ is countable. Set $D(F) = \{w_n, n = 1, 2, ...\}$ such that $\{|w_n|\}$ is nondecreasing sequence. The following factorization theorem is due to Titchmarsh:

THEOREM (Titchmarsh) 3.53. Let $m = 2\eta_F(0)$. If m = 0, then

$$F(w) = F(0)e^{-ic_f w} \prod_{n=1}^{\infty}(1 - w/w_n)^{\eta_F(w_n)}(1 - w/\bar{w}_n)^{\eta_F(\bar{w}_n)} \qquad (3.119)$$

and the infinite product is conditionally convergent. For $m > 0$

$$F(w) = \frac{1}{m!}F^{(m)}(0)w^m e^{-ic_f w} \prod_{n=1}^{\infty}(1 - w/w_n)^{\eta_F(w_n)}(1 - w/\bar{w}_n)^{\eta_F(\bar{w}_n)}. \qquad (3.120)$$

Proof. For the case m = 0 we have $F(0) \ne 0$ and since f belongs to $L_o^2(R)$, f is integrable and the product follows from Titchmarsh. For the case $m > 0$ set $g_1(x) = \int_{a_f}^t f(t)dt$. So $c_{g_1} = c_f$. Since $\int_{a_f}^{b_f}|g_1(x)|dx \le (b_f - a_f)\int_{a_f}^{b_f}|f(t)|dt$ we see that g_1 is integrable. We also have

$$F(w) = \int_{a_f}^{b_f}f(x)e^{-ixw}dx = g_1(x)e^{-iwx}|_{a_f}^{b_f} + iw\int_{a_f}^{b_f}g(x)e^{-iwx}dx. \quad (3.121)$$

Thus $F(w) = iwG_1(w)$. And $\eta_{G_1}(w) = \eta_F(w)$ for all nonzero w. Repeating this process m times gives

$$F(w) = (iw)^m G_m(w) \tag{3.122}$$

with $G_m(0) \neq 0$. By Titchmarsh's theorem for G_m we obtain

$$F(w) = Aw^m e^{-ic_J w} \prod (1 - w/w_n)^{\eta_F(w_n)} (1 - w/\bar{w}_n)^{\eta_F(\bar{w}_n)} \tag{3.123}$$

We leave it to the reader to check that $A = F^{(m)}(0)/m!$.

THEOREM (Titchmarsh) 3.54. Let $\gamma_F(w) = \eta_F(w)$ for w in $C\backslash R$ and let $\gamma_F(w) = 2\eta_F(w)$ for w in R. The the Hadamard factorization is

$$F(w) = Aw^m e^{aw} \prod_{n=1}^{\infty} (1 - w/w_n)^{\gamma_F(w_n)} e^{w/w_n \gamma_F(w_n)} \tag{3.124}$$

where the infinite product is absolutlely convergent and the series

$$\sum \gamma_F(w_n) Im(1/w_n) \tag{3.125}$$

is absolutely convergent.

By the standard conversion it follows that the product

$$\prod_{n=1}^{\infty} e^{iw\gamma_F(w_n) Im(1/w_n)} \tag{3.126}$$

is absolutely convergent. Thus the product

$$\prod_{n=1}^{\infty} (1 - w/w_n)^{\gamma_F(w_n)} e^{w\gamma_F(w_n) Re(1/w_n)} \tag{3.127}$$

is absolutely convergent.

Since $F^*(w)$ is also entire with zeros $\bar{w}_n$ of orders $\gamma_{F^*}(\bar{w}_n) = \gamma_F(w_n)$, a similar product can be developed with $\bar{w}_n$ replacing w_n. We have then

COROLLARY 3.55. Let β be an integer-valued function on C where $0 \leq \beta(w) \leq \eta_F(w)$ for all w in $C\backslash R$ and $0 \leq \beta(w) \leq 2\eta_F(w)$ on R; then the infinite product

$$\prod_{n=1}^{\infty} [(1 - w/\bar{w}_n)/(1 - w/w_n)]^{\beta(w_n)} \tag{3.128}$$

is absolutely convergent.

This result provides the zero flipping theorem of Walther:

THEOREM (Walther) 3.56. Let f belong to $L_o^2(R)$ and let $\{w_n\}$ be the distinct non-real zeros of the Fourier-Laplace transform F of f. Let β be as above and set

$$G(w) = F(w) \prod_{n=1}^{N} [(1 - w/\bar{w}_n)/(1 - w/w_n)]^{\beta(w_n)} \tag{3.129}$$

where N can equal ∞ and let g be the inverse transform of G. Then

$$|G(w)| = |F(w)| \tag{3.130}$$

for all w on R and $I(g) = I(f)$.

Proof. Set $\phi_m(w) = \prod_{n=1}^{m} [(1 - w/\bar{w}_n)/(1 - w/w_n)]^{\beta(w_n)}$. By the last corollary $\phi_\infty(w)$ is absolutely convergent and since $|\phi(u)| = 1$ for all real u we have $|G(u)| = |F(u)|$. Since F(u) is square integrable so are G and g. To show that $I(g) = I(f)$ we first check that $I(g) \subset I(f)$. Let $G_m(w) = \phi_m(w)F(w)$ and let g_m be the inverse transform of G_m. Let $\Gamma(w) = 1/(w_o - w)$ for an arbitrary non-real zero w_o of F. The inverse transform of Γ is $\gamma(x) = -\frac{1}{2}[sgn(v_o) + sgn(x)]e^{iw_o x}$ where $w_o = u_o + iv_o$ and

$$sgn(t) = \begin{cases} 1 & t > 0 \\ 0 & t = 0 \\ -1 & t < 0 \end{cases} \tag{3.131}$$

Set $H(w) = \frac{1 - w/\bar{w}_o}{1 - w/w_o} F(w)$. Then the inverse transform of H is

$$h(x) = \frac{w_o}{\bar{w}_o} f(x) - \frac{w_o}{\bar{w}_o}(w_o - \bar{w}_o)(\gamma \star f)(x). \tag{3.132}$$

The reader can check that $(\gamma \star f)(x) = 0$ for $x < a_f$ and $x > b_f$. Thus we have $I(\gamma \star f) \subset I(f)$ and it follows that $I(h) \subset I(f)$.

Since $|G(u) - g_m(u)|^2 \to 0$ as $m \to \infty$ for all real u, we have by the Lebesgue dominated convergence theorem that $\int_{-\infty}^{\infty} |G(u) - G_m(u)|^2 du \to 0$ as $m \to \infty$. Thus g_m converges to g in the L^2 norm. Since $I(g_m) \subset I(f)$ for m = 1,2,... it follows that $I(g) \subset I(f)$.

The reader is left to show that $I(f) \subset I(g)$.

For functions with disconnected support the following result generalizes Walther's theorem. Let I_n, n=1,...N, be a sequence of disjoint intervals which satisfy

$$(I_n - I_m) \cap (I_j - I_k) = 0 \tag{3.133}$$

for $j \neq k$ and ordered pair (n,m) not equal ordered pair (j,k). Set $A = \cup_{n=1}^{N} I_n$ and assume f,g in $L_o^2(R)$ have $S(f) \subset A$ and $S(g) \subset A$. Let $f_n(x) = f(x)$ for x in I_n and 0 otherwise and similarly for g. Let $B = \cap Z(F_n)$. We assume f_n are not identically zero for all n.

THEOREM (Crimmins-Fienup) 3.57. For these conditions, $|F(u)| = |G(u)|$ for all real u if and only if there is a real number θ and an integer-valued function α on C, $0 \leq \alpha(z) \leq min\eta_F(z)$ for $z \in C\backslash R$ and $0 \leq \alpha(z) < 2min_n \eta_{F_n}(z)$ for real u such that if

$$\phi(w) = e^{i\theta} \prod_{z \in B} [(1 - w/\bar{z}_n)/(1 - w/z_n)]^{\alpha(z)} \qquad (3.134)$$

then for $N \neq 2$

(1) $G_n(w) = e^{i(c_{f_n} - c_g)}\phi(w)F_n(w)$ and $c_{f_n} - c_{g_n} = c_f - c_g$ and for $N = 2$ either (1) holds or

(2) $G_n(w) = e^{i(c_f - c_g)}\phi(w)F_n(w)$ and $c_{f_n} - c_{g_n} = c_f + c + g$ for n = 1,2.

Proof. In the case $N = 1$ with $|F(u)| = |G(u)|$ we have $f \star \tilde{f} = g \star \tilde{g}, FF^* = GG^*, D(F) = D(G), \eta_F(0) = \eta_G(0)$. Using the product expansions and the results above we find that

$$G(w) = Ae^{idw}\psi(w)F(w) \qquad (3.135)$$

where $d = c_f - c_g, A = G^{(m)}(0)/F^{(m)}(0)$ and $\psi(w) = \prod_{z \in B}()^{\alpha(z)}$ with

$$\eta = \eta_F$$

$$\xi = \eta_G$$

$$b = 2\eta(0) = 2\xi(0)$$

$$\eta(w) - \xi(w) = \xi(\bar{w}) - \eta(\bar{w})$$

$$\beta(w) = \eta(w) - \xi(w)$$

$$\alpha(w) = \begin{cases} \beta(w) & \beta(w) > 0 \\ 0 & \text{otherwise} \end{cases}$$

So $0 \leq \alpha(w) \leq \eta(w)$ and $\alpha(w) = 0$ for w not in B.

We leave the more general case to the reader.

COROLLARY 3.58. If $B = 0$, then $|F(u)| = |G(u)|$ for all real u if and only if $f \sim g$.

COROLLARY (Akutowicz-Hofstetter-Walther) 3.59. Taking $N = 1$ and $B = Z(F)$ we have $|F(u)| = G(u)|$ for all real u if and only if there exists a real number θ and an integer valued function α on $Z(F)$, $0 \leq \alpha(z) \leq \eta_F(z)$ for all z in $Z(F)$ such that

$$G(w) = e^{i\theta}e^{i(c_f - c_g)w}F(w) \prod_{z \in Z(F)} ()^{\alpha(z)}. \qquad (3.136)$$

Flipping all the nonreal zeros of F yields:

$$F^*(w) = \bar{F}^{(m)}(0)/F^{(m)}(0)e^{2ic_f w}F(w)\prod_{z\in Z(F)}B(z)^{\eta_F(z)} =$$

$$\bar{F}^{(m)}(0)/F^{(m)}(0)e^{2ic_f w}F(w)\prod_{z\in W(F)}B(z)^{\sigma_F(z)}$$

where $m = 2\eta_F(0)$. Since F^* is the transform of $\tilde{f}$ and $\tilde{f}$ is equivalent to f we have:

THEOREM 3.60. The following are equivalent:
(1) F(u) is unambiguous
(2) F(w) has no non-real zeros or one non-real zero of order one
(3) $F(w)F^*(w)$ has no non-real zeros or two non-real zeros of order one.

EXAMPLE 3.61. The function $F(u) = sin(\pi u)/(\pi u)$ has only real zeros so F(u) is unambiguous.

THEOREM (Titshmarsh) 3.62. For f in $L_o^2(R)$ if $I(f) = [a_f, b_f]$, then $I(auto(f)) = [-d, d]$ where $d = b_f - a_f$.
Proof. Clearly $d \le b_f - a_f$. To show $d \ge b_f - a_f$, suppose $d < b_f - a_f$. Set g = Auto(f). So $g(y) = \int_{a_f}^{b_f - y}\bar{f}(x)f(x + y)dx = 0$ for almost all $y \ge c$. Let $\phi(x) = \bar{f}(x + a_f), \psi(x) = f(b_f - x), t = x - a_f, s = b_f - a_f - y, k = b_f - a_f - d$. Then $k > 0$ and $\int_0^s \phi(t)\psi(s - t)dt = 0$ for almost all $s \le k$. By a result of Titchmarsh it follows that there is a λ and μ such that $\phi(t) = 0$ almost everywhere in $(0, \lambda)$ and $\psi(t) = 0$ almost everywhere in $(0, \mu)$ and $\lambda + \mu \ge k$; this implies $\lambda > 0$ or $\mu > 0$ or both. However, $\lambda > 0$ implies f(x) = 0 a.e. in $(a_f, a_f + \lambda)$ which contradicts $S(f) = [a_f, b_f]$. Similarly for $\mu > 0$.

Crimmins and Fienup have shown that if there are no non-real zeros common to all the transforms of the segments I_n, then the function with these parts is unique among functions with the same support. And for non-negative functions having supports satisfying certain disconnection conditions, then if the transforms of the segments of the function have no non-real zeros in common, the function is unique among nonnegative functions.

3.18 Summary of Phase Retrieval in One Dimension

Let C be the class of all functions g in $L^2(R)$ for which
(1) $|g(x)| = m(x) \ne 0$ for all x in R

(2) $g = \mathcal{F}G$ where the support of G is contained in a bounded interval A of R. Then any two functions g,h in $\mathcal{C}$ are related by $h(z) = e^{i(a+bz)}B(z)g(z)$ where $B(z) = \prod_{l=1}^{\infty}(z - z_l^*)/(z - z_l)$ where z_l forms some subset of zeros of $g(z)$. The term $B_n(z) = (z - z_n)/(z - z_n^*)$ is the Blaschke factor and it has the effect of replacing a zero of $g(z)$ by its complex conjugate. On the real axis $|B(x)| = 1$.

THEOREM (Levin) 3.63. A necessary and sufficient condition for the infinite product B(z) to converge is that

$$\sum_{l=1}^{\infty} |Imz_l|/(1 + |z_l|^2) < \infty. \qquad (3.137)$$

A sufficient condition for the convergence of the infinite sum is that G(u) have only a finite number of jump discontinuities over A.

Combining these results we have the statement of Barakat and Newsam:

THEOREM (Barakat-Newsam) 3.64. Let A be a bounded interval in R. Let g,G be a solution of the 1-D phase retrieval problem. If $m(x) > 0$ and G(u) has only finite jump discontinuities and is nonzero in the neighborhood of the end points of its support A, then all other solution pairs h,H are given by

$$h(z) = e^{ia}B(z)g(z)$$
$$H(u) = (\mathcal{F}^{-1}h)(u)$$

where B(z) is any finite or infinite product of Blaschke factors and a is in R.

COROLLARY 3.65. g(z) admits the Hadamard factorization

$$g(z) = |g(0)|e^{i(a+bz)}\prod_{l=1}^{\infty}(1 - z/z_l)$$

which can be written as

$$g(z) = |g(0)|e^{i(a+bz)}\prod_{l\in\Lambda}(1 - z/z_l)\prod_{N\backslash\Lambda}(1 - z/z_l)$$

where Λ is a subset of the natural numbers. So every solution has the form

$$g(z) = e^{ic}g_1(z)g_2^*(z^*)$$

i.e., all possible solutions are in one to one correspondence with all possible factorizations of g(z).

Chapter 4

Homometric Distributions

4.1 Introduction

Let G be an abelian group. A finite distribution D on G has the form $\sum a(g)\delta(g)$ where $a(g)$ is in C and $a(g) \neq 0$ only finitely many times. The algebra of such distributions under convolution product is isomorphic to the group ring $C[G]$.

DEFINITION 4.1. The reflection D^* of D is the distribution $D^* = \sum \bar{a}(g)\delta(-g)$ where $\bar{a}(g)$ is the complex conjugate of $a(g)$.

Let Γ denote the group of all homomorphisms of R into the circle group $T = \{z \in C \mid |z| = 1\}$.

DEFINITION 4.2. The Fourier transform of $D = \sum a(g)\delta(g)$ in $C[G]$ is defined by $\hat{D}(\gamma) = \sum a(g)\gamma(g)$ for all γ in Γ.

Note that $\hat{D}^*(\gamma) = \bar{\hat{D}}(\gamma)$ for all γ.

THEOREM 4.3. If D_1, D_2 belong to $C[G]$ then the Fourier transform of the convolution gives $\widehat{D_1 \star D_2}(\gamma) = \hat{D}_1(\gamma)\hat{D}_2(\gamma)$.

COROLLARY 4.4. If $D \star D^* = E \star E^*$, then $|\hat{D}(\gamma)| = |\hat{E}(\gamma)|$ for all γ in Γ.

If A belongs to $C[G]$ and $\hat{A}(\gamma) = 0$ for all γ in Γ, then $A = 0$. Thus we have

THEOREM 4.5. For D,E in $C[G]$, then $D \star D^* = E \star E^*$ if and only if $|\hat{D}(\gamma)| = |\hat{E}(\gamma)|$ for all γ in Γ.

47

Patterson in 1939 introduced the concept of a homometric distribution:

DEFINITION 4.6. If D,E are in $C[G]$, then D and E are said to be homometric if $D \star D^* = E \star E^*$.

From the last result we see that D and E are homometric if and only if they have equivalent Fourier transform magnitudes. The study of homometric distributions arose in the study of x-ray crystallography. The question centers on determining D from $|\hat{D}|$ or equivalently the Patterson function $D \star D^*$. Rosenblatt cites the earliest reference to non-uniqueness causing a problem for x-ray crystallography in Pauling and Shappell (1930) who were studing the structure of bixbyite. Bixbyite is a solid solution of Mn_2O_3 and Fe_2O_3.

Let A be a multiset, i.e. a finite set with repetitions, in R^n and let ΔA denote the multiset of all vector differences $x - y$ with x, y in A. We say two multisets U, V in R^n are homometric if $\Delta U = \Delta V$. Clearly for multisets U, V in R^n the multisets

$$U + V = \{u + v | u \in U, v \in V\} \qquad (4.1)$$

$$U - V = \{u - v | u \in U, v \in V\}$$

are homometric. Not all homometric multisets arise this way. However, we have the following results:

THEOREM 4.7. The ring $K[R^n]$ is locally a UFD where the units in $K[R^n]$ are ux^v where v is in R^n and u is a unit of K.

For A in $K[R^n]$, say $A(x) = \sum a(v)x^v$ we set $A(x^{-1}) = \sum \bar{a}(v)x^{-v}$. Then we say $A_1(x)$ and $A_2(x)$ in $K[R^n]$ are homometric if

$$A_1(x)A_1(x^{-1}) = A_2(x)A_2(x^{-1}). \qquad (4.2)$$

Thus if A in R^n is a multiset and we associate with it the element A(x) in $Z[R^n]$ defined by $A(x) = \sum\{x^v | v \in A\}$ then $A(x)A(x^{-1}) = \sum\{x^{v-w} | v, w \in R^n\} = \sum\{x^v | v \in \Delta A\}$. So A_1, A_2 are homometric as sets if and only if $A_1(x)$ and $A_2(x)$ are homometric as ring elements.

THEOREM (Rosenblatt-Seymour) 4.8. Two elements $A_1(x), A_2(x)$ in $K[R^n]$ are homometric if and only if there exist $P(x), Q(x)$ in $K[R^n]$ and c in K of absolute value one such that $A_1(x) = P(x)Q(x)$ and $A_2(x) = cP(x)Q(x^{-1})$.

Proof. Certainly if $P(x), Q(x)$ belong to $K[R^n]$ then $P(x)Q(x)$ and $cP(x)Q(x^{-1})$ are homometric as long as $|c| = 1$. For the converse we use the fact that $K[R^n]$ is locally a UFD. We leave the details to the reader. One may also check that:

THEOREM 4.9. Two elements $A_1(x)$, $A_2(x)$ in $K[Z^n]$ are homometric if and only if there exist $P(x)$, $Q(x)$ in $K[Z^n]$, c in K of absolute value one and v_1, v_2 in A^n such that $A_1(x) = x^{v_1} P(x)Q(x)$ and $A_2(x) = cx^{v_2} P(x)Q(x^{-1})$.

DEFINITION 4.10. $A(x)$ in $K[R^n]$ is called symmetric if there exists v in R^n such that $A(x^{-1}) = x^v A(x)$. And A is called semisymmetric if $A(x^{-1}) = cx^v A(x)$ where c in K is of absolute value one. $A(x)$ is called reconstructible if where $B(x)$ in $K[R^n]$ is homometric to $A(x)$ we have $B(x) = cx^v A(x)$ or $B(x) = cx^v A(x^{-1})$, for some v in R^n and c in K of absolute value one.

THEOREM 4.11. If $A_1(x)$, $A_2(x)$ in $K[R^n]$ are homometric and both symmetric, then $A_1(x) = \epsilon x^v A_2(x)$ where $\epsilon \in \{1, -1\}$ and $v \in R^n$.

THEOREM (Rosenblatt-Seymour) 4.12. An element $A(x)$ in $K[R^n]$ is reconstuctible if and only if $A(x)$ has at most one prime factor counting multiplicities which is not semi-symmetric.

We refer the reader to the reference for details of the proof.

For any point set A in R^n we set $1_A = \sum \{\delta(x) | x \in A\}$ which is an $\{0, 1\}$ distribution on R^n.

DEFINITION 4.13. Two finite sets A,B in R^n are called homometric if 1_A and 1_B are homometric distributions.

Consider now G to be an abelian group with no elements of finite order. Let K be a ring, $Z \subset K \subset C$, which is closed under conjugation ($\bar{K} = K$), and such that $K[G]$ is a unique factorization domain. (Note that $C[G]$ is not a UFD.) Then any element in $K[Z^n]$ has a unique factorization (up to units) into prime factors relative to $K[Z^n]$. This lead to:

THEOREM (Rosenblatt) 4.14. If D,E in $K[G]$ are homometric, then there exists c_1, c_2 in K , $|c_1| = |c_2| = 1$, g_1, g_2 in G and A,B in K[G] such that $D = c_1 \delta(g_1) \star A \star B$ and $E = c_2 \delta(g_2) \star A \star B^*$.

COROLLARY 4.15. If K is a field and D,E are in K[G] are homometric, then there exists ϵ_1, ϵ_2 in $\{1, -1\}$ and A,B in $K[\frac{1}{2}G]$ such that $D = \epsilon_1 A \star B$ and $E = \epsilon_2 A \star B^*$.

Note that if G = R, then $\frac{1}{2}G = G$.

DEFINITION 4.16. If D is in $K[G]$, then D is said to be uniquely recoverable if whenever E is in $K[G]$ homometric to D, then $D = c\delta(g) \star E$ or $D = c\delta(g) \star E^*$ for some c in K, $|c| = 1$ and g in G.

DEFINITION 4.17. D in $K[G]$ is said to be semisymmetric if there exists a c in K, $|c| = 1$ and $D^* = c\delta(g) \star D$, for some g in G.

THEOREM (Rosenblatt) 4.18. If G is finitely generated, then D in $K[G]$ is uniquely retrievable if and only if at most one of its prime factors in $K[G]$ is not semi-symmetric.

In case G is a finite abelian group, then $K[G]$ is not a UFD. Let K be a field closed under conjugation as before. However, now a different approach is required.

DEFINITION 4.19. A unit U in $K[G]$ is called a spectral unit if $U \star U^* = \delta(0)$.

So U is a spectral unit if and only if U and $\delta(0)$ are homometric.

THEOREM (Rosenblatt) 4.20. If D,E are in $K[G]$, then D,E are homometric if and only if there exists a spectral unit U in $K[G]$ such that $U \star D = E$.

THEOREM (Rosenblatt) 4.21. If D,E in $K[Z^n]$, then D,E are homometric if and only if there exists ϵ_1, ϵ_2 in $\{-1, 1\}$, g_1, g_2 in Z^n such that $D = \epsilon_1 \delta(g_1) \star A \star B$ and $E = \epsilon_2 \delta(g_2) \star A \star B^*$.

Combining these two results we have for the general abelian group:

THEOREM (Rosenblatt) 4.22. Let G be an abelian group and suppose D,E are in $K[G]$ for some conjugation-closed field K in C. Then D and E are homometric if and only if there exists ϵ_1, ϵ_2 in $\{-1, 1\}$ and A,B in $K[\frac{1}{2}G]$ such that $D = \epsilon_1 A \star B$ and $E = \epsilon_2 A \star B^*$.

This factorization theorem does not extend to the non-discrete case. Let $\mathcal{S}_c(R^n)$ denote the set of Schwartz distributions with compact support.

DEFINITION 4.23. The involution distribution D^* is defined by $< D^*, \phi >= \overline{< D, \phi^* >}$ where $\phi^*(x) = \bar{\phi}(-x)$. And the conjugation distribution $\bar{D}$ is defined by $< \bar{D}, \phi >= \overline{< D, \phi >}$.

Using these definitions we have $\hat{D}^* = \bar{\hat{D}}$.

DEFINITION 4.24. If D,E are in $\mathcal{S}_c(R^n)$, then D and E are homometric if $|\hat{D}| = |\hat{E}|$ everywhere on R^n.

Clearly if A,B are in $\mathcal{S}_c(R^n)$ then $D = A \star B$ and $E = A \star B^*$ are homometric.

The Fourier transform $\hat{D}(z)$ of D in $\mathcal{S}_c(R)$ extend to complex function $\hat{D}(z)$ which are entire. The Paley-Wiener theorem then asserts that:

THEOREM (Schwartz-Paley-Wiener) 4.25. For D in $\mathcal{S}_c(R)$ there exists a constant $c > 0$, an integer $N \geq 1$ and a constant B such that

$$|\hat{D}| \leq c(1 + |z|)^N e^{B|Im(z)|} \tag{4.3}$$

for z in C. And any entire function satisfying the estimate (4.3) must be of the form $\hat{D}(z)$ for some D in $\mathcal{S}_c$ with support of D in [-B,B].

Thus we see that for any $\varepsilon > 0, |\hat{D}(z)| \leq ce^{(B+\varepsilon)|z|}$, i.e., the entire function is of order no larger than one. The type of $\hat{D}$ is the smallest B such that supp(D) is in [-B,B].

THEOREM 4.26. If $D \neq 0$ is in $\mathcal{S}_c$ and the support of D is not $\{0\}$, then D is of order one and normal type for its order.

Thus as long as $\hat{D}(z)$ is not a polynomial then $\hat{D}$ is of order one and we can express $\hat{D}$ by the Hadamard factorization theorem. If $\hat{D}$ has a finite number of zeros $z_1, ... z_N$ in C, then

$$\hat{D}(z) = e^{a+bz} z^k \prod_{n=1}^{N} (1 - z/z_n)e^{z/z_n}. \tag{4.4}$$

If there are infinitely many zeros, then

$$\hat{D}(z) = e^{a+bz} z^k \prod_{n=1}^{\infty} (1 - z/z_n)e^{z/z_n}. \tag{4.5}$$

As noted above, by a result of Titchmarsh if D belongs to $L^1(R) \cap \mathcal{S}_c(R)$, then there must be infinitely many zeros.

THEOREM 4.27. If $D \neq 0$ is in $\mathcal{S}_c$ and supp(D) is not a point, then $\hat{D}$ has infinitely many zeros.
Proof. $\hat{D}$ is of order one. If $\hat{D}$ has finitely many zeros, then by Hadamard $\hat{D}(z) = e^{a+bz} P(z)$ where P is a polynomial. And by (4.3)

$$e^a e^{xRe(b)} |P(x)| \leq c(1 + |x|)^N.$$

Letting x tend to $\pm\infty$, we have either $Re(b) = 0$, $lim_{x\to\infty} P(x) = 0$ or $lim_{x\to-\infty} P(x) = 0$. Since $D \neq 0$, neither limit is zero. Thus $Re(b) = 0$ and $Supp(D) = \{-Im(b)\}$.

COROLLARY 4.28. If $D \neq 0$ belongs to $\mathcal{S}_c(R)$ and supp(D) is not a point, then the exponent of convergence s of $\hat{D}$ satisfies $\frac{1}{2} \leq s \leq 1$.

THEOREM 4.29. For D,E in $\mathcal{S}_c(R)$ the following are equivalent:
(1) D and E are homometric
(2) $\hat{D}(z)\overline{\hat{D}(\bar{z})} = \hat{E}(z)\overline{\hat{E}(\bar{z})}$ for all z in C
(3) $D \star D^* = E \star E^*$.
Proof. The definition of (1) says that (2) holds for real z. Because products in (2) are analytic, (2) is equivalent to (1). The equivalence of (1) and (3) is obvious.

THEOREM (Rosenblatt) 4.30. D and E in $\mathcal{S}_c(R)$ are homometric if and only if there exist entire functions f_1, f_2 such that $\hat{D}(z) = f_1(z)f_2(z)$ and $\hat{E}(z) = f_1(z)\bar{f}_2(\bar{z})$.

A method of constructing all possible E with $|\hat{E}| = |\hat{D}|$ is given by Walther's theorem. Viz., let $\hat{D}(z)$ have the form (4.5). Then to flip a zero of $\hat{D}$ means to change $\hat{D}$ to

$$\hat{E}_1 = (1 - z/\bar{z}_n)e^{z/\bar{z}_n}(1 - z/z_n)^{-1}e^{-z/z_n}D(z).$$

One checks that zero flipping $z \to (z - \bar{a})/(z - a)$ leaves $\mathcal{S}_c$ invariant:

THEOREM 4.31. If D belongs to $\mathcal{S}_c$ with $\hat{D}(a) = 0$, then there exists E in $\mathcal{S}_c$ such that $E(z) = (z - \bar{a})D(z)/(z - a)$. And if D belongs to $C_c^\infty(R)$, then E belongs to $C_c^\infty(R)$.

Proof. This follows directly from checking the Paley-Wiener conditions

$$|\hat{D}(z)| \le c(1 + |z|^N)e^{B|Imz|}.$$

And for $|z| \ge 2|a| + 1$, we have

$$|(z - \bar{a})\hat{D}(z)/(z - a)| \le 2c(1 + |z|)^{N-1}|z - a|e^{B|Imz|} \le c_1(1 + |z|)^N e^{B|Imz|}.$$

COROLLARY 4.32. If D belongs to $\mathcal{S}_c$ and has support in [-B,B], then E in $\mathcal{S}_c$ also has support in [-B,B].

Walther's theorem in the present case states:

THEOREM (Walther) 4.33. If D,E are in $\mathcal{S}_c$, then D,E are homometric if and only if with $\hat{D}$ of form 4.5 there exists c in C with $|c| = 1$, d in R and a choice of $z_n^{e(n)} \in \{z_n, \bar{z}_n\}$ such that for all z in C

$$E(z) = ce^{idz}e^{a+bz}z^k \prod_{n=1}^\infty (1 - z/z_n)e^{z/z_n^{e(n)}}.$$

For a finite sequence of zeros flips this theorem follows from theorem 4.31. However, for an infinite sequence of zero flippings it is more subtle.

THEOREM (Rosenblatt) 4.34. Let D be in $\mathcal{S}_c$ and let $\{w_n\}$ be a choice of w_n in $\{z_n, \bar{z}_n\}$ of nonzero zeros of D. Then there exists E in $\mathcal{S}_c$ such that

$$E = e^{itz}e^{a+bz}z^k \prod(1 - z/w_n)e^{z/w_n}$$

and if $supp(D) \subset [-B, B]$ then $supp(E) \subset [-B, B]$; and if D belongs to $C_c^\infty(R)$ then E belongs to $C_c^\infty(R)$.

Proof. For E in S_c let D_n be the sequence in S_c defined by

$$\hat{D}_{n+1}(z) = (1 - z/w_n)e^{z/w_n}(1 - z/z_n)^{-1}e^{-z/z_n}\hat{D}_n(z)$$

with $\hat{D}_0(z) = \hat{D}(z)$. Letting

$$g(z) = \prod_n (1 - z/w_n)e^{z/w_n}(1 - z/z_n)^{-1}e^{-z/z_n}\hat{D}(z)$$

then $|g(x)| \le c(1 + |x|)^N$. So there is a tempered distribution $\hat{E}$ with $\hat{E}(x) = g(x)$. One checks that $\hat{D}_n$ converges to $\hat{E}$ uniformly on compact sets in R and so $\lim_n \hat{D}_n = \hat{E}$ on S_c.

The factorization theorem 4.22 does not hold for C_c^∞. Viz.,

THEOREM (Rosenblatt) 4.35. There exists D,E in C_c^∞ which are homometric such that for no A,B in S_c does $D = A \star B$ and $E = A \star B^*$.

Proof. This follows essentially by noting how to construct D in $C_c^\infty(R)$ such that $\hat{D}(z) = 0$ and $\hat{D}(\bar{z}) = 0$ cannot happen simultaneously.

Rosenblatt has also generalized the results of Titchmarsh on the distribution of zeros:

THEOREM (Rosenblatt-Titchmarsh) 4.36. Let D belong to S_c and let $supp(D) \subset [a, b]$. Let $z_n = r_n e^{i\theta_n}$ denote the zeros of $\hat{D}$. Then
(1) $\lim_{r \to \infty} n_D(r)\pi/(r(b - a)) = 1$
(2) $\sum_{n=1}^\infty 1/r_n = \infty$
(3) $\sum_{n=1}^\infty |sin(\theta_n)|/r_n < \infty$
(4) $\sum_{n=1}^\infty cos(\theta_n)/r_n$ converges.
Proof. The proof proceeds by taking the function h where $h \ge 0$,

$$\int_R h(x)dx = 1$$

where $h \in C_c^\infty(R)$ and $supp(h) \subset [-1, 1]$. Let $h_j(x) = jh(jx), j = 1, 2, 3....$ And define $D_j = h_j \star D$. Then $supp(D_j) \subset supp(h_j) + supp(D) = [-1/j, 1/j] + supp(D)$. The functions D_j converge pointwise on $C_c^\infty(R)$ to D. Let $r \ge 0$ and set $\hat{D}_j = \hat{h}_j \hat{D}$. Then

$$n_{h_j}(r) + n_D(r) = n_{D_j}(r).$$

But $h_j, D_j \in C_c^\infty(R)$ and Titchmarsh's results apply to these functions. In particular, $\lim_{r \to \infty} n_{h_j}(r)\pi/r(1/j) = 1$ and $\lim_{r \to \infty} n_{D_j}(r)/r(b_j - a_j) = 1$ where $[a_j, b_j]$ is the support of D_j. One can show that

$$limsup_{r \to \infty} n_D(r)\pi/r(b - a)$$

is no larger than

$$(1 - (1/j)/(b_j - a_j))(b - a + \epsilon_j)/(b - a)$$

where $|b_j - b| \leq \epsilon_j$, $|a_j - a| \leq \epsilon_j$ and $lim_{j \to \infty} \epsilon_j = 0$. Letting $j \to \infty$ shows that $limsup_{r \to \infty} n_D(r)\pi/r(b - a) \leq 1$. A similar argument holds for liminf. Thus $lim_{r \to \infty} n_D(r)\pi/r(b - a) = 1$ as asserted.

These facts now imply that $n_D(r_N) = n - r_N(b - a)/\pi$. Hence, the series $\sum_{N=1}^{\infty} 1/r_N$ can be compared to $\sum_{N=1}^{\infty} 1/N$ which diverges. This proves (2). Let $\{z_n^1\}$ be the zeros of $\hat{h}_1$. And let $\{z_n^2\}$ denote the zeros of $\hat{D}_1$. Set $z_n^j = r_n^j e^{i\theta_n^j}$. By Titchmarsh both series $\sum_{n=1}^{\infty} sin(\theta_n^j)/r_n^j, j = 1, 2$ converge absolutely. But z_n^2 is an interweaving of z_n^1 and the sequence of zeros $\{z_n\}$ of $\hat{D}$. Hence, $\sum_{n=1}^{\infty} sin(\theta_n)/r_n$ converges absolutely also. This proves (3). We leave the proof of (4) to the reader.

Chapter 5

Analytic Signals and Signal Recovery from Zero Crossings

5.1 Introduction

Analytic signals are described by

$$m(t) = h(t) + i\hat{h}(t) = f(t)e^{i\phi(t)} \tag{5.1}$$

where h(t) is a real signal and $\hat{h}(t) = H(h(t))$ is the Hilbert transform $\hat{h}(t) = \frac{1}{\pi}PV\int_{-\infty}^{\infty}\frac{h(s)}{t-s}ds$. The spectrum of h(t) is given by $S(f) = \mathcal{F}(h(t)) = \int_{-\infty}^{\infty} h(t)e^{-2\pi i f t}dt$ and it is related to the spectrum of $\hat{h}$ by $\mathcal{F}(\hat{h}(t)) = -isgn(f)S(f)$. The spectrum of m(t) is seen to be

$$F(f) = \begin{cases} 2S(f) & \text{for } f > 0 \\ S(0) & f=0 \\ 0 & f < 0. \end{cases} \tag{5.2}$$

5.2 Amplitude Modulation

A signal with amplitude modulation is conventionally defined as $s_{am}(t) = s(t)cos(\omega_o t)$ for $\omega_o \geq \pi W_o$ where ω_o is the carrier frequency and s(t) is a real modulating signal of bandwidth $\pm\omega_o/2$ hertz. In the analytic form we have $m_{am}(t) = s_{am}(t) + i\hat{s}_{am}(t) = s(t)e^{i\omega_o t}$.

55

5.3   Single Sideband Modulation

Single sideband modulation was proposed by Bedrossian to be given by
signals of the form

$$s(t) = Re(e^{i(\omega_c t + x(t) + i\hat{x}(t))}) = e^{-\hat{x}(t)} cos(\omega_c t + x(t)) \qquad (5.3)$$

where x(t) is the baseband signal, $\hat{x}(t)$ is the Hilbert transform of x(t) and
ω_c is the carrier frequency. This modulated signal, due to the relationship
between the amplitude modulation $e^{-\hat{x}(t)}$ and the phase modulation, has
no spectrum in the interval $(-\omega_c, \omega_c)$; in other words, the amplitude modulation removes the lower sideband. The complex form $m(t) = e^{-\hat{x} + i\omega_c t + ix}$
is analytic for all $\omega_c > 0$.

5.4   Exponential Single Sideband Modulation

Logan proposed a generalization of the exponential single sideband to the
analytic modulation signals of the form:

$$s(t) = Re(f(z(t))e^{i\omega_c t} \qquad (5.4)$$

for $\omega_c > 0$ where f(x) is an analytic function of the analytic signal z(t) = x(t)
+ iy(t). It is assumed that x(t) is bounded and band-limited with spectral
support $[-\Omega, \Omega]$. The Fourier transform of s(t) vanishes over $(-\omega_c, \omega_c)$ and
hence s(t) is called a "single sideband" signal.

   EXAMPLE 5.1. In the case f(z) = z, then s(t) is single sideband amplitude modulation. In this case the Fourier transform of s(t) vanishes outside
$[\omega_c, \omega_c + \Omega]$ and $[-\omega_c - \Omega, -\omega_c]$. So the bandwidth required for transmitting
s(t) is $2\Omega + \varepsilon$ for some small positive ε.

   Voelcker proposed a method for demodulating conventional single sideband signals via envelope detection. For single sideband amplitude modulation we assume the Fourier transform of z(t) vanishes outside $[0, \Omega]$ and
the received signal is $\sigma_R(t) = Re(e^{i\omega_c t} z(t))$. The envelope of the received
signal is $|z(t)|$.

   If z(t) is zero-free in the upper half plane (e.g. if $Re z(t) = x(t) > 0$)
and if $lim_{u \to \infty} z(t + iu) = 1$, then the imaginary part of log z(t) can be
determined from the real part; hence z(t) can be recovered from $|z(t)|$. In
particular if x(t) = 1+g(t) where $g(t) > -1$ and g(t) belongs to $L^p (1 \leq p < \infty)$
then $Log|z(t)|$ will belong to L^p and will have a Hilbert transform.

   EXAMPLE 5.2.   Exponential modulation $w = f(z) = e^z$ offers the
unique advantage of eliminating the need for preliminary in-phase and

quadrature detection of the received signal. In the present case $z = \log w$. If Log denotes the real part of the logarithm,

$$x(t) = Log|w(t)| \tag{5.5}$$

$$y(t) = arg(w(t))$$

then the transmitted signal is $s(t) = Re e^{i\omega_c t + z(t)}$ with envelope $|s(t)| = e^{x(t)}$. The instantaneous phase of s(t) is $A(t) = \omega_c t + y(t)$. Under perfect transmission conditions

$$x(t) = Log|s(t)| \tag{5.6}$$

and using an ideal discriminator or FM detector we have

$$y'(t) = \frac{dA(t)}{dt} - \omega_c. \tag{5.7}$$

As Logan has shown, if w_α is a filtered version of w(t), then if $w_\alpha(t)$ is zero free on the upper half plane, we see that $x_\alpha = Log|w_\alpha(t)|$ and x(t) agree over $(-\alpha, \alpha)$, the signal x(t) may be recovered by taking the Log of the envelope of the received signal and then filtering with an ideal low-pass filter on $[-\Omega, \Omega]$ which vanishes outside $[-\alpha, \alpha]$.

EXERCISE (Powers) 5.3. Consider the analytic signal $\psi(t) = s(t) + i\hat{s}(t)$ where $\hat{s}(t)$ is the Hilbert transform of s(t). Write $\psi(t)$ as $\psi(t) = \alpha(t)e^{i\phi(t)}$ and set $S(\omega)$ denote the spectrum of s(t), $S(\omega) = \int_{-\infty}^{\infty} s(t)e^{-i\omega t} dt$. Show that $\int_{-\infty}^{\infty} \alpha^2(t)e^{-i\omega t} dt = \frac{2}{\pi}\int_0^{\infty} S^*(\lambda)S(\lambda + \omega)d\lambda$ for $\omega > 0$. Assume s(t) is band-limited to bandwidth W. Show that the bandwidth of the square of the envelope of this single sideband signal is equal to the spectral width of th signal. Show that if $\alpha^2(t)$ is a nonnegative square integrable function for which $\int_{-\infty}^{\infty} \frac{|log\alpha^2(t)|dt}{1+t^2} < \infty$ and if the Fourier transform of $\alpha^2(t)$ vanishes for $|\omega| > 2\pi W$, then

$$\psi(t) = e^{\frac{i}{\pi}\int_{-\infty}^{\infty} \frac{log\alpha(u)}{t-u} du}$$

has a Fourier transform that vanishes outside $0 < \omega < 2\pi W$. Thus the transmitted signal has the same bandwidth as the original signal and distortionless reception of this signal is achieved through a square law detector.

5.5 Distortionless Demodulation

Zakai introduced the class $B(W, \delta)$ of functions in $L^2(R)$ which satisfy $f(t) = \int_{-\infty}^{\infty} f(s)k(t - s)ds$ for almost all t where $k(t) = \frac{2}{\pi\delta t^2}sin((W + \delta/2)t)sin(\delta t/2)$. The Fourier transform of k is

$$K(\lambda) = \begin{cases} 1 & |\lambda| \leq W \\ 1 - \frac{|\lambda| - W}{\delta} & W < |\lambda| < W + \delta \\ 0 & \text{otherwise.} \end{cases} \tag{5.8}$$

Thus $B(W, \delta)$ contains the standard class of band-limited functions CB(W). Masry and Cambanis characterized $B(W, \delta)$ by the following result:

THEOREM 5.4. Function f is in $B(W, \delta)$ if and only if f(t) = f(0)+tg(t) where g is in CB(W).

Thus $B(W, \delta)$ is actually independent of δ; we denote it by B(W). Functions in $B(W)$ but not in $CB(W)$ include periodic band-limited functions, e.g., $f(t) = \sum_{-N}^{N} a(n)e^{in\omega_p t}$ for $N\omega_p \leq W$, and the function $f(t) = \int_0^t \frac{sin(Ws)ds}{s}$.

As with CB(W), a function in B(W) has an analytic extension f(z) which is entire and satisfies $|f(z)| \leq K(1 + |z|)e^{W|Imz|}$.

DEFINITION 5.5. For f in B(W) with f(t) = f(0) + tg(t), g in CB(W), then the Hilbert transform of f is defined by $\tilde{f} = t\tilde{g}(t)$, where $\tilde{g}$ is the ordinary Hilbert transform of g.

EXAMPLE 5.6. Let f in B(W) be the periodic function

$$f(t) = \sum_{n=0}^{N} a(n)cos(n\omega_p t)$$

where $N\omega_p \leq W$. Then

$$g(t) = (f(t) - f(0))/t = -\sum a(n)(1 - cos(n\omega_p t))/t,$$

$$\tilde{g}(t) = \sum_{n=1}^{N} a(n)sin(n\omega_p t)/t$$

and

$$\tilde{f}(t) = t\tilde{g}(t) = \sum_{n=1}^{N} a(n)sin(n\omega_p t),$$

as is the custom.

The associated analytic signal to f(t) is then taken to be $f_a(t) = f(t) + i\tilde{f}(t)$. Consider the single sideband angle modulated signal

$$v(t) = e^{-\tilde{f}(t)}cos(\omega_c t + f(t))$$

where f(t) is the modulating signal, $\tilde{f}(t)$ is its Hilbert transform and ω_c is the carrier frequency. In the case where f(t) is band-limited with spectrum in [-W,W], Bernard gave an iterative algorithm to reconstruct f(t) from a band-limited version of v(t). Let H denote the bandpass filter and let $f_o(t)$ denote the output of this filter followed by a standard angle detector and finally by a low pass filter L. Under what conditions on f,H, and L will $f_o(t) = f(t)$, i.e., under what conditions is there distortionless domodulation?

Let H be characterized by

$$H = \begin{cases} 0 & |\lambda| \text{ not in } [\omega_c - \delta_1, \omega_c + W_1 + \delta_1] \\ 1 & |\lambda| \in [\omega_c, \omega_c + W_1] \end{cases} \tag{5.9}$$

where $W_1 \geq W, \delta_1 > 0$. And let

$$L = \begin{cases} 0 & |\lambda| > W + \delta \\ 1 & \lambda \in [-W, W]. \end{cases} \tag{5.10}$$

THEOREM (Masry) 5.7. Let f be in B(W) and let H,L be as above. Assume

(1) $|f_o(t)| < M$ for all t
(2) $\|h\|_{L_1}(e^{2M} - 2M - 1) < 1$
(3) $W_1 > W + \delta$, then the demodulation of v is distortionless, i.e., $f_o(t) = f(t)$ for all t.

The proof contains the following steps. Let $h(t) = \frac{1}{2\pi} \int_{-\infty}^{\infty} H(\lambda)e^{it\lambda} d\lambda$. Then the filtered analytic modulated signal is $v_o(t) = (v \star h)(t)$. Introducing the complex phase term q(t) we have

$$v_o(t) = Re(e^{i\omega_c t} e^{i(f_o(t) + q(t))}) \tag{5.11}$$

and the output r(t) of a standard angle detector is

$$r(t) = Re(f_a(t) + q(t)) = f_a(t) + Re(q(t)). \tag{5.12}$$

The final output of the demodulator is $f_o(t) = (r \star l)(t)$ where l is the Fourier transform of L. Since f is in $B(W)$, it follows that $f \star l = f$. And it can be shown that $(q \star l)(t) = 0$ for all t.

5.6   Modulation Reconstruction from Zeros

The modulation reconstruction problem is given by: can the zeros of a real modulating signal (or the signal itself) be recovered from the zeros of m(t)? The problem of determining complex zeros has led to an alternate problem of first converting the complex zeros into first order real zeros and then asking for modulation reconstruction from real-zeros.

DEFINITION 5.8. B-functions whose zeros are wholly, first-order real are called RZ functions.

RZ functions are described unambiguously by their zero crossings. One approach has been to find a mapping which converts general B-functions into RZ functions in order to determine the B-functions by the zeros of their RZ form. This approach has used the following theorem due to Hermite and Biehler:

THEOREM 5.9. A B-function $m(z) = f(z) + ig(z)$ with f and g real on the real line and with indicator diagram $[-ib,-ia]$ has no zeros in the lower half plane $Im(z) < 0$ and satisfies $a + b > 0$ if and only if $f(z) = p(z)f_1(z), g(z) = p(z)g_1(z)$ where all the zeros of p, f_1, g_1 are real the zeros of f_1, g_1 are single and interlaced.

Let f(z) be a band-limited function on $[-a,a]$, bounded and real on the real line and possessing a Hilbert transform $\hat{f}$. Then $m(z) = f(z) + i\hat{f}(t)$ is band-limited on $[0,a]$. By the last theorem f(z) is an RZ function if m(z) has no zeros in the closed lower half plane.

Let $m(z) = \sum_{k=0}^{n} c(k)e^{ik\Omega z}$. Then is is easy to see that

$$m_c(z) = \sum_{k=0}^{n} c(k)w^k + Aw^n$$

for $A > M = max|m(x)|$ has no zeros in the closed lower half plane. Thus $f_c(z) = Re(m(z)) + Acos(n\Omega z)$ is an RZ function.

Consider the analytic signal $m(z) = \sum_{k=0}^{n} c(k)e^{ik\Omega z}$ with zeros z_n. If all the zeros are in the lower half plane, then $|\alpha(n)| < 1$ for all n, where $\alpha(n) = e^{-i\Omega z_n}$. $m(z)$ is then said to be of minimum phase and the phase and log magnitude are Hilbert transforms.

If the zeros occur only in conjugate pairs and if n is even then $m(z) = s(z)e^{i\frac{n\Omega t}{2} + \theta_0}$ where s(z) is a real signal of period T and band-limited to $\pm nW/2$. And s can be factored as $s(t) = s_{cz}(t)s_{rz}(t)$ where

$$s_{cz} = \prod_{i=1}^{n_c} [cosh\Omega|\sigma_n| - cos\Omega(t - \tau_n)];$$

here $z_n = \tau_n + i|\sigma_n|$ and

$$s_{rz}(t) = \prod_{k,l=1}^{n_r} [cos\Omega(\tau_k - \eta)/2 - cos\Omega(t - \tau_k + \eta)].$$

5.7   Recovery from Zero Crossings

Consider a bandpass signal s(t) as in the last section. Then the real zeros portion may be recovered from sgn(s(t)). The following algorithm has been proposed by Voelcker and Requicha to be used to reconstruct band-limited functions from zero crossings:

(1) let $s_o(t) = B\,sgn(s(t))$

(2) set $s_{n+1}(t) = s_n(t) - c(B\Phi(s_n(t)) - s_o(t))$.

Here B is the bandpass operator and c is a positive constant. We return to this type of iterative algorithm in later chapters.

The study of reconstruction of signals from zero crossings rather than real zeros has been pursued by Logan.

THEOREM (Logan) 5.10. If $h(t) = Re(f(t)e^{i\omega_o t})$ is a real bandpass function where f(t) is band-limited and bounded, then h(t) is uniquely determined by its real zero-crossings within a multiplier if h(t) has no zeros in common with its Hilbert transform other than real simple zeros and if the bandwidth is less than an octave (i.e., $a > b/2$).

Logan's development goes as follows. Let $B_p(\lambda), 1 \le p \le \infty$, denote the L^p functions f(t) on the real line which extend as entire functions $f(\tau), \tau = t + iu$, of exponential type λ. Let $B_p(\alpha, \beta)$ denote the class of functions of the form $h(t) = f_1(t)e^{i\mu t} + f_2(t)e^{-i\mu t}$ where f_1, f_2 belong to $B_p(\lambda/2)$, $\lambda = \beta - \alpha, 0 < \alpha < \beta < \infty$ and $\mu = (\alpha + \beta)/2$. Functions in $B_p(\alpha, \beta)$ have Hilbert transforms

$$\hat{h}(t) = -if_1(t)e^{i\mu t} + if_2(t)e^{-i\mu t}.$$

If h(t) is real-valued, then $f_2(t) = \bar{f}_1(t)$; i.e., $h(t) = Re(f(t)e^{i\mu t})$ where f(t) = p(t)+iq(t), p,q in $B_\infty(\lambda/2)$, p,q real valued.

DEFINITION 5.11. A complex or real number z is said to be a free zero of h if $g(t) = \frac{at+b}{t-z}h(t)$ belongs to $B_\infty(\alpha, \beta)$ whenever h does.

THEOREM 5.12. A function h in $B_\infty(\alpha, \beta)$ has a free zero z if and only if $h(z) = 0$ and $\hat{h}(z) = 0$.

COROLLARY 5.13. If h is in $B_\infty(\alpha, \beta)$ and $h(z) = \hat{h}(z) = 0$, then g(t), given as above, belongs to $B_\infty(\alpha, \beta)$ and has Hilbert transform

$$\hat{g}(t) = \frac{at+b}{t-z}\hat{h}(t) \tag{5.13}$$

which is also in $B_\infty(\alpha, \beta)$.

COROLLARY 5.14. A real valued function h(t) of the form above has a free zero z if and only if $f(z) = f(\bar{z}) = 0$. And if f(z) is zero free in either the upper half plane or the lower half plane, then h has no free zeros.

THEOREM 5.15. If h_1 is a real-valued function in $B_\infty(\alpha, \beta)$ having a complex free zero or a multiple real free zero then there is a function h_2 in $B_\infty(\alpha, \beta)$ such that $sgn(h_1(t)) = sgn(h_2(t))$ and $h_2(t) \ne Ah_1(t)$ for a constant A and all t.

DEFINITION 5.16. Let $Z(\alpha, \beta)$ denote the set of all real-valued functions h(t) of the form $h(t) = Re(f(t)e^{i\mu t})$ where $0 < \alpha < \beta < \infty, \mu = (\alpha + \beta)/2$, f(t) belongs to $B_\infty(\lambda/2), \lambda = \beta - \alpha$, and h(t) has no pair of complex conjugate zeros and no real zeros that are not simple.

THEOREM 5.17. Let h_1, h_2 belong to $Z(\alpha, \beta)$. Then $sgn(h_1(t)) = sgn(h_2(t))$ for all t implies $h_1(t) = Ah_2(t)$ provided $\sigma(T)$, the number of sign changes of $h_1(t)$ in (0,T), satisfies

$$lim_{T \to \infty} sup\sigma(T)/T > \frac{\beta - \alpha}{\pi}. \tag{5.14}$$

We note that 5.14 is always satisfied if $h_1 \neq 0$ and $\alpha > \beta/2$.

We noted above that a band-limited function with a sufficiently large sinusoidal component at the upper edge of the band is an RZ function. Functions of the form

$$h(t) = Re(1 + x(t) - i\hat{x}(t))e^{i\beta t} \tag{5.15}$$

where $|x(t)| < 1$ and $x, \hat{x}$ are real and belong to $B_\infty(\lambda), 0 \leq \lambda < \beta$ are called full carrier, lower sideband signals. By the remarks above, such functions are RZ functions. Logan has generalized this fact to:

THEOREM (Logan) 5.18. Let f be a nonnull bounded band-limited function whose spectrum is confined to $[-\lambda, \lambda], 0 \leq \lambda < \infty$, but to no smaller interval – i.e., $e^{i\mu t}f(t)$ does not belong to $B_\infty(\lambda)$ for any non-zero μ. Assume f(z) is zero free in the closed lower half plane: $Im(z) \leq 0$. Then the zeros of h(t) given by $h(t, \mu) = Re(f(t)e^{i\mu t})$ are real and simple provided $\mu > 0$ or provided $\mu \geq 0$ if $f \neq$ constant.

THEOREM 5.19. If f satisfies the last theorem, then the functions $h_1(t, \mu) = Ref(t)e^{i\mu t}$ and $h_2(t, \mu) = Imf(t)e^{i\mu t}$ have all real, simple interlacing zeros for $\mu > 0$.

COROLLARY 5.20. A full carrier, lower sideband signal and its Hilbert transform have only real simple interlacing zeros.

The problem of actually recovering functions in $Z(\alpha, \beta)$ from their zero crossings appears to be difficult. In principle a full carrier lower sideband signal can be recovered by forming an infinite product having simple zeros z_k at the points of sign change: $h(z) = h(0) \prod_{k=1}^{\infty}(1 - z/z_k)$ where we have assumed $h(0) \neq 0$. The product converges conditionally if the zeros are ordered such that $|z_{k+1}| > |z_k|$. This is not a very practical approach.

Let S(b,c) denote the set of signals of the form $s(t) = g(t) + \cos(ct)$ where g(t) is real valued with spectrum confined to [-b,b], $0 < b < c < \infty$

and such that

$$(-1)^k s(k\pi/c) > 0 \qquad (5.16)$$

for $k = 0, \pm 1, \pm 2, \ldots$ This last condition implies s(t) has only real simple zeros. Alternatively we might assume $|g(t)| < 1$ for all t. The zeros $\{z_k\}$ of s(t) satisfy $\frac{k\pi}{c} < z_k < \frac{(k+1)\pi}{c}$ for $k = 0, \pm 1, \pm 2, \ldots$ i.e., they are interlaced with the zeros of sin(ct). Let J(t) be the jump function which increases by π at each zero t_k; let $J(0) = 0$. The function h(t) = J(t) -ct is called the fundamental function. The linear term -ct just offsets the growth of J(t) so that $-\pi < h(t) < \pi$ for all t. It can be shown that h(t) is a high-pass function, having no spectrum in the interval $(-\lambda, \lambda)$, where $\lambda = c - b$. Since h(t) is a bounded, high-pass function, a good estimate $\hat{h}_T(t)$ for the Hilbert transform $\hat{h}(t)$ can be made from knowledge of h on (t-T,t+T) – i.e., from the zeros of this interval.

The derivative function $h'(z) = \pi \sum \delta(z - z_k) - c$ can also be shown to be a high-pass distribution with no spectrum in $(-\lambda, \lambda)$. In the case $c > \mu$ we can think of s(t) as a lower sideband signal with full carrier cos(ct). The parameter $\lambda = c - b$ is the width of the guard band between the top frequency of g(t) and the frequency of cos(ct). It is called the gap frequency.

Logan has developed a reconstruction algorithm for functions s in S(b,c) determined only by the zeros of s in the interval (t-T,t+T).

THEOREM (Logan) 5.21. For s in S(b,c) the function

$$s_T(t) = \frac{1}{2} sgn(s(t)) e^{\hat{h}_T(t)} \qquad (5.17)$$

satisfies

$$|s(t) - s_T(t)| \leq 2e^{-\lambda T}/(1 - e^{-\lambda T})^2 |s(t)| \qquad (5.18)$$

for all t. Here $\hat{h}_T(t) = \sum L_T(t - z_k) + \frac{c}{\lambda}\mu(\lambda T)$ where

$$L_T(t) = \begin{cases} 0 & |t| > T \\ -\int_t^T \frac{f(x)}{x} dx & 0 < t < T \end{cases} \qquad (5.19)$$

$$f(T) = \frac{T sin\lambda \sqrt{T^2 - t^2}}{sinh(\lambda T)\sqrt{T^2 - t^2}}$$

$$\mu(x) = x I_o(x)/sinh(x) - \frac{2x}{sinh(x)\pi} \int_0^{\pi/2} e^{-x sin\theta} d\theta \sim \sqrt{2x/\pi}$$

as $x \to \infty$.

For the case $\lambda T = \pi/2$, $L_T(t) \sim log|t/T|$.

Logan has extended the signals designed for recovery after clipping to the case s(t) = cos(ct) + g(t) where g(t) is bounded real-valued function band-limited to $[-b, b], 0 \leq b < c$, and such that s(z) has only real zeros z_k with multiplicities m_k. For this class of functions Logan shows that:

(1) every interval of length $> 2\pi/\lambda$ contains at least one zero of s(t)

(2) if $N_T(x)$ denotes the number of zeros of s(z) counted according to multiplicity in the closed interval [x,x+T], then

$$N_T(x) \leq \frac{cT}{\pi} + \frac{2c}{\lambda} \qquad (5.20)$$

for all x with equality if and only if $\frac{2c}{\lambda} = m$, m an integer and $s(t) = \frac{(-1)^n}{2}\{2cos(\lambda t/2) + n\pi/m\}^m$, n an integer and x,x+T are zeros

(3) $|s(t)| \leq 2^{(2c/\lambda)-1}$.

5.8   A Zero-Crossing Spectrum Analyzer

A conventional digital spectrum analyzer may consist of an antialiasing filter to band limit the signal, an A/D converter, and an FFT. However, as we see in this section, a spectrum analyzer can be built in a somewhat simpler fashion based on a zero-crossing detector.

Let s(t) be a periodic band-limited signal observed over the interval $-T/2 \leq t < T/2$. Let W be the single sided bandwidth in hertz. We represent s(t) as

$$s(t) = \sum_{m=-M}^{M} c(m)e^{im\omega_o t}$$

where $\omega_o = 2\pi/T, c(-m) = c(m)^*$ and $M = 2\pi W/\omega_o = WT$. Let $z_k = e^{i\omega_o(t_k+is_k)}$. Then s(t) can be represented by the 2M degree polynomial $s'(z) = \sum_{-M}^{M} c(m)z^m$; the zero crossings of s(t) correspond to the real zeros of $s'(z)$.

We note that s(t) has 2M = 2WT zeros in period T. So a zero count will give 2WT. If $s'(z)$ has only real zeros, then s(t) could be reconstructed by

$$s(t) = 2^{2M}|c_M| \prod_{k=1}^{2M} sin(\omega_o(t - t_k)/2). \qquad (5.21)$$

The Fourier coefficients can be realized also in this case since

$$s'(z)/c(M) = z^M + \frac{c(M-1)}{c(M)}z^{M-1} + ... + \frac{c(-M)}{c(M)}z \qquad (5.22)$$

$$= z^{-M} \prod(z - z_k)$$

$$= z^M \prod_{k=1}^{2M}(z - e^{-i\omega_o t_k}).$$

The discrete Fourier transform of length $N = 2M+1$ of $s(n\Delta)$ is given by

$$S(k) = \sum_{n=-M}^{M} s(n\Delta)e^{-i2\pi kn/N} \qquad (5.23)$$

$k = -M,...M$, $N = 2M + 1 = 2WT + 1 = T/\Delta$. Then the $S(k)$ are related to the $c(k)$ by

$$S(k) = Nc(k) \qquad (5.24)$$

$k=-M,...,M$. Since the zero crossings in this case give $c(k)$, this also determines the DFT spectrum.

To convert the complex zeros of $s(t)$ we form

$$x(t) = s(t) + A\cos 2\pi(W + \frac{1}{T}) \qquad (5.25)$$

where $A > max_{-T/2 \leq t \leq T/2}|s(t)|$. As we have seen above, $s(t)$ now has real zeros. In terms of its Fourier series

$$x(t) = \sum_{-(M+1)}^{M+1} c(m)e^{im\omega_o t} \qquad (5.26)$$

where $c(M+1) = A/2$. The $2M+2$ zero crossings of

$$x'(z) = c(M + 1)z^{-(M+1)}z^{-(M+1)} \prod_{k=1}^{2(M+1)} (z - e^{i\omega_o t_k}) \qquad (5.27)$$

allow the DFT coefficients to be constructed as outlined above.

Here $x(t)$ has one zero crossing in each subinterval of length $T_1 = 1/2(W + 1/T)$. Note $x(0) > 0$ but $x(T_1) = s(T_1) - A < 0$. So there is at least one zero crossing of $s(t)$ on $[0, T_1]$. But the added cosine term is periodic with period $2T_1$. So there is a minimum of $2(WT+1)$ zeros in the interval $[0,T)$. But $x'(z)$ is of degree $2(WT+1)$ and can have no more than $2(WT+1)$ roots. Thus all the zeros of $x(t)$ are real.

The following recursive algorithm to compute the DFT coefficients has been suggested by Kay and Sudhaker. Let z_i denote the zero crossing. Then define $a_i^{(k)}$, the coefficients of $P_k(z) = \prod_{i=1}^{k}(z - z_i)$, by

$$a_o^{(k)} = 1 \qquad (5.28)$$

$$a_i^{(k+1)} = a_i^{(k)} - z_{k+1}a_{i-1}^k \quad i = 1,...k$$

$$= -z_{k+1}a_k^{(k)} \quad i = k + 1$$

This continues until k = L = 2(M+1) and $P_L(z)$ is the desired polynomial with coefficients a_i which are related to c(M) by $c(M) = a_{M+1-i}$, i=-(M+1),...M+1. Kay and Sudhaker have also shown that for a clock rate of $(2W)2^B$ hertz where B is the number of bits used for quantizing the zero crossing time, the zero crossing spectrum analyzer with this clock rate has an output SNR equivalent to a standard A/D FFT spectrum analyzer with a B-bit A/D.

Chapter 6

Signal Representation by Fourier Phase and Magnitude in One Dimension

6.1 Introduction

Shitz and Zeevi have shown the following result based on Logan's work:

THEOREM 6.1. Let f(t) be a complex finite time signal whose Fourier transform F has no complex conjugate zeros. The signal f(t) is uniquely determined (within a positive multiplicative constant) by its Fourier phase $\theta_f(s)$. If F(s) has no conjugate zeros at all (neither real nor complex), then f(t) is uniquely defined (within a real constant) by $tan(\theta_f(s))$.

Proof. Set $y(t) = .5(x(t - T_f) + x^*(t - T_f))$. Since $y(t) = y^*(-t)$, the Fourier transform Y(s) of y(t) is real:

$$Y(s) = .5(X^*(s)e^{isT_f} + X(s)e^{-isT_f}) \qquad (6.1)$$

$$= ReX(s)e^{-isT_f}$$

$$= R_f(s)cos(T_f - \theta_f(s)).$$

Both F(s) and Y(s) are EFETs. The Fourier transform of F(-s) is zero outside $[0, \lambda]$ where $\lambda = T$. And z_o is a common zero of Y(s) and $\hat{Y}(s)$ if $F(z_o) = F(z_o^*) = 0$. If $T_f > T$, then the one octave bandwidth requirement of Logan is satisfied. Thus by Logan's results if there are no free zeros except simple real zeros, then F(s), y(t) and hence x(t) are uniquely determined within a multiplicative constant by real zero crossings of Y(s). These are uniquely determined by $\theta_f(s)$.

Consider now the discrete signal $x(t) = \sum x(n)e^{-ist_n}$ $0 \leq t_n < T$. The Fourier transform is $F(s) = \sum x(n)e^{-ist_n}$. If $\sum |x(n)| < \infty$, then $|F(s)|$ is bounded and we can apply Logan's theory. If we set $z = e^{is}$, then is $z_o = e^{is_o}$, $s_o = w_o = iu_o$ is a zero of F(s), then so is $z_1 = e^{is_o^*} = e^{u_o}e^{iw_o}$. Clearly $z_1 = z_o^{*-1}$. Since by theorem 5.18 simple real free zeros are permitted, so $u_o \neq 0$. In terms of the z-plane this means simple zeros on the unit circle are permitted. Specifying $\theta_x(\omega)$ implies more than specifying the real zeros of $Y(\omega)$, since $\theta_x(\omega)$ also determines the sign of $Y(\omega)$. Thus, if $\theta_x(\omega)$ is specified and Logan's conditions are satisfied then $Y(\omega)$ is determined within a positive constant. In summary:

THEOREM (Hayes-Lim-Oppenheim) 6.2. Let x(n) and y(n) be two finite length real sequences whose z-transforms have no zeros in reciprocal conjugate pairs or on the unit circle. If $\theta_x(\omega) = \theta_y(\omega)$ for all ω, then $x(n) = \beta y(n)$ for some constant β. If $tan(\theta_x(\omega)) = tan(\theta_y(\omega))$ for all ω, then $x(n) = \beta y(n)$ for some real constant β. Here $\theta_x(\omega)$ and $\theta_y(\omega)$ are the phases of the Fourier sequences x(n) and y(n).

This theorem also follows from the following result of Poggio et al.

THEOREM 6.3. Let h(t) be a real periodic bandpass function

$$h(t) = \sum_{n=-N}^{N} c(n)e^{int} \qquad (6.2)$$

$$c(n) = c(-n)^*.$$

Then h(t) is uniquely determined by its real zeros within a multiplicative constant provided h(t) and $\hat{h}(t)$, where

$$\hat{h}(t) = \sum_{n=-N}^{N} \hat{c}(n)e^{int} \qquad (6.3)$$

$$\hat{c}(n) = -isgn(n)c(n),$$

have no zeros in common except real simple zeros and c(n) = 0 for $|n| \leq N/2$ (or (N+1)/2 for N odd).

Let d(n) be a complex sequence which is zero outside [0,N-1]. The discrete Fourier transform is given by

$$D(w) = \sum_{n=0}^{N-1} d(n)e^{-inw} = R_D(w)e^{i\theta_D(w)}. \qquad (6.4)$$

Set v = w+iu. Define the bandpass in time sequence

$$\alpha(n) = .5(d(n - M) + d^*(-n - M)) \tag{6.5}$$

for $M > 0$ with Fourier transform

$$A(v) = \sum_{n=-(M+N-1)}^{M+N-1} \alpha(n)e^{-inv}. \tag{6.6}$$

A(w) is real since $d(n) = d^*(-n)$. The Hilbert transform $\hat{A}(v)$ of A(v) is given by $\hat{A}(v) = \sum \hat{\alpha}(n)e^{-inv}$ where $\hat{\alpha}(n) = isgn(n)\alpha(n)$. The one octave bandwidth requirement is satisfied by choosing $M > (N - 1)$. One finds that if v_o is a free zero of A(-v) then both the zero and its conjugate v_o^* are zeros of $N_A(v)$ where $A(v) = M_A(v) + N_A(v)$ where $M_A(v) = .5D^*(v)e^{ivM}$, and $N_A(v) = .5D(v)e^{-ivM}$.

Let $A_1(v)$ be the Fourier sequence of a time bandpass sequence $\alpha 1_n$ which is zero for $|n| \in [M, M + (N - 1)]$ and given by 6.6 when $d(n) = d1_n$. It can be checked that

$$A(w)\hat{A}_1(w) - \hat{A}(w)A_1(w) = R_D(w)R_{D_1}sin(\theta_D - \theta_{D_1}) = \sum_{-(N-1)}^{N-1} c(n)e^{-inw} \tag{6.7}$$

and the Vandermonde determinant associated with the real roots of 6.7 is not equal to zero if Logan's conditions are satisfied. So if $\theta_D(v) = \theta_{D_1}(v)$ for at least 2N-1 distinct frequencies on $(0, 2\pi)$, then e(n) are all zero. Therefore, $sin(\theta_D(w) - \theta_{D_1}(w)) = 0$ for all w; or $tan(\theta_D(w)) = tan(\theta_{D_1}(w))$. In summary we have:

THEOREM (Hayes) 6.4. Let x(n) and y(n) be complex sequences which are zero outside (0,N-1) with z-transforms which have no zeros in conjugate reciprocal pairs or on the unit circle. If $M \geq 2N - 1$ and $\theta_x(w) = \theta_y(w)$ at M distinct frequencies in $[0, 2\pi)$, then $x(n) = \beta y(n)$ for some positive number β. If $M \geq 2N - 1$ and $tan(\theta_x(w)) = tan(\theta_y(w))$ at M distinct frequencies on $[0, 2\pi)$, then $x(n) = \beta y(n)$ for some real number β.

6.2 Arrival Time Determination

Li and Kurkjian have proposed the following arrival time determination technique. Let two receivers collect the signals $x_1(t) = s(t) \star \sum_{i=1}^{p} \gamma_i \delta(t - \xi_i)$ and $x_2(t) = s(t) \star \sum_{i=1}^{q} \beta_i \delta(t - \tau_i)$. The object is to determine s(t), p,q, and $\gamma_i, \xi_i, \beta_i, \tau_i$. The Fourier transforms of the received signals are $X_1(f) = S(f) \sum_{i=1}^{p} \gamma_i e^{-i2\pi f \xi_i}$ and $X_2(f) = S(f) \sum_{i=1}^{q} \beta_i e^{-i2\pi f \tau_i}$. The cross spectrum is then

$$G_{x_1 x_2} = X_2(f)^* X_1(f) = S(f)S(f)^* \sum_{i=1}^{p} \sum_{k=1}^{q} \gamma_i \beta_k e^{-i2\pi f(\tau_k - \xi_i)}. \quad (6.8)$$

We rewrite this as

$$G_{x_1 x_2}(f) = G_{ss}(f) = \sum^{N} \alpha_n e^{-i2\pi D_n}. \quad (6.9)$$

Define now $A(f), \phi(f)$ and y_m by

$$A(f)e^{i\phi(f)} = \sum_{n=1}^{N} \alpha_n e^{-i2\pi f D_n} = \sum_{m=0}^{M-1} y_m e^{-i2\pi km/M}$$

where $k = Mf\Delta t$ and Δt is small enough that it divides each D_n an integral number of times. It is assumed that no two D_n values are precisely equal. Since $G_{ss}(f)$ is a positive real-valued function, the phase $\phi(f)$ is the same as the phase of sequence y_m. Using the results from Hayes-Lim-Oppenheim, if a sequence is finite length and has no zero phase (i.e., even) convolutional components, then the signal can be uniquely reconstructed from the phase of its DFT. The reconstruction of y_m comes from the iterative algorithm which applies the phase constraint $\phi(f)$ in the DFT domain and the support constraints on y_m based on a priori knowledge on bounds on the inter-reciever delays D_n.

Once the sequence y_m is reconstructed, one determines values of p,q such that N = pq. One makes a guess at mapping i and k into n to give values for $\tau_k - \xi_i$ and $\gamma_i \beta_k$. One peels off the answer by assuming that $\xi_1 = 0$ and $\gamma_1 = 1$. If the choices are correct, the deconvolved wavelets s(t) will be the same for x_1 and x_2 and the wavelets will contain a delay corresponding to the absolute arrival time of the first arrival at the first receiver. This search through p,q is continued throughout all possible combinations. The correct solution will be eventually obtained.

6.3 Fourier Representation by Fourier Magnitude

From Walther's theorem we have concluded:

THEOREM 6.5. If x(t) is a complex finite time signal and its Fourier transform has no zeros in the upper half plane or lower half plane, then x(t) is uniquely determined by its Fourier magnitude.

It follows directly from this result that:

THEOREM (Hayes-Lim-Oppenheim) 6.6. Let x(n), y(n) be two real sequences whose z-transforms contain no reciprocal pole-zero pairs and which have all poles, not at $z = \infty$, inside the unit circle. If the magnitude of the Fourier transforms of x(n) and y(n) are equal, then $x(n) = \pm y(n + m)$ for some integer m.

6.4 Representation by Fourier Signed Magnitude

Let x(t) be a complex signal which vanishes outside [0,T] with x(0) being real. The Fourier transform X(w) of s(t) is analytic, in fact, EFET. We write

$$X(w)e^{-i\alpha} = P_x^\alpha(w) + iQ_x^\alpha(w) = R_x(w)e^{i(\theta_x(w)-\alpha)}. \tag{6.10}$$

Here $P_x^\alpha(w) = P_x(w)\cos\alpha + Q_x(w)\sin\alpha = R_x(w)\cos(\theta_x(w) - \alpha)$ is a real EFET with $Q_x^\alpha(w) = -P_x^\alpha(w)$. Note that P_x^α, Q_x^α are Fourier transforms of signals with finite extent

$$P_x^\alpha = .5\mathcal{F}(x(t)e^{-i\alpha} + x(t)^*e^{i\alpha}) \tag{6.11}$$

$$Q_w^\alpha = .5\mathcal{F}(x(t)e^{-i\alpha} - x^*(-t)e^{i\alpha}). \tag{6.12}$$

The signed Fourier magnitude G_x^α is defined by

$$G_x^\alpha(w) = S_x^\alpha(w)|R_x^\alpha(w)| \tag{6.12}$$

where

$$S_x^\alpha = \begin{cases} 1 & \alpha - \pi \le \theta_x(w) < \alpha \\ -1 & \text{otherwise} \end{cases} \tag{6.13}$$

The real zero-crossings of $Q_x^\alpha(w)$ are determined by $S_x^\alpha(w)$. Since $Q_x^\alpha(w)$ is a real EFET, the constant in the Hadamard factorization is real, k = 0 and all zeros come in conjugate pairs. The real zeros of $Q_x^\alpha(w)$ determine it up to some positive function. Since the magnitude of $Q_x^\alpha(w)$, viz., $R_x(w)$, is known, $Q_x^\alpha(w)$ is uniquely specified. In summary we have:

THEOREM (Shitz-Zeevi) 6.7. Let x(t) be a complex time limited signal with x(0) real. Then the signal x(t) is uniquely determined (within x(0) for $\alpha = 0$ or π) by the Fourier signed magnitude $G_x^\alpha(w)$ for any $0 \le \alpha \le \pi$ provided the Fourier transform X(v) contains no multiple real zeros.

COROLLARY (Hayes-Lim-Oppenheim) 6.8. Let x(n) and y(n) be two real causal (or anti-causal) finite extent sequences with z-transforms which have no zeros on the unit circle. Then if $G_x^\alpha(w) = G_y^\alpha(w)$ for all w and $0 < \alpha < T$, then x(n) = y(n). If $\alpha = \pi$ or 0, and $G_x^\alpha(w) = G_y^\alpha(w)$ for all w, and x(0) = y(0) = 0, then x(n) = y(n).

6.5 Representation by One Bit of Fourier Phase

Let x(t) be a complex signal which is zero outside [0,T) with x(0) real. Consider the analytic signal $X(v)e^{-i\alpha}$. By Logan's theorem

$$Re\{(2\pi)^{-1}X(v)e^{-i\alpha-iTv/2+i\mu v}\}$$

has only real simple zeros if $\mu > 0$ and $(2\pi)^{-1}x(-v)e^{-i(iTv/2+\alpha)}$ is zero free in the lower half plane: $v = w + iu, u \leq 0$. Taking $\mu = T/2$, we see

$$P_x^\alpha(v) = Re(X(v)e^{-i\alpha}) \tag{6.14}$$

has only simple real zeros if $X(v)$ is zero free in the closed, upper half plane $u \geq 0$. As noted above, P_x^α is an EFET which is real on $v = w$ and it determines x(t) there except for x(0). If the zeros of $X(v)$ are all in the closed upper half plane, then $P_x^{\alpha+\pi/2}(w)$ has only simple real zeros and it is determined by $S_x^\alpha(w)$ within a multiplicative constant. In summary we have:

THEOREM (Shitz-Zeevi) 6.9. Let x(t) be a complex signal of finite extent and x(0) real. Then x(t) is determined (within a positive, real multiplicative constant for $\alpha \neq 0, \pi$) (or within a real constant for $\alpha = 0, \pi$) by the one bit of Fourier phase $S_x^\alpha(w)$, provided the Fourier transform $X(v)$ is zero free in the closed, upper half plane.

COROLLARY (Oppenheim-Lim-Curtis) 6.10. Let x(n), y(n) be two real finite extent causal (or anitcausal) sequences with all zeros outside (inside) the unit circle. If $S_x^\alpha(w) = S_y^\alpha(w)$ for all w, $\alpha = 0, \pi$ then $x(n) = \beta y(n)$. For $\alpha = \pi/2, \beta$ is positive.

Chapter 7

Recovery of Distorted Band-Limited Signals

7.1 Introduction

We begin this chapter with a discussion of signal companding. First we characterize the space of signals that we are treating.

THEOREM 7.1. The set of signals (i.e., f(t) in $L^2(-\infty, \infty)$) band-limited to $(-\Omega, \Omega)$ is a closed subspace of L^2.

Proof. Suppose a sequence of signals, f_n, with band $(-\Omega, \Omega)$, converges in norm to f. Then f belongs to L^2 and by Plancherel's theorem

$$\int_{-\infty}^{\infty} |f_n(t) - f(t)|^2 dt = \int_{-\infty}^{\infty} |F_n(w) - F(w)|^2 dw. \qquad (7.1)$$

Thus, F(w) must vanish for $|w| > \Omega/2$.

DEFINITION 7.2. Let B denote this subspace of band-limited signals.

DEFINITION 7.3. A compander is a function $\phi(x)$ such that $\phi(f(t))$ is in L^2 if f(t) is in L^2.

DEFINITION 7.4. Let χ denote the characteristic function

$$\chi(w) = \begin{cases} 1 & 1\,|w| \le \Omega \\ 0 & \text{otherwise.} \end{cases} \qquad (7.2)$$

Define the functional equation

$$S(w) = \chi(w)\mathcal{F}\phi(f(t)) \qquad (7.3)$$

where $\phi(x)$ is a monotonic increasing compander.

THEOREM 7.5. S(w) is in L^2 and equation 7.3 has at most one solution f(t) in B.

Proof. If f_1, f_2 are two solution of 7.3, then by Plancherel's theorem

$$\int (f_1 - f_2)(\phi(f_1) - \phi(f_2))dt = \int (F_1(w) - F_2(w))\overline{(\mathcal{F}\phi(f_1) - \mathcal{F}\phi(f_2))}dw. \tag{7.4}$$

The first integrand on the right vanishes for $|w| > \Omega$ and the second vanishes for $|w| < \Omega$ by definition. Based on monotony of ϕ we have $f_1 = f_2$.

THEOREM (Landau-Miranker) 7.6. If $\phi(x)$ is differentiable and such that $m \leq \phi' \leq M$, then 7.3 has a solution f(t) in B for each S(w) in L^2.

Proof. The proof is based on the construction of a sequence of approximations f_n to f. The iteration formula arises from the equation

$$f = c\mathcal{F}^{-1}S(w) + \mathcal{F}^{-1}\chi\mathcal{F}(f - c\phi(f)) \tag{7.5}$$

where c is a constant. Since $\mathcal{F}^{-1}\chi\mathcal{F}$ is the identity map on B we have

$$0 = c\mathcal{F}^{-1}S(w) - \mathcal{F}^{-1}\chi\mathcal{F}c(\phi f) \tag{7.6}$$

and a solution of 7.6 is a solution of 7.3 Set

$$f_{n+1} = c\mathcal{F}^{-1}S(w) + \mathcal{F}^{-1}\chi\mathcal{F}(f_n - c\phi(f_n)). \tag{7.7}$$

Then we must show that f_n converges to a solution of 7.7 in B, provided f_o belongs to B and c is a suitably chosen. If f_n is a Cauchy sequence then the limit of f_n belongs to B since B is closed. A direct calculation and the mean value theorem show that

$$\|f_{n+1} - f_n\| =$$

$$\|\mathcal{F}^{-1}\chi\mathcal{F}[f_n - f_{n-1} - c(\phi(f_n) - \phi(f_{n-1}))\| \leq max_x|1 - c\phi'(x)|^2\|f_n - f_{n-1}\|^2. \tag{7.8}$$

Thus f_n is a Cauchy sequence if $\theta^2 = max|1 - c\phi'(x)|^2 < 1$. By the assumptions $\phi'(x) \leq M$, thus c is chosen to be a positive constant less than 2/M.

In the case of logarithmic companding Logan has shown the following explicit recovery scheme. Assume $f(x) = \frac{1}{2}log(1 + x)$ for $x > -1$ (i.e., we require $f(t) > -1$). Let $\lambda_\Omega(t) = \int_{-\infty}^{\infty} \phi(f(t))sin\Omega(t - s)ds$ then $\lambda_\Omega(t)$ belongs to L^p and it has the Hilbert transform

$$\phi_\Omega(t) = \int \lambda_\Omega(s)\frac{1 - cos\Omega(t - s)ds}{\pi(t - s)} = \int \phi(f(t))\frac{1 - cos\Omega(t - s)ds}{\pi(t - s)}. \tag{7.9}$$

Set $w_\Omega(t) = \lambda_\Omega(t) + i\phi_\Omega(t)$. Then

$$w_\Omega(t) = \int \phi(f(t))K_\Omega(t - s)ds \tag{7.10}$$

$$K_\Omega(t) = (e^{i\Omega} - 1)/i\pi t.$$

Let B_Ω be the band-limiting operator in L^p given by

$$B_\Omega h(t) = \int_{-\infty}^{\infty} h(s) \frac{\sin\Omega(t-s)ds}{\pi(t-s)}. \tag{7.11}$$

Then

$$\lambda_\Omega(t) = .5B_\Omega log(1 + g(t)) \tag{7.12}$$

for g in $B_p(\Omega), g > -1$; and

$$g(t) = |B_\Omega e^{\lambda_\Omega(t) + i\hat\lambda_\Omega(t)}|^2 - 1 \tag{7.13}$$

THEOREM (Logan) 7.7. Given a function $\lambda_\Omega(t)$ in $B_p(\Omega)1 \le p < \infty$. Then equation 7.12 has a solution in $B_p(\Omega)$ satisfying $g(t) > -1$ if and only if

$$z(t) = B_\Omega e^{\lambda_\Omega + i\hat\lambda_\Omega(t)} \tag{7.14}$$

extends as a function zero-free in the upper half plane. The solution of 7.12 in this case is given by 7.13.

Proof. Since $1+g(t)$ is a positive band-limited function, by the Boas-Kac theorem $1 + g(t) = \gamma(t)\gamma(t)^*$ where the Fourier transform of $\gamma(t)$ vanishes outside $[-\Omega/2, \Omega/2]$ and $\gamma(t)$ is zero free in the upper half plane. Then $z(t) = \gamma(t)e^{i\Omega t/2}$ is a function with spectral support $[0, \Omega]$ and $z(t)$ is zero free in the upper half plane. Clearly $1 + g(t) = |z|^2$. Since $g(t) \to 0$ as $t \to \pm\infty$, we may assume that $lim_{t\to\pm\infty} z(t) = 1$. Thus, $z(t) = e^{w(t)}$ where $w(t) = \frac{i}{2}\int log(1 + g(s))ds$.

7.2   Landau's Convolution Equation

For f in $L^2 = L^2(-\infty, \infty)$ with the Fourier transform

$$F(w) = (2\pi)^{-1/2}\int f(t)e^{-iwt}dt$$

and for the open set S of the real line define the subspaces of L^2

$$\mathcal{D}(S) = \{f \in L^2 | f(t) = 0, t \in S\} \tag{7.15}$$

$$\mathcal{B}(S) = \{f \in L^2 | F(w) = 0, w \in S\}.$$

If S is a bounded set and f belongs to $\mathcal{B}(S)$, then writing f(t) as the inverse Fourier transform of F(w) we have seen that f can be viewed as the restriction to the reals of an EFET, f(t+iu). If $\|f\| = 1$, then one sees that f(t+iu) is bounded in any given horizontal strip by a constant depending only on the strip. We have shown in the last section that $\mathcal{B}(S)$ is a closed subspace. Similarly, it is checked that $\mathcal{D}(S)$ is closed. Let D and B denote the orthogonal projection operators of L^2 onto $\mathcal{D}(S)$ and $\mathcal{B}(S)$ respectively. Viz., set

$$D_S f(t) = \chi_S(t) f(t) \tag{7.16}$$

and

$$B_S f(t) = (2\pi)^{-1/2} \int_{-\infty}^{\infty} \chi_S(w) F(w) e^{iwt} dw = \mathcal{F}^{-1} \chi_S \mathcal{F} f(t). \tag{7.17}$$

Since projections are bounded by 1, self-adjoint and idempotent, we have

$$(B_S D_Q B_S f, f) = (D_S^2 B_S f, B_S f) = \|D_Q B_S f\|^2 \le \|f\|^2 \tag{7.18}$$

where S,Q have finite measure. Thus, $B_S D_Q B_S$ is bounded by one, self-adjoint and positive. Let $\mathcal{F}(h) = \chi_S(w)$. Then we can write explicitly

$$B_S D_Q f = (2\pi)^{-1/2} \int \chi_Q(u) h(t-u) f(u) du. \tag{7.19}$$

By Parseval's theorem

$$\int_{-\infty}^{\infty} \int_{-\infty}^{\infty} |\chi_Q(u) h(t-u)|^2 = \int |\chi_Q(u)|^2 du \int |\chi_S(w)|^2 dw. \tag{7.20}$$

Thus, the operator $B_S D_Q$ is completely continuous. It follows that the operators $B_S D_Q B_S$ and $D_Q B_S D_Q$ are also completely continuous.

We write $A \sim B$ if A and B are completely continuous operators which have the same nonzero eigenvalues, including multiplicities. If $B_S D_Q B_S f = \lambda(S,Q) f$ with $\lambda(S,Q) \ne 0$, then $f = B_S f$ since B_S is a projection, and $\|D_Q B_S f\| \ne 0$. Applying D_Q to the equation we have

$$D_Q B_S D_Q (D_Q B_S f) = \lambda D_Q B_S f. \tag{7.21}$$

Thus, λ is also an eigenvalue of $D_Q B_S D_Q$. These arguments show that $B_S D_Q B_S \sim D_Q B_S D_Q$.

Since the spectrum is real, $D_Q B_S D_Q \sim C D_Q B_S D_Q C$ where C denotes complex conjugation. Note that C commutes with χ and $C \mathcal{F} C = \mathcal{F}^{-1}$. Thus, $C D_Q B_S D_Q C = \chi_Q \mathcal{F} \chi_S \mathcal{F}^{-1} \chi_Q$. Since $\mathcal{F}$ is unitary we have $\chi_Q \mathcal{F} \chi_S \mathcal{F}^{-1} \chi_Q \sim \mathcal{F}^{-1} \chi_Q \mathcal{F} \chi_S \mathcal{F}^{-1} \chi_Q \mathcal{F} = B_Q D_S B_Q$; in summary

$$B_S D_Q B_S \sim B_Q D_S B_Q.$$

In terms of the eigenvalues we have $\lambda_k(S,Q) = \lambda_k(Q,S)$. Similarly, one shows that

$$\lambda_k(S,Q) = \lambda_k(S+a, Q+b) = \lambda_k(\alpha S, \alpha^{-1} Q) \tag{7.22}$$

for a,b in R and $\alpha > 0$. One has the standard result

$$\sum_k \lambda_k(S,Q) = (2\pi)^{-1/2} \int_{-\infty}^{\infty} \chi_Q(x)h(0)ds = (2\pi)^{-1}m(S)m(Q) \quad (7.23)$$

where m(S) is the Lebesgue measure of set S.

We gather these and related results in the following theorem.

THEOREM (Landau) 7.8. If S,Q are sets in R with finite measure, then $B_S D_Q B_S$ is a bounded self-adjoint positive completely continuous operator. If $\lambda_k(S,Q)$ denotes its eigenvalues arranged in nonincreasing order k = 0,1,2,..., then for all k we have

(1) $\lambda_k(S,Q) = \lambda_k(S+a, Q+b) = \lambda(\alpha S, \alpha^{-1}Q)$ for a,b in R and $\alpha > 0$.

(2) $\lambda_k(S,Q) = \lambda_k(Q,S)$

(3) $\sum_k \lambda_k(S,Q) = (2\pi)^{-1}m(S)m(Q)$

(4) $\sum_k \lambda_k^2(S,Q) \geq \sum_k \lambda_k^2(S,Q_1) + \sum_k \lambda_k^2(S,Q_2)$ where $Q = Q_1 \cup Q_2$ and Q_1 and Q_2 are disjoint;

(5) $\sum_k \lambda_k^2(S,Q) \geq \{(2\pi)^{-1}sq - \pi^{-2}log(sq) - 1\}$ if S and Q are lines of length s,q;

(6) $\lambda_k(S,Q_1) \leq \lambda_k(S,Q_2)$ if $Q_1 \subset Q_2$.

7.3 Time-Limited and Band-Limited Functions

Let f belong to $L^2(R)$ and have Fourier transform $F(w) = \int_{-\infty}^{\infty} f(t)e^{-iwt}dt$. Then Parseval's theorem states that

$$\int_{-\infty}^{\infty} f(t)g(t)dt = \frac{1}{2\pi} \int_{-\infty}^{\infty} F(w)G(w)dw. \quad (7.24)$$

Let $B(\Omega)$ denote the subclass of $L^2(R)$ of those functions f(t) whose Fourier transforms vanish for $|w| > \Omega$. Here $\Omega = 2\pi W$. Thus, if f belongs to $B(\Omega)$, then f can be written as a finite Fourier transform

$$f(t) = \frac{1}{2\pi} \int_{-\Omega}^{\Omega} F(w)e^{iwt}dw.$$

DEFINITION 7.9. B is called the class of band-limited functions. For F and f in L^2 define the function Bf in B by

$$Bf(t) = \frac{1}{2\pi} \int_{-\Omega}^{\Omega} F(w)e^{iwt}dw. \quad (7.25)$$

B is the band-limiting operator, i.e., Bf(t) results from passing f through an ideal low-pass filter with cut off frequency Ω.

DEFINITION 7.10. Let $\mathcal{D}$ denote the subclass of functions f in $L^2(R)$ which vanish for $|t| > T/2, T > 0$. $\mathcal{D}$ is called the class of time-limited functions. The projection of f in L^2 onto a function Df(t) in $\mathcal{D}$ is given by

$$Df(t) = \begin{cases} f(t) & t \le T/2 \\ 0 & |t| > T/2. \end{cases} \tag{7.26}$$

From these definitions it follows that

$$BDf(t) = \frac{1}{2\pi} \int_{-\Omega}^{\Omega} dw e^{iwt} \int_{-T/2}^{T/2} ds f(s) e^{-isw} = \int_{-T/2}^{T/2} \rho_\Omega(t - s) f(s) ds \tag{7.27}$$

where $\rho_\Omega(t) = \frac{sin(\Omega t)}{\pi t} = \frac{1}{2\pi} \int_{-\Omega}^{\Omega} dw e^{iwt}$. By Parseval's theorem we have

$$\int_{-\infty}^{\infty} \rho_\Omega(t - u) \rho_\Omega(u - s) du = \rho_\Omega(t - s). \tag{7.28}$$

Using this equation one has

$$\|BDf(u)\|^2 = \int_{-\infty}^{\infty} du [BDf(u)] \overline{[BDf(u)]} \tag{7.29}$$

$$= \int_{-\infty}^{\infty} du \int_{-T/2}^{T/2} dt \int_{-T/2}^{T/2} ds \rho_\Omega(t - u) \rho_\Omega(u - s) f(t) \bar{f}(s) =$$

$$\int_{-T/2}^{T/2} \int_{-T/2}^{T/2} ds \rho_\Omega(t - s) f(t) \bar{f}(s)$$

since ρ_Ω is real and even.

Consider the eigenvalue problem

$$BDf(t) = \lambda f(t) \tag{7.30}$$

for $|t| \le T/2$, i.e.,

$$\int_{-T/2}^{T/2} \rho_\Omega(t - s) f(s) ds = \lambda f(t). \tag{7.31}$$

We turn in the next section to a study of this eigenvalue equation.

7.4 Prolate Spheroidal Functions

The differential equation

$$(1 - t^2)u'' - 2tu' + (\chi - c^2t^2)u = 0 \qquad (7.32)$$

has continuous solutions for $|t| \leq 1$ for certain discrete real positive values of $\chi(c)$. The solutions of this equation are the angular prolate spheroidal function (PSWFs) $S_{on}(c, t)$. These functions satisfy the equation

$$\frac{2c}{\pi}(R_{on}^{(1)}(c, 1))^2 S_{on}(c, t) = \int_{-1}^{1} \frac{sinc(t - s)}{\pi(t - s)} S_{on}(c, s) ds \qquad (7.33)$$

where n = 0,1,2,..., $|t| \leq 1$. The eigenvalues are given by

$$\lambda_n(c) = \frac{2c}{\pi}(R_{on}^{(1)}(c, 1))^2 \qquad (7.34)$$

where $R_{on}^{(1)}(c, t)$ are the radial spheroidal functions. The angular prolate spheroidal functions are orthogonal on (-1,1) and are complete on $L^2(-1, 1)$. We note that $S_{on}(c, t)$ has exactly n zeros on (-1,1) and $S_{on}(c, 0) = P_n(0)$, the nth Legendre polynomial.

Since the kernel of

$$\lambda f(t) = \int_{-1}^{1} \rho_c(t - s) f(s) ds \qquad (7.35)$$

is positive definite it follows by Bochner's theorem that the eigenvalues are strictly positive. If we set

$$(u_n(c))^2 = \int_{-1}^{1} S_{on}(c, t)^2 dt \qquad (7.36)$$

and let

$$\psi_n(c, t) = \sqrt{\lambda_n(c)} S_{on}(c, 2t/T)/u_n(c) \qquad (7.37)$$

then 7.35 becomes

$$\lambda_n \psi_n(t) = \int_{-T/2}^{T/2} \frac{sin\Omega(t - s)}{\pi(t - s)} \psi_n(s) ds \qquad (7.38)$$

for n = 0,1,2... Clearly ψ_n belongs to $\mathcal{B}$. Since $\psi_n(c, \omega T/2\Omega)$ are complete on $|w| \leq \Omega$ we see that $\psi_n(t)$ are complete in $\mathcal{B}$.

We summarize the following properties of PSWFs:

THEOREM 7.11. The PSWFs are continuous solutions on the interval [-1,1] of the eigenvalue equation

$$\int_{-1}^{1} \frac{sin(c(w - w'))}{\pi(w - w')} \psi_m(c, w') dw' = \lambda(c) \psi_m(c, w). \qquad (7.39)$$

The $\psi_m(c, w)$ can be uniquely extended to entire functions. The eigenvalues form a decreasing sequence $1 > \lambda_0 > \lambda_1 > .. > 0$ and they have a step like behavior: viz., they are approximately unity for values of $m < 2c/\pi$ (which is called the Shannon number) and thereafter they fall to zero exponentially.

(2) The PSWFs are orthogonal over the infinte interval:

$$\int_{-\infty}^{\infty} \psi_m(c, w)\psi_n(c, w)ds = \delta_{m,n}.$$

(3) The PSWFs form a basis in the space of square integrable functions whose Fourier transforms have compact support.

(4) The PSWFs are orthogonal over [-1,1] where

$$\int_{-1}^{1} \psi_m(c, w)\psi_n(c, w)dw = \delta_{m,n}\lambda_n(c)$$

(5) The functions $u_{m,c}(x) = (\lambda_m(c))^{-1/2}\psi_m(c, x)$ form a basis in the space $L^2(-1, 1)$.

(6) $\psi_m(c, w)$ is the Fourier transform of a function with compact support; viz.,

$$\int_{-\infty}^{\infty} \psi_m(c, w)e^{-iwx}dw = (-i)^m \sqrt{2\pi/c\lambda_m}\psi_m(2, x/c)\chi(x/c)$$

where

$$\chi(s) = \left\{ \begin{array}{ll} 1 & |s| \leq 1 \\ 0 & |s| > 1. \end{array} \right.$$

7.5 Extrapolation by PSWFs

Given F on (-1,1) then $F(w) = \sum_{m=0}^{\infty} a(m)\psi_m(w)$, where

$$a(m) = \frac{1}{\lambda_m} \int_{-1}^{1} F(w)\psi_m(w)dw. \tag{7.40}$$

then

$$F(w) = \sum_{m=0}^{\infty} \lambda_m^{-1}\psi_m(w) \int_{-1}^{1} F(w)\psi_m(w)dw. \tag{7.41}$$

This provides an extrapolation formula which only requires knowledge of F(w) over (-1,1), the PSWFs over $(-\infty, \infty)$ and the eigenvalues λ_m. This method requires only numerical evaluation of the integrals. However, for large m, λ_m are very small.

The eigenvalues $\lambda_n(c)$ satisfy

$$\lambda_n(c) = e^{i\pi n/2}|\lambda_n(c)|. \tag{7.42}$$

We set $\sigma_n(c) = |\lambda_n(c)|$; i.e., $\sigma_n(c)$ is the singular value. It can be shown that for a fixed c and increasing n

$$\sigma_n(c) \sim 1 - \alpha_n e^{-\beta_n c^2}. \tag{7.43}$$

EXERCISE 7.12. For

$$\frac{d(1-x^2)}{dx}\frac{df_n}{dx} + (\chi_n - c^2 x^2)f_n = 0 \tag{7.44}$$

or

$$\lambda_n f_n(x) = \int_{-1}^{1} sin(c(x-y))f_n(y)dy \tag{7.45}$$

show that for $n << c$ we have

$$1 - \lambda_n \sim \sqrt{\pi} 2^{3n+2} e^{-2c} c^{n+1/2} (n!)^{-1} \tag{7.46}$$

and for n large, $n \sim c$

$$(n + 1/2)\frac{1}{2}\pi \sim c + \phi(b) - \frac{1}{2}bln(4c) \tag{7.47}$$

$b = 1/2(c - \xi/c)$ and $1 - \lambda_n \sim (1 + e^{\pi b})^{-1}$. Here $\psi(b)$ is given by $\Gamma(1/2 + ib/2) = (\frac{\pi}{cosh(\pi b/2)})^{1/2}e^{i\psi(b)}$.

THEOREM (Landau-Pollak-Slepian-Widom) 7.13. If D,E are bounded subsets of R^N with volumes $|D|$ and $|E|$ and surface areas $|\partial D|$ and $|\partial E|$, then for large c the number $n(c, \alpha)$ of singular values of $P_{cD}\mathcal{F}P_{cE}$ greater that α is given by

$$n(c, \alpha) \sim |D||E|c^{2N} - \gamma|\partial D||\partial E|c^{2N-2}log(\alpha^{-1} - 1)log(c) + \mathcal{O}(c^{2N-3}) \tag{7.48}$$

where cD is the set $\{cd | d \in D\}$ and γ is a constant independent of c, D, and E.

COROLLARY 7.14. For

$$\int_{-1}^{1} sin f_n(t)dt = f_n(s)\lambda_n \tag{7.49}$$

if $n(c, \alpha)$ is the number of eigenvalues λ_n such that $\lambda \geq \alpha$, then

$$n(c, \alpha) = 4c + 2log(\alpha^{-1} - 1)logc + \mathcal{O}(c^{2N-3}) \tag{7.50}$$

and if b is fixed and n,c chosen so that

$$n = 4c + \frac{2b}{\pi^2}log(2\sqrt{2\pi}c) \tag{7.51}$$

then $lim_{n\to\infty}\lambda_n = (1 - e^b)^{-1}$.

Thus, $\lambda_n \sim 1$ if $n < 4c$ and ~ 0 if $n > 4c$ with the change from unity to zero occuring in a strip of width ($\sim logn$) centered at n = 4c.

The Shannon number S(K) is defined by

$$S(K) = \sum \lambda_n = Tr(K). \tag{7.52}$$

For a symmetric kernel $k^*(t) = k(-t)$ we have

$$S(K) = \int_{-1}^{1} k(s-s)ds = 2k(0). \tag{7.53}$$

7.6   Riemann Zeta Function

The Riemann zeta function is defined by

$$\zeta(s) = \sum_{n=1}^{\infty} \frac{1}{n^s} = \prod_p \frac{1}{1 - p^{-s}} \tag{7.54}$$

for $Re(s) > 1$ where the product is taken over all prime numbers p. As an analytic function $\zeta(s)$ can be continued to the left of Re(s) = 1 so that $\zeta(s)$ is analytic on all of C except for a simple pole at s = 1. The Riemann zeta function satisfies the functional equation

$$\zeta(s) = \chi(s)\zeta(1-s) \tag{7.55}$$

where $\chi(s) = \pi^{s-\frac{1}{2}}\Gamma(\frac{1-s}{2})/\Gamma(s/2)$. We see immediately that $\zeta(-2n) = 0$ for $n \geq 1$. These are the so called trivial zeros. The nontrivial zeros are those located in the region $0 \leq Re(s) \leq 1$. We let them be denoted by ρ. The Riemann hypothesis states that $Re(\rho) = 1/2$ for all such ρ.

Write ρ as $\rho = 1/2 + i\gamma$. A study of the distribution of the nontrivial zeros has led to the Montgomery pair correlation conjecture that

$$\frac{|\{(\gamma,\gamma')|0 < \gamma, \gamma' \leq T, 2\pi\alpha(logT)^{-1} \leq \gamma - \gamma' \leq 2\pi\beta(logT)^{-1}\}|}{T(logT)/2\pi} \sim$$

$$\int_{\alpha}^{\beta} (1 - (\frac{sin\pi u}{\pi u})^2)du \tag{7.56}$$

for fixed $0 < \alpha < \beta < \infty$. It says that the function $(1 - (sin\pi u/\pi u)^2)$ is the pair correlation function of the nontrivial zeros of the zeta function. We note that the Gaussian unitary ensemble (GUE) has the same pair correlation function. Here the GUE consists of nxn complex Hermitian matrices of the form $A = (a_{jk})$ where $a_{jj} = \sqrt{2}\sigma_{jj}, a_{jk} = \sigma_{jk} + i\eta_{jk}, j <$

$k, a_{jk} = \hat{a}_{kj}, j > k$; here σ_{jk} and η_{jk} are independent standard normal variables.

Let P(k,u) denote the probability density of the kth neighbor spacing in the GUE – i.e., the probability that the normalized differences between an eigenvalue of the GUE and the (k+1)st smallest eigenvalue of those that are larger than it lies in the interval (a,b) is $\int_a^b p(k, u)du$. Let λ_n be the eigenvalues of the equation

$$\lambda f(x) = \int_{-1}^{1} \frac{sin(\pi t(x - y)/2)}{\pi(x - y)} f(y) dy \qquad (7.57)$$

for t in R^+. Order the λ_n so that $1 \geq \lambda_1 \geq \lambda_2 ... 0$.

EXERCISE 7.15. Show that

$$p(0, t) = \frac{d}{dt} \prod_n (1 - \lambda_n). \qquad (7.58)$$

7.7 Recovery of Stochastic Processes

In section 7.1 we discussed Beurling's result on the recovery of band-limited signals that have been companded. Viz., if A is the monotonic function representing the compander and f is band-limited to (-W,W), then knowledge of the spectrum of A(f(t)) in the band (-W,W) is sufficient to determine f(t) uniquely. The iterative formula of Landau demonstrates a procedure of recovery for a class of companding functions A(x).

In 1973 Masry showed that an analogue of Beurling's result holds for a class of stationary band-limited stochastic processes. In particular he was able to show that within the class of jointly stationary and jointly Gaussian band-limited processes, a Gaussian process is uniquely determined by its zero crossings. Furthermore, if A is monotonic, an iterative algorithm based on Landau's method provides for the recovery of the input process.

Consider an input process $\{X(t, w)\}$ on the real line R which is a real second-order, mean-square, continuous stationary band-limited stochastic process (on the $\sigma-$ algebra of Lebesgue measurable sets on R). Let $S(\lambda)$ denote the spectral density of the process and C(t) its covariance function, so

$$C(t) = \int_{-W}^{W} e^{it\lambda} dS(\lambda) \qquad (7.59)$$

where (-W,W) is the process bandwidth. We are assuming X(t) is normalized to have zero mean and unit variance.

Let $X = (X_1, X_2)$ where we assume X_1, X_2 are jointly stationary. Let B denote the class of processes with these properties. We note that if

$S = \{S_{X_i X_j}\}$ is the spectral distribtuion of $X = (X_1, X_2)$ for X_1, X_2 in $\mathcal{B}$, then

$$|S_{X_i X_j}(B)|^2 \le S_{X_i X_i}(B) S_{X_j X_j}(B) \qquad (7.60)$$

for any Borel set on R. Thus, the processes are jointly band-limited on (-W,W).

Let A be a Borel measurable function on R and define the output process $Y(t,w) = A(X(t,w))$ for t in R. This process is stationary. We also assume that A satisfies

(1) $E(A^2(x(t))) < \infty$ for all t
(2) $\{A(x(t))^2\}$ is a mean-square, continuous process
(3) $E(X(0)A(X(0))) \ne 0$.

Let $\mathcal{A}$ denote the class of companding functions for which (1)-(3) hold. We note that Y(t) is mean-square, continuous if A is Lipshitz, i.e.,

$$|A(y) - A(x)| < M|x - y| \qquad (7.61)$$

for x,y in R. However, A does not have to be Lipshitz. E.g., if A(x) is the hardlimiter $A(x) = sgn(x)$, then

$$R_{YY}(t) = \frac{2}{\pi} sin^{-1} C_{XX}(t) \qquad (7.62)$$

so $Y(t) = sgnX(t)$ is a stationary second order, mean-square, continuous process.

The desired output is the low pass version of Y. Let L be an ideal low pass filter with transfer function

$$H(i\lambda) = \left\{ \begin{array}{ll} 1 & |\lambda| \le W \\ 0 & \text{otherwise.} \end{array} \right. \qquad (7.63)$$

Then the measured output is Z(t) where

$$Z(t) = L(A(X))(t) \qquad (7.64)$$

which we abbreviate to $Z(t) = T(X)(t)$. Note that Z(t) is a second order, mean-square, continuous stationary band-limited process for A in $\mathcal{A}$.

THEOREM (Masry) 7.16. Let $X_1(t), X_2(t)$ be two jointly Gaussian processes in $\mathcal{B}$. Let $Z_i(t) = T(X_i)(t)$. Then $Z_1(t) = Z_2(t)$ a.s. for some $t = t_o$ implies $X_1(t) = X_2(t)$ a.s. for all t.

Proof. Define the functional

$$J = E((X_1(t_o) - X_2(t_o))(Z_1(t_o) - z_2(t_o))). \qquad (7.65)$$

J vanishes by hypothesis. However,

$$J = R_{X_1 Z_1}(0) - R_{X_1 Z_2}(0) - R_{X_2 Z_1}(0) + R_{X_2 Z_2}(0) \qquad (7.66)$$

$$= \int_{-\infty}^{\infty} H(i\lambda) d(S_{X_1 Y_1}(\lambda) - S_{X_1 Y_2}(\lambda) - S_{X_2 Y_1}(\lambda) + S_{X_2 Y_2}(\lambda)$$

$$= \int_{-W}^{W} d(S_{X_1 Y_1} \ldots)$$

and $R_{X_i Y_j}(t) = C_{X_i Y_j}(t)$ since the input processes have zero means. Since Gaussian processes have the cross correlation property we have $C_{X_i Y_j} = a_{ij} C_{X_i X_j}$ where $a_{ij} = E(X_i(t) A(X_j(t)))$ and a_{ij} is independent of t. Thus we have

$$J = \alpha E |X_1(t) - X_2(t)|^2 \qquad (7.67)$$

for all t where $\alpha = E(X(t) Z(X(t)))$. Since $J = 0$ and $\alpha \neq 0$ we see that $X_1(t) = X_2(t)$ a.s. for all t; so the processes $X_1(t)$ and $X_2(t)$ are equivalent.

We note that monotonicity is not required for this result. Also the low pass need not be ideal (see Masry for this result).

Consider now the two-level quantizer $A(x) = \text{sgn}(x)$. As we noted above A belongs to $\mathcal{A}$. The output Y is a stationary binary process associated with the zero crossing of $X(t)$ in $\mathcal{B}$. Since $X(t)$ has analytic sample paths, all the level crossings are genuine in the sense that they consist of up crossings and down crossings with no tangencies.

THEOREM (Masry) 7.17. Let $\mathcal{G}$ be the class of real zero-mean jointly stationary and jointly Gaussian band-limited processes. Then every process $X(t)$ in $\mathcal{G}$ is uniquely determined up to a multiplicative constant within $\mathcal{G}$ by the band-limited version of $Y(t)$.

This theorem asserts that if X_1, X_2 belong to $\mathcal{G}$ and we have the equality $\text{sgn} X_1(t) = \text{sgn} X_2(t)$ a.s. for all t, then $X_1(t) = b X_2(t)$ a.s. for all t, where b is a real constant. Since $\text{sgn}(X(t))$ is uniquely determined by the zero crossings of $X(t)$ in $\mathcal{G}$ up to multiplicative $\pm$ sign, it follows that no two processes in $\mathcal{G}$ can have the same zero crossings unless they are a constant multiple of each other. In other words:

COROLLARY 7.18. Every process $X(t)$ in $\mathcal{G}$ is uniquely determined up to a multiplicative constant within $\mathcal{G}$ by its zero crossings.

Consider now the case of real zero-mean second order, mean-square, continuous band-limited processes with A monotonic, increasing and continuous:

$$u(x_1 - x_2) \leq A(x_1) - A(x_2) \leq U(x_1 - x_2) \qquad (7.68)$$

for all x_1, x_2 in R. Let $\mathcal{B}'$ denote this class of processes.

THEOREM (Masry) 7.19. Let X_1, X_2 be real jointly stationary processes in $\mathcal{B}'$. Assume L is an ideal filter. Then $Z_1(t) = Z_2(t)$ a.s. for all t implies $X_1(t) = X_2(t)$ a.s. for all t.

Masry finally developed the following analogue of Landau's iterative reconstruction:

THEOREM (Masry) 7.20. Assume L is ideal and X is in B'. Then

$$X_{n+1}(t) = X_n(t) + c\{Z(t) - T(X_n)(t)\} \qquad (7.69)$$

converges in the mean-square sense to the input process $X(t)$ for all $0 < c < 2/U$.

LEMMA 7.21. Let X_1, X_2 be two jointly stationary processes in B'. Let $K(X)(t) = X(t)\text{-}cT(X)(t)$. Then

$$\|K(X_1)(t) - K(X_2)(t)\| \le d\|X_1(t) - X_2(t)\| \qquad (7.70)$$

$d < 1$ for all $0 < c < 2/U$.

Proof. Define the process $r(t)$ by

$$r(t) = X_1(t) - X_2(t) - c(AX_1(t) - AX_2(t)).$$

Then $Lr = X_1 - X_2 - c(T(X_1) - T(X_2))$. If $S(t)$ is the spectral distribution of $r(t)$, then by the spectral representation theorem

$$\|Lr(t)\|^2 = \int_{-\infty}^{\infty} |H(i\lambda)|^2 dS_{rr}(\lambda) =$$

$$\int_{-W}^{W} dS_{rr}(\lambda) \le \int_{-\infty}^{\infty} dS_{rr}(\lambda) = \|r(t)\|^2 = \|(X_1 - X_2)(1 - \frac{c(AX_1 - AX_2)}{X_1 - X_2}\|^2.$$

Let $d = sup_{x_1,x_2} |1 - \frac{c(A(x_1) - A(x_2))}{x_1 - x_2}|$. Then $\|Lr\|^2 \le d^2\|X_1 - X_2\|^2$, where $d < 1$. Since $Lr = K(X_1) - K(X_2)$ then the lemma is shown.

The theorem follows from the lemma by noting that

$$\|X - X_{n+1}\| \le d\|X - X_n\| \le d^n\|X - X_1\|$$

so $\|X(t) - X_{n+1}(t)\| \to 0$ as $n \to \infty$.

Chapter 8

Compact Operators, Singular Value Analysis and Reproducing Kernel Hilbert Spaces

8.1 Introduction

We have seen in the course of proving theorem 7.8 that $B_S D_Q B_S$ is a self-adjoint, completely continuous or compact operator. The spectrum of a compact self-adjoint operator K is particularly simple. Viz., the eigenspace associated to each nonzero eigenvalue λ is finite dimensional and the eigenvalues form a sequence $\lambda_1, \lambda_2, \ldots$ which (if infinite) converges to zero. We turn our attention to certain topics on compact operators and signal representation in this chapter.

Let X be a Hilbert space with inner product $(\ ,\ )$. If S is a subset of X, we let $\bar{S}$ denote the norm closure of S and $S^\perp$ will denote the orthogonal complement of S, i.e.,

$$S^\perp = \{y \in X | (x,y) = 0 \text{ for all } x \in S\}. \tag{8.1}$$

Let T be a continuous linear operator from Hilbert space X into Hilbert space Y; then the adjoint of T is the continuous linear operator $T^* : Y \to X$ defined by

$$(Tx, y) = (x, T^*y) \tag{8.2}$$

for all x in X and y in Y. The range $R(T)$ and null space $N(T)$ of a linear operator with domain $D(T)$ are defined by

$$R(T) = \{Tx | x \in D(T)\} \tag{8.3}$$

$$N(T) = \{x \in D(T) | Tx = 0\}.$$

THEOREM 8.1. If $T : X \to Y$ is a continuous linear operator, then $R(T)^{\perp} = N(T^*)$ and $N(T)^{\perp} = \overline{R(T^*)}$. And we have the orthogonal decompositions

$$X = N(T) \bigoplus N(T)^{\perp} \tag{8.4}$$

$$Y = \overline{R(T)} \bigoplus R(T)^{\perp}.$$

Consider now the operator equation $Tf = g$. Then four exclusive cases arise:

(a) $R(T)$ is dense in Y (so $N(T^*) = 0$) and g is in $R(T)$

(b) $\underline{R(T)}$ is dense in Y and g is not in $R(T)$

(c) $\overline{R(T)}$ is a proper subspace of Y and g is in $R(T) + R(T)^{\perp}$

(d) $\overline{R(T)} \neq Y$ and g is not in $R(T) + R(T)^{\perp}$.

Let Q denote the projection of Y onto $\overline{R(T)}$. We consider now the projection equation

$$Tf = Qg \tag{8.5}$$

for g in Y. This is different from the original equation only if $R(T)$ is not dense in Y. However, for any g in $R(T) \bigoplus R(T)^{\perp}$ 8.5 is solvable and the set S of all solutions is a closed convex subset of X; hence, it contains a unique element of minimal norm. This mapping of y to the unique solution of minimal norm is called the generalized inverse.

DEFINITION 8.2. Let $T : X \to Y$ be a bounded linear operator. The generalized inverse $T^{\dagger}$ of T is the mapping whose domain is $D(T^{\dagger}) = R(T) \bigoplus R(T)^{\perp}$ with $T^{\dagger}y = v$ in X where v belongs to $S = \{x \in X | Tx = Qy\}$ with $\|v\| < \|u\|$ for all u in S, $u \neq v$. Here v is called the least squares solution of minimal norm of $Tf = g$.

THEOREM 8.3. The following conditions on f in X are equivalent:

(1) $Tf = Qg$

(2) $\|Tf - g\| \leq \|Tx - g\|$ for any x in X

(3) $T^*Tf = T^*g$.

THEOREM 8.4. The generalized inverse is a linear operator and

$$N(T^{\dagger}) = R(T)^{\perp} \tag{8.6}$$

and

$$R(T^\dagger) = N(T)^\perp.$$

If $R(T)$ is closed, then $D(T^\dagger) = Y$ and $T^\dagger$ is a bounded linear operator. We leave this for the reader to verify. In general $T^\dagger$ is linear but not necessarily bounded. In fact $T^\dagger$ is bounded if and only if $R(T)$ is closed.

DEFINITION 8.5. A bounded linear operator $T : X \to Y$ is called compact if for each bounded set U in X, the closure of the set $T(U)$ is compact in the norm topology of Y.

EXAMPLE 8.6. Let $Tf(s) = \int_a^b k(s,t)f(t)dt$ for s in [a,b] and $k(s,t)$ in $L^2([a,b],[a,b])$. Then T is a compact operator. If the kernel k is finite rank, i.e., $k(s,t) = \sum_{i=1}^m g_i(s)h_i(t)$ for certain g_i, h_i, then R(T) is finite dimensional and the solution of 8.5 reduces to solving a linear algebraic system.

THEOREM 8.7. If T is a compact linear operator then $R(T)$ is closed if and only if $R(T)$ is finite dimensional.

Returning to the cases (a)-(d) cited above, we see that in case (a) $T^\dagger g$ gives the minimal-norm solution of $Tf = g$. In case (c) equation $Tf = g$ has a least-square solution (which is unique if and only if $N(T) = \{0\}$). Cases (b) and (d) are pathological since in both cases g does not belong to $D(T^\dagger)$.

DEFINITION 8.8. Equation $Tf = g$ is well-posed in the least squares sense if for each g in Y, the equation has an unique least-squares solution of minimal norm which depends continuously on g.

It follows by the open mapping theorem that:

THEOREM 8.9. The following statements are equivalent:
(1) $Tf = g$ is well-posed in the least squares sense
(2) $R(T)$ is closed
(3) $T^\dagger$ is bounded.

For compact operators we have the following summary:

THEOREM 8.10. Let A be a compact operator $A : X \to X$ on Hilbert space X. Then (1) for each $\lambda \neq 0$ the operator $A - \lambda I$ has a bounded generalized inverse $(A - \lambda)^\dagger$ defined on X and $\hat{x} = (A - \lambda)^\dagger y$ is the unique best approximate solution of minimal norm of $Ax - \lambda x = y$ for each y in X. If $\lambda \neq 0$ and y is in $R(A - \lambda I)$, then $\hat{x} = (A - \lambda I)^\dagger y$ is the unique solution of minimal norm. And if $\lambda \neq 0$ is in the resolvent of A then $(A - \lambda I)^\dagger = (A - \lambda)^{-1}$ and $Ax - \lambda x = y$ has a unique solution for each y in X.

(2) The operator A has a generalized inverse $A^\dagger$ defined on $R(A) \bigoplus R(A)^\perp$. The equation $Ax = y$ has a unique least squares solution for each y in $D(A^\dagger)$. If y in $N(A^*)^\perp$ and $\sum \frac{1}{\mu_n} | < y, \phi_n > |^2 < \infty$ where $AA^* \phi_n = \mu_n \phi_n$, then $\hat{x} = A^\dagger y$ is the unique solution of $Ax = y$ of minimal norm. Finally, the set of all least-squares solutions of $Ax = y$ for y in $R(A) \bigoplus R(A)^\perp$ is given by $A^\dagger y \bigoplus N(A)$.

Let $A : X \to Y$ be a compact operator, which is symmetric (i.e., $A = A^*$) and nonnegative definite, i.e., $(Ax, x) \geq 0$ for all x in X. Then A has a finite or countably infinite number of eigenvalues λ_n. Let ϕ_n denote the corresponding orthonomal eigenvectors. We assume $A \neq 0$ so $\lambda_1 > 0$. Let $\mathcal{E}$ be the set of subscripts for which $\lambda_n \neq 0$. Let $\sigma(A)$ denote the nonzero spectrum of $A = \{\lambda_i | A\phi_i = \lambda_i \phi_i, \lambda_i > 0, \phi_i \neq 0\} = \{\lambda_n | n \in \mathcal{E}\}$. The set $B = \{\phi_n | n \in \mathcal{E}\}$ is a Schauder basis for $\overline{R(A)}$, i.e., $\overline{spanB} = \overline{R(A)}$. Thus, every element in R(A) can be expanded in terms of ϕ_n:

$$Ax = \sum_{n \in \mathcal{E}} (Ax, \phi_n)\phi_n = \sum_{n \in \mathcal{E}} \lambda_n (x, \phi_n)\phi_n. \tag{8.7}$$

It is easy to check that

THEOREM 8.11. B is a complete orthonormal system for X if and only if $N(A) = 0$.

We apply these results to $A = KK^*$ where $K : X \to Y$ is a compact linear operator. A pair of nonzero vectors ϕ in X and ψ in Y which satisfy $\phi = \mu K\psi$ and $\psi = \mu K^*\phi$ are said to be a pair of singular functions of K with singular value μ. And in this case ϕ, ψ are eigenfunctions of KK^* and K^*K respectively:

$$\psi = \mu^2 KK^*\psi \tag{8.8}$$
$$\phi = \mu^2 KK^*\phi.$$

Set $\sigma(K^*K) = \{\lambda_n | n \in \mathcal{E}\}$ and let $\mu_n = \lambda_n^{-1/2}$ for n in $\mathcal{E}$. Set $U = \{u_n | KK^*u_n = \lambda_n u_n, n \in \mathcal{E}\}$ and $V = \{v_n | K^*Kv_n = \lambda_n v_n, n \in \mathcal{E}\}$. The set (u_n, v_n, λ_n) is called a singular system for compact operator K.

THEOREM 8.12. (1) $\{u_n\}$ is a complete orthnormal set for $\overline{R(K)} = N(K^*)^\perp$

(2) $\{v_n\}$ is a complete orthonormal set for $R(K^*) = N(K)^\perp$.

Picard's necessary and sufficient condition for the existence of a solution of $Ax = y$ is: given y in H there is a vector x such that $Ax = y$ if and only if $\sum(y, \phi_n)^2/\lambda_n^2 < \infty$ and $(y, u) = 0$ for all u such that $Au = 0$. We can combine Picard's condition and least-squares solutions as follows:

THEOREM 8.13. Let (u_n, v_n, λ_n) be a singular system for compact operator $K : X \to Y$, let y belong to Y, and let $K^\dagger$ be the generalized inverse of K. Then the following statements are equivalent:

(1) y belongs to $D(K^\dagger) = R(K) \oplus R(K)^\perp$

(2) $P_{\overline{R(K)}}y$ belongs to $R(K)$

(3) $\sum_{n \in \mathcal{E}} \mu_n^2 |(y, u_n)|^2 < \infty$.

COROLLARY 8.14. $Kx = y$ has a solution if and only if (3) holds and y belongs to $N(K^*)^\perp$, or if and only if $\sum |(y, u_n)|^2/\lambda_n^2 < \infty$ (where if $\lambda_n = 0$ we set $(y, u_n) = 0$).

If y satisfies Picard's criterion, then the solution with minimal norm of $Kx = y$ is given by $K^\dagger y$. And the set of all solutions is $\sum \mu_i(y, u_i)v_i + (spanV)^\perp$. The solution is unique if and only if $\overline{spanV} = X$, i.e., V is a Schauder basis for X.

For the compact operator

$$(Kx)(s) = \int_0^1 k(s,t)s(t)dt \quad 0 \le s \le 1 \tag{8.9}$$

where k belongs to $L^2([0,1]x[0,1])$. Then we can express the kernel $k(s,t)$ as $k(s,t) = \sum_{n \in \mathcal{E}} u_i(s)v_i(t)/\mu_i$ in terms of the singular system. In particular we have the trace formula

$$\sum_{n \in \mathcal{E}} 1/\mu_n^2 = \int_0^1 \int_0^1 k^2(s,t)dsdt. \tag{8.10}$$

8.2 Singular Value Analysis

The general one-dimensional imaging equation can be written

$$g(t) = \int_{-\infty}^{\infty} f(s)k(t,s)ds \tag{8.11}$$

where f is the scalar complex amplitude of the one-dimensional, object function, $g(t)$ is the scalar complex amplitude of the image function and $k(y,x)$ is the system point-spread function. With coherent illumination, for an ideal diffraction-limited, optical instrument, assuming f has finite support, say $[-X/2, X/2]$, and setting the magnification equal to one, then the point spread function is given by

$$k(t) = \frac{1}{2\pi} \int_{-c}^{c} e^{iwt} dw \tag{8.12}$$

and 8.11 is given by

$$g(x) = \int_{-X/2}^{X/2} \frac{sin(c(x-y))}{\pi(t-s)} f(y)dy \quad |x| \le X/2. \tag{8.13}$$

Here $[-c, c]$ is the passband of the system; i.e., the Fourier transform of the analytic continuation of $g(x)$ is zero outside $[-c, c]$.

If we introduce the new variables $t = 2x/X$ and $c = X\Omega/2$, then equation 8.13 becomes

$$g(t) = \int_{-1}^{1} \frac{sin(c(t-s))}{\pi(t-s)} f(s)ds \quad |s| \leq 1. \tag{8.14}$$

EXERCISE 8.15. Show that $\int k(x-y)k(y)dy = k(x)$ for point spread function $k(x)$.

DEFINITION 8.16. The problem of object restoration in diffraction-limited imaging can be stated as: find a function f in $L^2(-1, 1)$ such that

$$g(t) = (Af)(t) = \int_{-1}^{1} \frac{sin(c(t-s))}{\pi(t-s)} f(s)ds$$

for $-\infty < t < \infty$. The Shannon number is defined by $S = 2c/\pi = X\Omega/\pi$ and the Rayleigh resolution distance is defined by $R = X/S = \pi/\Omega = X/2c$.

We have seen in the last section that A maps $L^2(-1, 1)$ into itself (i.e., the geometric image) and has the following properties:

THEOREM 8.17. The integral operator A on $L^2(-1, 1)$ is self-adjoint, injective (i.e., the only solution of $Af = 0$ is $f = 0$), nonnegative, compact and has trace $Tr(A) = S$. The eigenvalues of A are $u_k(t) = \lambda_k^{-1/2} \psi_k(c, t)$ where $\psi_k(c, t)$ are the linear prolate spheroidal functions; here $Au_k = \lambda_k u_k$.

The functions u_k form a basis in $L^2(-1, 1)$. We write the solution to 8.14 as

$$f(x) = \sum g_n u_n(x)/\lambda_n \tag{8.15}$$

$$g_n = \int_{-1}^{1} g(x)u_n(x)dx. \tag{8.16}$$

However, since the λ_n tend to zero exponentially fast for $n > S = 2c/\pi$ this series converges only for a very special class of images.

For normal objects much larger than the wavelength of light the small diffraction spread outside the geometrical image contributes little to the problem. However, for objects with sizes of the order of R, the entire image may be considered for inversion.

Let K denote the extension of A from $L^2(-1, 1)$ to $L^2(-\infty, \infty)$, i.e.,

$$(Kf)(t) = \int_{-1}^{1} sin(c(t-s))f(s)ds \tag{8.17}$$

for $-\infty < t < \infty$. Again the solution to the inversion problem is equivalent to the solution of the integral equation $Kf = g$. Let $\mathcal{B}$ denote the space of band-limited functions on $[-c, c]$. Let B denote the projection operator to $\mathcal{B}$:

$$(Bg)(t) = \int_{-\infty}^{\infty} \frac{\sin(c(t-s))}{\pi(t-s)}g(s)ds = \frac{1}{2\pi}\int_{-c}^{c}\hat{g}(w)e^{iwt}dw. \qquad (8.18)$$

Let D denote the operator

$$Dg(t) = \left\{ \begin{array}{ll} g(t) & |t| \leq 1 \\ 0 & |t| > 1. \end{array} \right. \qquad (8.19)$$

Thus, for an image g, Dg is the restriction of g to the geometrical image region.

Let K^* denote the adjoint of the operator K, i.e.,

$$(K^*g)(t) = \int_{-\infty}^{\infty} k^*(s-t)g(s)ds \quad |t| \leq 1. \qquad (8.20)$$

Since the kernel of K is real, we have $K^* = K^T$. The range of K is dense in $\mathcal{B}$, so $BK = K$. It follows that $K^* = K^*B$ and so the null space of K^* is the orthogonal complement of $\mathcal{B}$. If we set

$$(K^*Kf)(t) = \int_{-1}^{1} h(t-s)f(s)ds \quad |t| \leq 1, \qquad (8.21)$$

then by Parseval's equality we have

$$h(t) = \int_{-\infty}^{\infty} k(s-t)k(t)^*ds = \frac{1}{2\pi}\int_{-c}^{c} e^{-iwt}dw = \sin(ct)/\pi t. \qquad (8.22)$$

In summary, we have:

THEOREM 8.18. (1) $K^T = DB$, i.e.,

$$(K^Tg)(t) = \int_{-\infty}^{\infty} \sin(c(t-s))g(s)ds \quad |t| \leq 1.$$

(2) The null space of K^T is the subspace of L^2 functions whose Fourier transforms are zero over $[-c, c]$.

(3) $K^TK = DBK = DK = A$.

If we introduce the functions

$$v_k(t) = \psi_k(c, t) \quad -\infty < t < \infty \qquad (8.23)$$

which form an orthonormal system in $\mathcal{B}$, then

$$Ku_k = \sqrt{\lambda_k} v_k \tag{8.24}$$

$$K^T v_k = \sqrt{\lambda_k} u_k.$$

COROLLARY 8.19. The system (u_k, v_k, λ_k) forms a singular value system for K.

Any image $g(t)$ can thus be expressed as

$$g(t) = \sum_{k=0}^{\infty} g(k) v_k(t) \tag{8.25}$$

where $g(k) = \int_{-\infty}^{\infty} g(t) v_k(t) dt$. This uses the entire image. The solution to 8.14 is given by

$$f(t) = \sum_{k=0}^{\infty} g(k) u_k(t) / \sqrt{\lambda_k}. \tag{8.26}$$

EXERCISE 8.20. If we express the restriction of g to the geometrical region as

$$(Dg)(t) = \sum_{k=0}^{\infty} g'(k) u_k(t) \tag{8.27}$$

where $g'(k) = \int_{-1}^{1} g(t) u_k(t) dt$, then show that $g'(k) = \sqrt{\lambda_k} g(k)$; and so the solution of $Af = Dg$ is given by

$$f(t) = \sum g'(k) u_k(t) / \lambda_k \tag{8.28}$$

and is equivalent to 8.26.

8.3   The Effect of Noise on Inversion

The impact of noise on inversion has been discussed by Bertero and Pike. If f is an object on $L^2(-1, 1)$ which has a basis $\{u_k\}$, then

$$f(t) = \sum a(k) u_k(t) \quad |t| \leq 1 \tag{8.29}$$

where $a(k) = \frac{1}{\sqrt{\lambda_k}} \int g(t) v_k(t) dt$ if singular functions are used and $a(k) = \frac{1}{\sqrt{\lambda_k}} \int_{-1}^{1} g(t) u_k(t) dt$ if the eigenfunction method is used.

If there is noise $n(t)$ present, we set

$$g_n(t) = g(t) + n(t) \quad -\infty < t < \infty. \tag{8.30}$$

If we resolve $n(t)$ into in band and out of band components, then

$$n(t) = \sum b(k)v_k + \frac{1}{2\pi} \int_{|w|>c} \hat{n}(w)e^{itw}dw \tag{8.31}$$

where $b(k) = \int_{-\infty}^{\infty} n(t)v_k(t)dt$. The effect of noise on the coefficients of the reconstructed object in the singular value method is

$$a(k)^{(s)} = \frac{1}{\sqrt{\lambda}} \int_{\infty}^{\infty} g(t)v_k(t) = a(k) + b(k)/\sqrt{\lambda_k} \tag{8.32}$$

while for the eigenfunction method

$$a(k)^{(e)} = \frac{1}{\lambda_k} \int_{-1}^{1} g(t)u_k(t)dt = a(k) + \frac{b(k)}{\sqrt{\lambda_k}} + \frac{c(k)}{\lambda_k} \tag{8.33}$$

where $c(k) = \frac{(-1)^k}{\sqrt{2\pi c}} \int_{|w|>c} \hat{n}(w)\psi_k(c,\omega/c)dw$.

EXERCISE 8.21. If the object and image noise are white noise uncorrelated processes with power spectrum E and $\mathcal{E}$ respectively – i.e.,

$$< f^*(t)f(t') > = E^2\delta(t - t') \tag{8.34}$$

$$< n^*(t)n(t') > = \mathcal{E}^2\delta(t - t')$$

show that $< |a(k)^{(s)}|^2 > = E^2 + \mathcal{E}^2/\lambda_k$ and $< |a(k)^{(e)}|^2 > = E^2 + \mathcal{E}^2/\lambda_k^2$. Thus, in the inversion process we can estimate only those components such that $\lambda_k \geq (\mathcal{E}/E)^2$ for the singular function method and components such that $\lambda_k \geq \mathcal{E}/E$ for the eigenfunction method.

8.4 Laplace Transform Inversion

Numerical solution of integral equations of the form

$$g(t) = \int_0^{\infty} K(vt)p(v)dv \quad 0 \leq t < \infty \tag{8.35}$$

have been studied by McWhirter, Pike, Bertero and others. The eigenvalues of the equation

$$\int_0^{\infty} K(vt)\phi_s(v)dv = \lambda_s\phi_s(t) \tag{8.36}$$

are given by the Mellin transform of K:

$$\tilde{K}(s) = \int_0^\infty x^{s-1} K(x) dx. \tag{8.37}$$

Viz., $\phi_s^\pm(v) = \sqrt{\tilde{K}(s)} v^{-s} \pm \sqrt{\tilde{K}(1-s)} v^{s-1}$ where $\lambda_s^\pm = \pm\sqrt{\tilde{K}(s)\tilde{K}(1-s)}$.
K must satisfy $\int_0^\infty |K(x)| x^{-1/2} dx < \infty$ for these transforms to exist.

We set $s = \frac{1}{2} + iw$ where w is real and write the eigenvalues as $\lambda_w^\pm = \pm|\tilde{K}(\frac{1}{2} + iw)|$.

Consider the special case

$$g(t) = \int_0^\infty e^{-\alpha vt} cos(\beta vt) p(v) dv \tag{8.38}$$

where α, β are real. The Mellin transform becomes

$$\tilde{K}(\frac{1}{2} + iw) = \int_0^\infty z e^{-\alpha z} cos(\beta z) dz \tag{8.39}$$

$$= \Gamma(\frac{1}{2} + iw) cos(\frac{1}{2} + iw) tan^{-1}(\frac{\beta}{\alpha}) / (\alpha^2 + \beta^2)^{1/4 + iw/2}.$$

In the limit $\beta \to 0$, 8.39 becomes the Laplace transform and $\tilde{K}(1/2 + iw) \to \alpha^{-1/2 - iw} \Gamma(1/2 + iw)$ and the eigenvalue spectrum is given by

$$|\lambda_w^\pm|^2 = \frac{1}{\alpha} |\Gamma(1/2 + iw)|^2 = \frac{\pi}{\alpha cosh(\pi w)} \tag{8.40}$$

which decays as $(\pi/\alpha) e^{-\pi w}$ for large w.

The Laplace transform for a function of compact support on $[a, b]$ can be viewed as a linear operator $K : L^2(a, b) \to L^2(0, \infty)$ given by

$$(Kf)(p) = \int_a^b e^{-pt} f(t) dt \quad 0 \le p < \infty. \tag{8.41}$$

It is easy to show that K is continuous and $Kf = 0$ has only the trivial solution, i.e., K is injective. The adjoint operator K^* is given by $K^*(g)(t) = \int_0^\infty e^{-tp} g(p) dp, a \le t \le b$, and one can check that K^* is a compact linear operator from $L^2(0, \infty)$ into $L^2(a, b)$. Since K is injective, the range of K^* is dense in $L^2(a, b)$. K^* is also injective and, therefore, the range of K is dense in $L^2(0, \infty)$.

The Laplace operator K thus admits a singular system (u_k, v_k, α_k) where

$$Ku_k = \alpha_k v_k \tag{8.42}$$

$$K^* v_k = \alpha_k u_k$$

$$KK^*u_k = \alpha_k^2 u_k.$$

From the results above, the compact self-adjoint operator KK^* is given by

$$KK^* = \int_a^b f(s)ds \quad a \le t \le b. \tag{8.43}$$

The operator KK^* is of trace class and

$$Trace(KK^*) = \sum_{k=0}^\infty \alpha_k^2 = \int_0^\infty dt/zt = log(\gamma)/2 \tag{8.44}$$

where $\gamma = b/a$.

The set $\{u_k\}$ is an orthogonal basis of $L^2(a,b)$ and $\{v_k\}$ is an orthogonal basis of $L^2(0,\infty)$. Thus, if g is in the range of K, then $f = K^{-1}g$ is given by

$$f(t) = \sum g(k)u_k(t)/\alpha_k \tag{8.45}$$

where $g(k) = \int_0^\infty g(p)v_k(p)dp$. This solution does not depend continuously on the data since the singular values α_k tend to zero, in fact, exponentially fast by the results of Hille and Tamarkin. In the presence of noise this expansion generally is not convergent.

The most simple regularization technique is to truncate this series:

$$\tilde{f}(t) = \sum_{k=0}^{K_s-1} g(k)u_k(t)/\alpha_k \tag{8.46}$$

which is a smoothed version of f(t). This can be seen by using

$$g(k) = \alpha_k \int_a^b f(s)u_k(s)ds \tag{8.47}$$

to get

$$\tilde{f}(t) = \int_a^b M_s(t,s)f(s)ds \tag{8.48}$$

where $M_s(t,s) = \sum_{k=0}^{K_s-1} u_k(t)u_k(s)$.

In the presence of noise on g(p) 8.46 has an additional noise component. However, McWhirter and Pike have shown that this noise contribution does not exceed 10 per cent if K_s is chosen as the greatest of the values of the index k such that $\alpha_k \ge \mathcal{E}/E$ where $\mathcal{E}^2$ is the power spectrum of the noise and E^2 is the power spectrum of the signal. So in the presence of noise, it is possible to extract from the given Laplace transform the projection of f(t) on the subspace spanned by the singular function $u_0(t), ... u_{K_s-1}$. However, this space is very small; e.g., in case $\gamma = 2$, there are only three singular values greater than 10^{-3}. So Laplace transform inversion is a severely ill-posed problem.

8.5   Resolution with Incoherent Illumination

The problem of object restoration for incoherent illumination is equivalent
to the inversion of the integral operator

$$(Af)(x) = \int_{-X/2}^{X/2} \frac{\sin^2(\Omega(x-y))f(y)dy}{\pi\Omega(x-y)^2} \quad |x| \leq X/2. \tag{8.49}$$

The function

$$\hat{S}(w) = \int_{-\infty}^{\infty} \frac{\sin^2(\Omega x)e^{-iwx}dx}{\pi\Omega x^2} = (1 - \frac{|w|}{2\Omega})rect(w/2\Omega) \tag{8.50}$$

is the normalized optical transfer function. Here

$$rect(t) = \left\{ \begin{array}{ll} 1 & |t| \leq 1 \\ 0 & \text{otherwise.} \end{array} \right. \tag{8.51}$$

Thus, if we continue the image $g = Af$ over the whole image space, then
its Fourier transform vanishes outside $[-2\Omega, 2\Omega]$; i.e., the bandwidth for
incoherent imaging is twice the bandwidth for coherent imaging. Thus, the
Nyquist interval for g is one-half the Rayleigh resolution distance $R = \pi/\Omega$.

Making the change of variables $t = 2x/X$ and $c = X\Omega/2$, we rewrite
8.49 as

$$(Af)(t) \int_{-1}^{1} \frac{\sin^2(c(t-s))f(s)ds}{\pi c(t-s)^2} \quad |t| \leq 1. \tag{8.52}$$

8.6   Reproducing Kernel Hilbert Spaces

The reproducing kernel Hilbert space which was studied in detail by Aronszajn has appeared in several areas of modern engineering analysis. We
propose to use this concept below in the context of iterative solutions to
integral equations of the type which arise in the image restoration problem.

DEFINITION 8.22. A reproducing kernel Hilbert space (RKHS) H is a
Hilbert space of functions defined on a set T such that there exists a unique
function $K(s,t)$ on T x T such that
(1) $K(.,t)$ is in H for all t in T
(2) $x(t) = (x, K(.,t))$ for all t in T and all x in H.

Clearly $K(s,t)$ is nonnegative definite symmetric kernel. Conversely, by
the Aronszajn-Moore theorem every nonegative definite symmetric function
K on T x T determines a unique Hilbert function space H_K for which K
is the reproducing kernel.

THEOREM 8.23. Let B be the class of $L^2(-\infty, \infty)$ functions such that their Fourier transforms vanish everywhere outside the interval $(-\pi, \pi)$. Then B is an RKHS on $T = R$, where $K(s,t) = sin(\pi(t - s))/\pi(t - s)$.

COROLLARY 8.24. Let H be the class of $L^2(0, \infty)$ functions whose sine-transforms

$$F(w) = lim_{A \to \infty} \frac{2}{\pi} \int_{-A}^{A} sin(wt)f(t)dt \qquad (8.53)$$

vanish almost everywhere outside $(0, \pi)$. Then H is an RKHS on $T = (0, \infty)$ with kernel

$$K_H(s,t) = \frac{1}{\pi}(\frac{sin\pi(t - s)}{t - s} - \frac{sin\pi(t + s)}{t + s}). \qquad (8.54)$$

THEOREM 8.25. For an abstract RKHS H with kernel $K(s,t)$ on T, let $\{t_1, ...t_n\}$ be n sampling points in T with real finite sample values $\{M_1, ...M_n\}$. Let H_o denote the subspace of f in H such that $f(t_i) = M_i$, $i = 1,...n$. Then for any g in H the unique element in H_o for which

$$min_{f \in H_o} \|f - g\|^2 = \|f_o - g\|^2 = \sum_{i=1}^{n} d_i^2 \qquad (8.55)$$

is given by $f_o(s) = g(s) + \sum_{i=1}^{n} d_i D_i(s)$ where

$$d = \frac{1}{(G_{i-1}G_i)^{1/2}} \begin{vmatrix} K(t_1,t_1) & ... & K(t_n,t_1) \\ .. & .. & .. \\ K(t_1,t_{n-1}) & ... & K(t_n,t_{n-1}) \\ M_1 & ... & M_n \end{vmatrix} \qquad (8.56)$$

$m_i = M_i - g(t_i), G_o = 1, G_i = det K(t_j, t_k), j, k = 1, ...i$

$$D_i(s) = \frac{1}{(G_iG_{i-1})^{1/2}} \begin{vmatrix} K(t_1,t_1) & ... & K(t_n,t_1) \\ .. & .. & .. \\ K(t_1,t_{n-1}) & ... & K(t_n,t_{n-1}) \\ K(s,t_n) & ... & K(s,t_n) \end{vmatrix}. \qquad (8.57)$$

COROLLARY 8.26. The signal f in H with f(t) = M and minimum $\|f\|^2$ is given by f(s) = MK(s,t)/K(t,t).

DEFINITION 8.27. A class Ω of functions on T is said to possess a sampling expansion for $\{t_i \in T\}$ if there exist a set of sampling functions $\psi(s, t_i)$ in Ω for which

(1) $\psi_i(t_j, t_i) = \begin{cases} 1 & i = j \\ 0 & i \neq j \end{cases}$

(2) for any f in Ω there is an uniformly convergent expansion given by $f(s) = \sum f(t_i)\psi_i(s, t_i)$ for s in T.

THEOREM 8.28. Given an RKHS H with kernel K(s,t) on T, let $\{\phi_i(s, t_i)|t_i \in T\}$ be a complete orthonormal (CON) system in H. If there are constants c_i such that $\phi_i(s, t_i) = c_i K(s, t_i)$ and $|K(t, t)| \leq c < \infty$ for t in T then the CON expansion of any f in H is given by

$$f(s) = \sum a(i)\phi_i(s, t_i) \tag{8.58}$$

for s in T, $a(i) = (f, \phi_i)$, and $\psi(s, t_i) = c_i\phi_i(s, t_i)$ is a sampling expansion.

Since $\phi_i = sin(\pi(s - i)/$ is complete and orthonormal in B we have

COROLLARY 8.29. In the RKHS B any f in B possesses a sampling expansion given by

$$f(s) = \sum f(i)sin(\pi(s - i))/\pi(s - i)$$

for s in R.

A finite Shannon sampling expansion is given by

$$g(s) = c \sum f(t_i)sin(s - t_i). \tag{8.59}$$

In the case $c = 1, t_i = i, i = 1, ...n$, f in B satisfies the two properties:
Interpolation: $g(t_i) = f(t_i), i = 1, ...n$
Minimum Energy: $\|g\|^2 = min_{h \in B|h(t_i)=f(t_i)i=1,...n}\|h\|^2$.
In general, these two properties do not hold for an arbitrary set of sample points $\{t_i\}$. Conditions for these two properties to hold are given by:

THEOREM (Yao) 8.30. A necessary and sufficient condition for a finite expansion of the form 8.59 with arbitrary distinct real sample points $\{t_i\}$ to satisfy the Interpolation and Minimum Energy properties is that $c = 1/\lambda_n, f(t_i) = \theta_{n,i}$ where λ_n is the largest eigenvalue of $K\theta = \lambda\theta$ with eigen vector θ_n where

$$K = [\frac{sin\pi(t_i - t_j)}{\pi(t_i - t_j)}]_{i,j=1,...n} \tag{8.60}$$

$$\theta^T = (\theta_1, ..., \theta_n)$$

One reason for using RKHS in approximation theory is that strong convergence in H implies pointwise convergence, since

$$|f(t) - f_n(t)| = |(f - f_n, K_t)_K| \leq \|f - f_n\|\sqrt{Q(t,t)}. \tag{8.61}$$

Nashed has used RKHS to study the convergence of approximations to minimal norm solutions to integral equations of the form

$$Kx = \int_0^1 K(.,t)s(t)dt = y. \tag{8.62}$$

THEOREM 8.31. Let K(s,t) be a continuous reproducing kernel and let λ_n, ϕ_n be the eigenvalues and eigenfuctions of K. Then

$$K(s,t) = \sum \lambda_n \phi_n(s)\phi_n(t) \tag{8.63}$$

converges pointwise and

$$\sum \lambda_n^2 = \int \int K^2(s,t)dsdt \tag{8.64}$$

$$H_K = \{f | \sum_{n=1}^{\infty} (f,\phi_n)/\lambda_n < \infty\}$$

where if $\lambda_n = 0$ we have $(f,\phi_n) = 0$. And the inner product of H_K is given by

$$< f,g >= \sum_{n=1}^{\infty} (f,\phi_n)(g,\phi_n)/\lambda_n = (K^{-1/2}f, K^{-1/2}g) \tag{8.65}$$

for g,f in H_K.

Landweber in 1951 proposed an iterative method for solving integral equations of the first kind

$$\int_{-1}^1 K(x,y)f(y)dy = g(x) \tag{8.66}$$

where K(x,y) is an L^2-kernel and g belongs to $L^2(-1,1)$. If we write the equation as

$$Tf = g, \tag{8.67}$$

then the Landweber iteration method is

$$f_n = f_{n-1} + T^*(g - Tf_{n-1}) \ n=1,2... \tag{8.68}$$

where T^* is the adjoint operator to T. This is given by

$$T^*g = \int_{-1}^1 K(x,y)g(x)dx \tag{8.69}$$

for g in $L^2(-1,1)$. Clearly for f in $L^2(-1,1)$

$$< Tf, g >= \int \int K(x,y)f(y)g(x)dydx = \int_{-1}^1 f(y)dy \int_{-1}^1 K(x,y)g(x)dx$$

(8.70)

$$=< f, T^*g > .$$

We return to this iterative method in the next chapter.

We relate now RKHS and Picard's condition. Let T be the integral operator

$$(Tf)(s) = \int_{-1}^1 K(s,t)f(t)dt = g(s)$$

and define

$$H_Q = \{g \in L^2(-1,1)| \sum |(g,u_n)|^2/\lambda_n < \infty\} \qquad (8.71)$$

where $\lambda_n = \mu_n^{-2}$ and the sum is taken over all non zero λ_n. Define the inner product $(f,g)_Q = \sum \frac{1}{\lambda_n}(f,u_n)(g,u_n)$ on H_Q. Then H_Q is a RKHS. Viz., set $Q(s,s') = K(s,t)\widetilde{K}(s',t)dt$ and $Q_s(s') = Q(s,s')$. Then H_Q is an RKHS with kernel Q. The spaces H_Q and $L^2(-1,1)$ are related by $H_Q = Q^{-1/2}L^2(-1,1)$ with $\|g\|_Q = \|Q^{-1/2}g\|_{L^2}$.

Assume $R(T)$ is dense in H_Q. Then $N(T^*) = R(T) = H_Q$ implies $N(T^*) = 0$. And, hence, $N(T) = 0$ since T is self-adjoint. By the open mapping theorem we have

THEOREM 8.32. $T : L^2 \to H_Q$ is onto and $T^{-1} : H_Q \to L^2$ is bounded.

DEFINITION 8.33. The superdirective ratio for a finite line source is given by

$$\gamma(f) = \frac{\int_{-\infty}^{\infty} |g(u)|^2 du}{\int_{-1}^1 |g(u)^2 du}. \qquad (8.72)$$

For a function g in H_Q a solution (i.e., an aperture distribution) f in $L^2(-1,1)$ exists. The problem of antenna synthesis is given by the constrained least squares problems:

$$inf\{\|g - Tf\| | f \in \Gamma\}$$

where $\Gamma = \{f \in L^2(-1,1)|\gamma(f) \leq \gamma_0\}$.

EXERCISE 8.34. Show that $\gamma(f) = 2\pi\|Q^{-1/2}g\|/c\|g\|$ where Q is the integral operator on $L^2(-1,1)$ with kernel

$$Q(x,s) = \int_{-1}^1 K(x,t)K(s,t)dt = \int_{-1}^1 e^{ict(x-s)}dt = \frac{sin(c(x-s))}{c(x-s)}.$$

Chapter 9

Kaczmarz Method, Landweber Iteration, Gerchberg-Papoulis and Regularization

9.1 Introduction

The Kaczmarz (1937) method (or K-method) has been revived in the study of ART (algebraic reconstruction technique) in tomography. The K-method always converges to a solution provided that at least one solution exists.

Consider the equation

$$Ax = b \qquad (9.1)$$

where $A : H_1 \to H_2$ with $dim H_1 = m$ and $dim H_2 = n$. Let N(A) be the null space of A and $N(A)^\perp$ its orthogonal complement. Assume b is in $R(A)$ – i.e., assume that 9.1 has at least one solution in H_1. Then we have:

THEOREM 9.1. Under the conditions above, the K-method converges to

$$x = \hat{u} + P_{N(A)}x \qquad (9.2)$$

where $\hat{u}$ is the unique solution of 9.1 in $N(A)^\perp$ and $P_{N(A)}$ is the projection from H_1 into N(A).

Let $(e_1, ... e_n)$ be an orthonormal basis of H_2. Then the K-method is given by

$$x_{k+1} = x_k - \frac{(Ax_k - b, e_{i_k})A^* e_{i_k}}{\|A^* e_{i_k}\|}. \tag{9.3}$$

Without loss of generality one may take $\|A^* e_{i_k}\| = 1$, $i = 1,...,n$.

The selection of $\{i_k\}$ represents the strategy. The optimal strategy is to compute for every fixed k the defects $|(Ax_k - b, e_k)|$, i=1,...n and then project x_k onto the particular plane for which this defect is maximal.

9.2 The Cimmino Method

An almost always convergent method which uses the weighted average of the defects was introduced by Cimmino in 1938. The Cimmino method is given by

$$x_{k+1} = x_k - 2 \sum_{i=1}^{n} w_i \frac{(Ax_k - b, e_i)A^* e_i}{\|A^* e_i\|^2} \tag{9.4}$$

or

$$x_{k+1} = x_k - 2 \sum w_i (Ax_k - b, e_i)A^* e_i \tag{9.5}$$

assuming the normalization $\|A^* e_i\| = 1$. Here w_i are weights, $w_i > 0$, with $\sum_{i=1}^{n} w_i = 1$.

THEOREM 9.2. If b belongs to R(A), then 9.4 always converges, provided $rank(A) > 1$.

A modification of the Cimmino method has been presented by Ansorge. Viz., let

$$x_{k+1} = x_k - 2 \sum_{i=1}^{n} w_i^{(k)} (Ax_k - b, e_i)A^* e_i \tag{9.6}$$

where

$$w_i^{(k)} = |(Ax_k - b, e_i)|^\gamma (1 - w_o) / \sum_{j=1}^{n} |(Ax_k - b, e_j)|^\gamma \tag{9.7}$$

where $\gamma > 0$.

THEOREM (Ansorge) 9.3. If b belongs to R(A), then 9.6 converges to a solution of Ax = b.

We summarize the results on the Cimmino method in the following:

THEOREM (Cimmino-Ansorge-Votruba) 9.4. If $rank(A) > 1$ and $m_i = \|r_i\|^2$, where $r_1,...r_n$ are the rows of A, $1 \leq i \leq n$, then

$$x^{(k)} = x^{(k-1)} - \frac{2}{\mu} \sum_{i=1}^{n} m_i (r_i, x^{(k-1)} - y_i)r_i / \|r_i\|^2 \tag{9.8}$$

where $\mu = \sum_{k=1}^{n} m_i$ converges to $(I-P)x^{(o)} + A^{\dagger}y$ where P is the orthogonal projection on $N(A)^{\perp}$.

9.3 Cimmino for Fredholm

The method of Cimmino has been extended to Fredholm equations by Kammerer and Nashed. Let

$$Tx = \int_a^b K(.,t)x(t)dt = y \qquad (9.9)$$

for y in $L^2(a,b)$. Let H_s denote the hyperplane

$$H_s = \{x \in L^2(a,b)| \int_a^b K(s,t)x(t)dt = y(s)\} \qquad (9.10)$$

for almost every s in [a,b]. Then the orthogonal projection of x_o in $L^2(a,b)$ onto H_s is given by

$$z_s = x_o + \lambda(s)K(s,.) \qquad (9.11)$$

where

$$\lambda(s) = \frac{y(s) - \int_a^b K(s,r)x_o(r)dr}{\int_a^b |K(s,r|^2 dr}. \qquad (9.12)$$

If we let $m(s) = \int_a^b |K(s,t)|^2 dt$ denote the mass density, and if we let $\beta = \int_a^b \int_a^b |K(s,t)|^2 dt ds$ denote the total mass, then the generalized Cimmino method is

$$x_{n+1} = x_n - \frac{2}{\beta}T^*Tx_n(t) + \frac{2}{\beta}(T^*y)(t). \qquad (9.13)$$

First we develop a generalization of the results on the iterative methods of Fridman and Bialy.

THEOREM 9.5. For $0 < \alpha < 2/\|T\|^2$, the sequence

$$x_n = (I - T^*T)x_{n-1} + \alpha T^*y \qquad (9.14)$$

converges for any initial approximation to $T^{\dagger}y + (I - P)x_o$, which is the best approximate solution of Tx = y.

This is based on Petryshyn's result:

THEOREM (Petryshyn) 9.6. Let $T : X \to Y$ be a bounded linear operator on Hilbert spaces X,Y. Assume $N(T) \neq 0$ and let α be any positive real number. Then $(I - \alpha T^*T)^n$ converges to $I - P_{N(T)}^{\perp} = I - P$ in operator norm if and only if $R(T)$ is closed and $0 < \alpha < 2/\|T\|^2$; the error estimate is

$$\|(I - \alpha T^*T)^n - (I - P)\| \le (\frac{c^2 - 1}{c^2 + 1})^n \qquad (9.15)$$

where $c = \|T\|\|T^*\|$ is the pseudo-conditional number of T.

If T is closed, there is a Neumann-type series expansion

$$T^\dagger = \sum_{n=0}^{\infty}(I - \alpha T^*T)^n \alpha T^* \qquad (9.16)$$

for $0 < \alpha < 2/\|T\|^2$.

Using the successive approximations

$$x_n = (I - \alpha T^*T)x_{n-1} + \alpha T^*y \qquad (9.17)$$

and the relationship $T^*TT^\dagger y = T^*Qy = T^*y$ one finds that

$$x_n - T^\dagger y = (I - \alpha T^*T)^n(x_o - T^\dagger y). \qquad (9.18)$$

Thus,

$$lim_{n\to\infty}(x_n - T^\dagger y) = \qquad (9.19)$$

$$lim_{n\to\infty}(I - \alpha T^*T)^n(x_o - T^\dagger y) = (I - P)(x_o - T^\dagger y) = (I - P)x_o$$

since $T^\dagger y$ belongs to $R(T^*)$, which completes the proof of 9.6

What about the case when $R(T)$ is not necessarily closed? In this case

$$\sum_{k=0}^{m}(I - \alpha T^*T)^k \alpha T^*y \qquad (9.20)$$

for $0 < \alpha < 2/\|T\|^2$ converges in norm monotonically to $T^\dagger y$ for any y in $D(T^\dagger) = R(T) \bigoplus R(T^\dagger)$. In summary, we have:

THEOREM (Kammerer-Nashed) 9.7. Let $T : X \to Y$ be a bounded operator with $R(T)$ not necessarily closed. Then 9.14 starting with $x_o = 0$ converges in norm monotonically to $T^\dagger y$ whenever y belongs to $D(T^\dagger)$ and α is as above. If Qy belongs to $R(TT^*)$, then

$$\|x_n - T^\dagger y\| \le \frac{\|T^\dagger y\|\|(TT^*)^\dagger y\|^2}{\|(TT^*)^\dagger y\|^2 + n\alpha(2 - \alpha)\|T\|^2)\|T^\dagger y\|^2}. \qquad (9.21)$$

Fridman considered the case

$$Tx = \int_a^b K(.,t)x(t)dt \qquad (9.22)$$

where $K(s,t)$ is a symmetric positive definite kernel. He showed

THEOREM 9.8. If $Tx = y$ has a solution, then for any α where $0 < \alpha < 2\alpha_1$ where α_1 is the smallest eigenvalue of $K(s,t)$ then

$$x_{n+1}(s) = x_n(s) + \alpha(y(s) - Tx_n(t)) \qquad (9.23)$$

converges in $L^2(a,b)$ norm to the solution starting from any x_o in $L^2(a,b)$.

Bialy generalized Fridman's results to the following case.

DEFINITION 9.9. A bounded linear operator A on Hilbert space H is said to be nonnegative in $(Ax, x) \geq 0$ for all x in H.

THEOREM (Bialy) 9.10. Let A be a nonnegative bounded linear operator $A : H \to H$. Consider the iterative process

$$x_{n+1} = x_n + \alpha(y - Ax_n) \qquad (9.24)$$

where $0 < \alpha < 2\|A\|^{-1}$. Then $Ax_n \to Qy$ where Q is the orthogonal projection on $R(A)$ and x_n converges if and only if $Ax = y$ has a solution in which case $x_n \to (I - P)x_o + \hat{x}$ where $\hat{x}$ is the solution of minimal norm.

If the initial approximation is $x_o = 0$, then $\{x_n\}$ converges to the minimum norm solution. Bialy's result really applies to any operator since the minimum norm least squares solution (if it exists) is the minimum norm solutuion of $A^t Ax = A^t y$ where $A^t A$ is nonnegative. Thus, we have the following result due originally to Landweber:

THEOREM (Landweber) 9.11. Let $A : H_1 \to H_2$ be a bounded linear operator; for y in $R(A) \oplus R(A)^{\perp}$ consider the iterative equation

$$x_o = 0 \qquad (9.25)$$

$$x_n = x_{n-1} + \alpha A^t(y - Ax_{n-1}, \ n \geq 1$$

for $0 < \alpha < 2/\|A^t A\|$. Then $\{x_n\}$ converges to the minimum norm least squares solution $A^\dagger y$.

In summary, the result of Kammerer and Nashed, theorem 9.7, generalizes the results of Fridman and Bialy to the setting of best approximate solutions and expresses the results and error bounds in terms of the generalized inverse.

Kammerer and Nashed have also generalized Cimmino's method as described above. From theorem 9.7 and a study of integral operators they have shown:

THEOREM (Cimmino Generalized) 9.12. If

$$Tx = \int_a^b K(.,t)x(t)dt = y \qquad (9.26)$$

for y in $L^2(a, b)$ and $R(T) > 1$, then the generalized method of Cimmino converges monotonically to a best approximate solution of minimal norm of 9.25 starting from $x_o = 0$ for any y in $D(T^\dagger) = R(T) \bigoplus R(T^\dagger)$ and

$$\|x_n - T^\dagger y\| \leq \frac{\|T^\dagger y\|^2 \|(TT^*)^\dagger y\|^2}{\|(TT^*)^\dagger y\|^2 + \frac{4n}{\beta^2}(\beta - \|T\|^2)\|T^\dagger y\|^2} \quad (9.27)$$

where $\beta = \int_a^b \int_a^b K(u, s) du ds$.

Note here that if $\lambda_1 \geq \lambda_2...$ are the eigenvalues of T^*T and since $R(T^*T) > 1$, then the key observation is that λ_1, which equals the spectral radius of T^*T, is strictly less than β.

9.4   Filtering and Landweber's Iteration

For the integral equation

$$(Kf)(x) = \int_a^b K(x, y)f(y) dy = g(x) \quad (9.28)$$

where $K(x, y)$ is an L^2-kernel, Landweber proposed the iteration

$$f_k = f_{k-1} + K^*(g - Kf_{k-1}). \quad (9.29)$$

In the presence of noise the equation to solve has the form $Kf = g + \varepsilon$ where g belong to $R(K)$ and ε is an error term. Using the singular function system $\{u_n, v_n, \mu_n\}$ we might approximate a solution by

$$f_N = \sum_{n=1}^N (g, u_n)\mu_n v_n + \sum_{n=1}^N (\varepsilon, u_n)\mu_n v_n. \quad (9.30)$$

One approach is to develop an operator D which will cut off the higher order terms above μ_N^2. This was developed by Strand where the Landweber iteration is replaced by

$$f_k = f_{k-1} + DK^*(g - Kf_{k-1}) \, k = 1, 2,... \quad (9.31)$$

Let $Dv_n = p_n v_n$ where $0 < p_n \lambda_n < 2$ for all n, and the sequence $\{p_n\}$ is bounded. Then we have:

THEOREM (Strand) 9.13.   Under the above conditions on D and assuming g belongs to $R(K)$ let $\bar{g} = P_U \bar{g} + g^\perp$ for f_o in L^2 and let $f_o = P_V f_o + f^\perp = f + f^\perp$ where f belongs to V and $f^\perp$ belongs to $V^\perp$. Then

$$lim_{k \to \infty} f_k = \tilde{f} + f^\perp \quad (9.32)$$

where $\tilde{f}$ is the least squares solution of minimum norm for $Kf = \bar{g}$.

Of course the case $D = I$ gives Landweber's original result.

EXAMPLE (Twomey-Tikhonov solution) 9.14. For the matrix equation $Af = g$ if we set $D = (A^T A + \gamma)^{-1}$ for $\gamma > 0$, one finds $p_n = \frac{1}{\lambda_n + \gamma}$. This gives the iterated Twomey-Tikhonov solution.

9.5 Image Restoration by the Projections

The image restoration problem is stated as follows: an ideal image $f(x, y)$ is degraded by a system operator D and random noise $n(x, y)$ is added. The final image is

$$g(x, y) = Df(x, u) + n(x, y). \tag{9.33}$$

In the digitized version we have

$$a_{11}f_1 + a_{12}f_2 + \dots + a_{1N}f_N = g_1 \tag{9.34}$$

$$\cdot\cdot$$

$$a_{M1}f_1 + a_{M2}f_2 \dots + a_{MN}f_N = g_M$$

where g_i are samples of the degraded image. The problem of restoration is thus equivalent to the solution of (9.34) for f_i. The projection or K-method which has been applied by Huang et al. gives for an initial estimate $f^{(o)} = (f_{o1}, \dots, f_{oN})$

$$f_1^{(1)} = f^{(o)} - \frac{(f^{(o)}A_1 - g_1)A_1}{A_1 A_1} \tag{9.35}$$

where $A_1 = (a_{11}, a_{12} \dots a_{1N})$.

Alternate algorithms involving orthogonalization of the sets of equation and prefiltering by Bernstein polynomials have been proposed by Ramakrishnam et al. Improved results are demonstrated in this reference.

9.6 Van Cittert Deconvolution

One of the earliest iterative deconvolution algorithms was given by Van Cittert. The problem involves the integral operator equation

$$A : L^2(S) \rightarrow L^2(T) \tag{9.36}$$

$$f \rightarrow Af(t) = \int_S h(t-s)f(s)ds \text{ t in T.}$$

We assume $S = T$, $\int_{-\infty}^{\infty} |h(t)|^2 dt < \infty$ and $\hat{h}(w) \geq 0$ for all w where $\hat{h}$ denotes the Fourier transform.

THEOREM 9.15. Under these conditions A is a nonnegative bounded linear operator.

To see that A is nonnegative one checks that

$$(Af, f)_{L^2(S)} = \int_S (Af)(s)f(s) = \int_R (h \star J_S f)(s)J_S(s)ds \qquad (9.37)$$

where J_S is the projection onto S and $\star$ denotes convolution. By Parseval's equality we obtain

$$(Af, f)_{L^2(S)} = \int_{-\infty}^{\infty} \hat{h}(w)|(J_S f)(w)|^2 dw \qquad (9.38)$$

which is always nonnegative.

DEFINITION 9.16. The deconvolution problem states: given the output $g(s)$, s in S, of the system described by operator A, recover the input $f(s)$ for s in S.

Van Cittert's iterative approach to this problem was to take:

$$f_o = 0 \qquad (9.39)$$

$$f_{n+1} = f_n + a(g - h \star J_S f_n)$$

that is

$$f_{n+1}(s) = f_n(s) + \alpha(g(s) - \int_S h(s-t)f_n(t)dt \quad s \text{ in S} \qquad (9.40)$$

where $0 < \alpha < 2/\|A\|$. The discrete version of the Van Cittert algorithm is as follows: for $A : l_2(B) \to l_2(B)$ where $l_2(B) = \{a(m)m \in B \subset A^n, \sum_{m \in B} |a(m)|^2 < \infty\}$ and $(Ax)(k) = \sum_{m \in B} h(k-m)x(m)$. As above, we assume $\sum_{m \in B} |h(m)|^2 < \infty$ and $\hat{h}(w) = \sum_m h(m)e^{-2wim} \geq 0$ for all w. Then the discrete Van Cittert algorithm is

$$x_m(k) = x_{m-1}(k) + \alpha(g(k) - h \star J_B x_{m-1}(k)) \text{ k in B.} \qquad (9.41)$$

This will converge by Bialy's theorem to the minimum norm solution of $g(m) = (h \star J_B f)(m)$ for m in B provided at least a solution exists for $0 < \alpha < 2/\|A\|$ (e.g., $0 < \alpha < 2/sup|\hat{h}|$). If we let $\delta : Z \to C$ where $\delta(0) = 1$ and $\delta(k) = 0$ for $k \neq 0$, we can rewrite 9.41 as

$$x_m(k) = \alpha g(k) + \{(\delta - \alpha h) \star J_B x_{m-1}(k)\} \text{ k in B.} \qquad (9.42)$$

DEFINITION 9.17. Let $A : H \rightarrow H$ be a linear operator. If $\|Aa - Ab\| \leq r\|a - b\|$ for a,b in some closed subspace of H and $r < 1$, then A is said to be a contraction operator on that subspace.

THEOREM (Contraction Mapping Theorem) 9.18. If A is a contraction on some subspace, then A has a unique fixed point in that subspace and every sequence of successive approximations

$$x_{k+1} = Ax_k \qquad (9.43)$$

converges to the fixed point x for every starting signal x_o in that subspace; furthermore,

$$\|x - x_k\|_H \leq \frac{r^{k+1}}{1-r}\|x - x_o\|_H. \qquad (9.44)$$

For the integral operator A we take $A = I - \lambda T$ where T is the integral operator 9.40.

EXERCISE 9.19. Show that $(\delta - \alpha h) \star J_B$ is a contraction operator if B is bounded and $|1 - \alpha\hat{h}(w)| < 1$ whenever $\hat{h}(w) \neq 0$. Note that if $\hat{h}(w)$ is real and positive, then this condition means that α must satisfy

$$0 < \alpha < 2/max_w\hat{h}(w). \qquad (9.45)$$

Note that if B is not bounded then the contraction property mapping principle cannot be used to prove convergence. However, Bialy's theorem still can be used. In the continuous deconvolution problem even when S is bounded, the operator $(I - \alpha h) \star J_S$ will not be a contraction mapping in general.

In summary then, Kammerer-Nashed or Bialy iteration provides a non-trivial extension of the Van Cittert algorithm. We note that an improved version of the Van Cittert algorithm with positivity constraints has been studied by Thomas.

9.7 Gerchberg-Papoulis Algorithm

The extrapolation problem can be stated as follows. Let B denote the class of signals that are band-limited to $[-W/2, W/2]$. Let $g_o(s)$ represent a known segment of f in B over the interval $[-T/2, T/2]$.

DEFINITION 9.20. The extrapolation problem is that given $g_o(s)$ as described above, determine f for all x.

Define the projection operators

$$G_W(f) = \begin{cases} 1 & |f| \leq W/2 \\ 0 & \text{otherwise} \end{cases} \qquad (9.46)$$

$$G_T(s) = \begin{cases} 1 & |s| \leq T/2 \\ 0 & \text{otherwise.} \end{cases}$$

DEFINITION 9.21. The Gerchberg-Papoulis extrapolation algorithm is given by

$$g_n = g_o + (1 - G_T)\mathcal{F}^{-1}G_W\mathcal{F}g_{n-1}. \qquad (9.47)$$

We set $B = \mathcal{F}^{-1}G_W\mathcal{F}$ which is the low pass filtering or band-limiting operator.

9.8   Extrapolation by Gori's Algorithm

The iterative technique for achieving super resolution is essentially the Burger Van Cittert algorithm. Viz., let f have known support on the interval $A = (-a,a)$ and assume the Fourier transform $\hat{f}$ of f is specified on $(-\Omega/2, \Omega/2)$. If g is the function where

$$\hat{g}(w) = \begin{cases} \hat{f} & |w| \leq \Omega/2 \\ 0 & \text{otherwise} \end{cases} \qquad (9.48)$$

then g is related to f by

$$g(x) = \int_{-a}^{a} f(y)S(x - y)dy \qquad (9.49)$$

where $S(x)$ is the perfect low pass filter $\Omega sinc(\Omega x)$. Perfect restoration of the object f from the image g involves solving 9.49 for f given g and S.

The Fredholm equation of the first kind 9.49 can be converted into a Fredholm equation of the second kind by setting

$$S(x) = \delta(x) - K(x) \qquad (9.50)$$

viz.,

$$g(x) = f(x) - \int_A f(y)K(x - y)dy. \qquad (9.51)$$

We define K to be the operation of truncation to domain A followed by convolution with $K(x)$:

$$Kh(x) = \int_A h(y)K(x - y)dy. \qquad (9.52)$$

Then 9.51 is f = g+Kf.

The following successive approximations are proposed:

$$f_1 = g \qquad (9.53)$$

$$f_n = g + K f_{n-1}.$$

If we set $f(x) = \sum d(m)\psi_m(x)$ and $g(x) = \sum b(m)\psi_m(x)$ where $A\psi$ is the restriction of ψ to A, then, in general, we have

$$K\psi_m = A\psi_m - \lambda_m\psi_m. \qquad (9.54)$$

If we restrict x to the range of A, then $K\psi_m = (1 - \lambda_m)\psi_m$. Repeated applications of K gives $K^n\psi_m = (1 - \lambda_m)\psi_m$. It follows that

$$K^n g = \sum_{m=0}^{\infty} b(m)(1 - \lambda_m)^n \psi_m. \qquad (9.55)$$

Thus,

$$f_n = \sum_{m=0}^{\infty} b(m)(1 + (1 - \lambda_m) + ... + (1 - \lambda_m)^{n-1})\psi_m = \qquad (9.56)$$

$$\sum_{m=0}^{\infty} b(m)\frac{\lambda_m}{1 - (1 - \lambda_m)^n}\psi_m = \sum_{m=0}^{\infty} d(m)p_n(m)\psi_m.$$

Since λ_m are positive and less than one, $p_n(m)$ converges to unity as n tends to infinity. Thus, we have

$$lim_{n\to\infty} f_n = \sum_{m=0}^{\infty} d(m)\psi_m; \qquad (9.57)$$

i.e., the Neumann series converges to the solution of 9.51.

9.9 Gerchberg's Method of Extrapolation

The Gori algorithm can be described as follows:

(1) restrict the current approximation $f_{n-1}(x)$ to the range (-a,a):

$$f_n^{(1)}(x) = f_{n-1}(x)P(2ax)$$

where

$$P(2ax) = \left\{ \begin{array}{ll} 1 & |x| < a \\ 0 & \text{otherwise} \end{array} \right.$$

(2) convolve $f_n^{(1)}(x)$ with $\delta(x) - \Omega sinc(x\Omega)$ to get $f_n^{(2)}$

(3) add the assigned approximation g(x):

$$f_n(x) = g(x) + f_n^{(2)}(x)$$

(4) let n = n+1 and go to (1).

The Gerchberg algorithm for superresolution is given by

(1) take the inverse Fourier transform of the current approximation to the Fourier transform

$$f_n^{(1)}(x) = \mathcal{F}^{-1}\hat{f}_{n-1}(w)$$

(2) restrict this function to (-a,a)

$$f^{(2)}(x) = f_n^{(1)}(x)P(2ax)$$

(3) take the Fourier transform

$$\hat{f}_n^{(3)}(w) = \hat{f}_{n-1}(w) \star \mathcal{F}(P(2ax))$$

(4) set the transform to zero in the region where it was originally known

$$\hat{f}_n^{(4)}(w) = \hat{f}_{n-1}(w) \star \mathcal{F}(P(2ax)) - \hat{f}_{n-1}(w) \star \mathcal{F}(P2ax))P(\Omega w)$$

(5) add the original known part of the Fourier transform

$$\hat{f}_n(w) = \hat{g}(w) + \hat{f}_n^{(4)}(w)$$

(6) set n = n+1 and go to (1).

If we take the inverse Fourier transform of (2), we have

$$f_n(x) = g(x) + f_{n-1}P(2ax) - f_{n-1}(x)P(2ax) \star \Omega sinc(x\Omega)$$

which is precisely Gori's algorithm. So, in summary, the Papoulis-Gerchberg iteration can be described by

$$\hat{f}_o = 0 \qquad\qquad (9.58)$$

$$f_n = sinc_A \star (J_\Omega \hat{f} + (I - J_\Omega)\hat{f}_{n-1}).$$

9.10 Gerchberg-Papoulis Again

The problem dealt with by Papoulis and Gerchberg was to determine the Fourier transform F(w) of a finite energy Ω-band-limited function f(t) when only a finite segment g(t) of f(t) is known, say on $|t| < T$. So we have $F(w) = 0$ for $|w| > \Omega$. The algorithm is as follows:

(1) let $g_o(t) = g(t)$

(2) form $G_o(w) = \int_{-T}^{T} g(t)e^{-iwt}\,dt$

(3) $F_n(w) = G_{n-1}(w)P_\Omega(w)$

(4) $f_n(t) = \int_{-\Omega}^{\Omega} F(w)e^{iwt}dw = g_{n-1}(t) \star \frac{sin(\Omega t}{\pi t}$

(5) $g_n(t) = f_n(t) + (f(t) - f(t))P_T(t) = g(t)$

(6) $G_n(w) = \int_{-\infty}^{\infty} g_n(t)e^{-iwt}dt$

(7) replace n by n+1 and go to (3).

Gerchberg refers to the algorithm as the energy reduction algorithm since by Parseval's formula we have

$$\frac{1}{2\pi}\int_{-\infty}^{\infty}|F(w)-G_{n-1}(w)|^2dw > \int_{-\infty}^{\infty}|f(t)-f_n(t)|^2dt > \int_{-\infty}^{\infty}|f(t)-g_n(t)|^2dt$$

$$\tag{9.59}$$

$$= \frac{1}{2\pi}\int_{-\infty}^{\infty}|F(w) - G_{n-1}(w)|^2dw.$$

To show convergence we expand f(t) in terms of prolate spheroidal wave functions

$$f(t) = \sum_{k=0}^{\infty} a(k)\phi_k(t). \tag{9.60}$$

Note that if $f(t) = \phi_k(t)$, then

$$f_1(t) = 1 - (1 - \lambda_k)\phi_k(t). \tag{9.61}$$

Suppose that $f_n(t) = A_n\phi_k(t)$ where $A_n = 1 - (1 - \lambda_k)^n$ for some $n \geq 1$. Then by the algorithm we have

$$g_n(t) = A_n\phi_k(t) + (1 - A_n)\phi_k(t)P_T(t) \tag{9.62}$$

and since

$$\int_{-T}^{T} \phi_k(s)\frac{sin\Omega(t - s)ds}{\pi(t - s)} = \lambda_k\phi_k(t) \tag{9.63}$$

and

$$\int_{-\infty}^{\infty} \phi_k(s)\frac{sin\Omega(t - s)ds}{\pi(t - s)} = \phi_k(t) \tag{9.64}$$

we get

$$f_{n+1}(t) = A_n\phi_k(t) + (1 - A_n)\lambda_k\phi_k(t) = A_{n+1}\phi_k(t) \tag{9.65}$$

where $A_{n+1} = A_n + (1 - A_n)\lambda_k$. Thus, we have shown:

THEOREM 9.22.

$$f_n(t) = f(t) - \sum_{k=0}^{\infty} a(k)(1 - \lambda_k)^n \phi_k(t). \tag{9.66}$$

THEOREM 9.23. For all t

$$f_n(t) \rightarrow f(t) \tag{9.67}$$

as $n \rightarrow \infty$.

Proof. Let $e_n(t) = f(t) - f_n(t)$ denote the error term. Then by the last result

$$e_n(t) = \sum_{k=0}^{\infty} a(k)(1 - \lambda_k)^n \phi_k(t). \tag{9.68}$$

The mean square error is

$$E_n = \int_{-\infty}^{\infty} e_n^2(t) dt = \sum a(k)(1 - \lambda_k)^{2n}. \tag{9.69}$$

Since the eigenvalues λ_k are less than one and tend to zero monotonically one can check that $E_n \rightarrow 0$ as $n \rightarrow \infty$. Since $|e_n(t)| \leq \sqrt{E_n \Omega / \pi}$ the result of the theorem follows.

Note that $G(w) = B \sum_{k=0}^{\infty} a(k) \sqrt{\lambda_k} \phi_k(bw)$ where $B = \sqrt{2\pi T/\Omega}, b = T/\Omega, F_n(w) = BP_\Omega$, and

$$G_n(w) = F_n(w) + B \sum_{k=0}^{\infty} a(k) \sqrt{\lambda_k} (1 - \lambda_k)^n \phi_k(bw). \tag{9.70}$$

It follows directly from the last theorem that

COROLLARY 9.24. $F_n(w)$ converges to F(w) as $n \rightarrow \infty$.

9.11   Landweber's Iteration

In this section we show that the Gerchberg-Papoulis algorithm is given by a Landweber iteration scheme. Let

$$(Kg)(x) = \int_{\Omega} g(w) e^{-2\pi i w x} dw$$

for x in A. Theorem 9.5 gives the recursion

$$f_o = 0 \tag{9.71}$$

$$f_m = f_{m-1} + \alpha K^t(g - K f_{m-1})$$

where $0 < \alpha < 2/\|K^t K\|$. The adjoint integral operator is given by

$$K^t(h)(w) = \int_A h(x) e^{2\pi i w x} dx \tag{9.72}$$

for w in Ω. Clearly $\|K^tK\| \leq 1$, so $\alpha = 1$ is an admissible value. Thus 9.71 will converge in the energy norm to the unique solution f of the integral equation $(Kf)(x) = g(x)$ for x in A.

The inverse Fourier transforms $f_m^\vee$ and $f^\vee$ will also converge. Since f_n, f are supported on Ω we have

$$f_n^\vee(x) = \int_\Omega f_n(w)e^{-2\pi i x w}\,dw \qquad (9.73)$$

and

$$f^\vee(x) = \int_\Omega f(w)e^{-2\pi i w x}\,dw. \qquad (9.74)$$

Applying the inverse Fourier transform to 9.71 gives

$$f_o^\vee = 0 \qquad (9.75)$$

$$f_m^\vee = f_{m-1}^\vee + \{\int_A (g(x) - \int_\Omega f(w)e^{-2piixw}\,dw)e^{2\pi i x}\,dx\}^\vee.$$

Setting $g_m = f_m^\vee$ we have

$$g_o = 0 \qquad (9.76)$$

$$g_m = g_{m-1} + sinc_\Omega \star (J_A g - J_A g_{m-1})$$

However, since g is Ω-band-limited we have

$$g_m = sinc_\Omega \star (J_A g + (I - J_A)g_{m-1}) \qquad (9.77)$$

which is the Papoulis-Gerchberg algorithm.

Sanz and Huang have extended 9.71 to the case

$$(Ks)(x) = \int_\Omega \hat{h}^{1/2}(w)e^{2\pi i w x}s(w)\,dw \qquad (9.78)$$

for x in A where $\hat{h}(w)$ is a nonnegative bounded function which is zero for w not in Ω and such that $\int_\Omega \frac{|\hat{g}(w)|^2 dw}{\hat{h}(w)} < \infty$. The Landweber iteration is defined as above for this K. Letting

$$g_m(x) = \int_\Omega f_m(w)\hat{h}^{1/2}(w)e^{-2\pi i w x}\,dw \qquad (9.79)$$

we have the more standard form

$$g_o = 0 \qquad (9.80)$$

$$g_m = g_{m-1} + \alpha h(-z) \star J_A(g - g_{m-1})$$

for $0 < \alpha < 2/sup|\hat{h}(w)|$. Here $h(z) = \int_\Omega h(w)e^{2\pi i w z}\,dz$.

9.12   Discrete-Continuous Iterative Extrapolation

Suppose y(m) for m in Z is a sequence of complex numbers which are band-limited to Ω, i.e.,

$$\sum_{m \in Z} y(m)e^{2\pi imw} = 0 \qquad (9.81)$$

for w not in $\Omega \subset (-1/2, 1/2)$. The discrete-continuous extrapolation problem assumes a piece of y(m) is given for m in A (where A is a finite subset of Z), then find z(m) where a(m) = y(m) for m in A and z is band-limited to Ω. If we set $K : L^2(\Omega) \to l_2(A)$ to be

$$(Kf)(m) = \int_\Omega f(w)e^{2\pi imw}\,dw \qquad (9.82)$$

for m in A, then K is a bounded linear operator and the extrapolation problem is to find f in $L^2(\Omega)$ such that (Kf)(m) = z(m) for m in A. By Bialy's iteration we have

$$f_o = 0 \qquad (9.83)$$

$$f_m(w) = f_{m-1}(w) + \alpha \sum_{k \in A}(z(k) - \int_\Omega f(z)e^{2\pi izk}\,dz)e^{-2\pi ikw}$$

for w in Ω; and f_m converges to the minimum norm solution. Setting $y_m(k) = \int_\Omega f_m(w)e^{2\pi ikw}\,dw$ we get the more standard form of iteration:

$$y_o = 0 \qquad (9.84)$$

$$y_m(k) = y_{m-1}(k) + \alpha \sum sinc_\Omega(k-j)(z(j) - y_{m-1}(j))$$

for $n \geq 1$ which will always converge to y(k), the minimum energy solution, for $0 < \alpha < 2$. In other words this algorithm will always converge but it will not distinguish signal from noise.

9.13   Discrete-Discrete Extrapolation

The discrete-discrete problem is given by a periodic sequence $\alpha(m)$ of period N = 2M+1 which is band-limited to $[-k_o, k_o]$, i.e.

$$\sum_{m=0}^{N-1} \alpha(m)e^{2\pi imk/N} = 0 \qquad (9.85)$$

for $M \geq |k| > k_o$. The extrapolation problem states that given a piece of α, say $\alpha(m)$ for m in[-L,L], find $\beta(m), |m| < M$ where $\beta(m) = \alpha(m)$ for

$|m| \le L$ and β band-limited to $[-k_o, k_o]$. Let $K : C^{2k_o+1} \to C^{2k_o+1}$ be the inverse discrete Fourier transform

$$K\beta(n) = \frac{1}{N} \sum_{-k_o}^{k_o} \beta(k)e^{2\pi ikn/N} \tag{9.86}$$

for $|n| \le L$ where $k_o \ge L$. Again the extrapolation problem is equivalent to findng a vector β in C^{2k_o+1} such that $(K\beta)(k) = \alpha(k)$ for $|k| \le L$. We can apply the Bialy iterative technique to find the minimum-norm solution β:

$$\beta_o(k) = 0 \; |k| \le k_o \tag{9.87}$$

$$\beta_n = \beta_{n-1} + \alpha K^t(\alpha - K\beta_{n-1}) \; n \ge 1.$$

Taking the inverse discrete Fourier transform of this equation gives

$$y_o = 0 \tag{9.88}$$

$$y_n(k) = \frac{1}{N} \sum_{-k_o}^{k_o} e^{2\pi ihk/M} \sum_{m=-N}^{N} e^{-2\pi imh/M}(J_L + (I - J_L)y_{n-1}(m)).$$

To connect the DFT implementation of the Papoulis-Gerchberg algorithm and the continuous extrapolation, we recall:

THEOREM (Schlebusch-Splettstosser) 9.24 Let $g : [-T, T] \to C$ be a piece of an Ω−band-limited function f and let Δ be a real positive number such that $M = 2\pi/(\Omega\Delta^2)$ is an odd integer. If

$$h(j) = \frac{1}{M} \sum_{k=-k_o}^{k_o} z(k)e^{2\pi ijk/M}$$

is a solution to the discrete extrapolation problem where

$$x(j) = g(j\Delta)$$

for $-L \le j \le L = [T/\Omega]$ and $k_o = [\Omega M\Delta/2\pi] = L$, then the trigonometric polynomials

$$\phi_\Delta(t) = \frac{1}{M} \sum_{k=-k_o}^{k_o} z(k)e^{2\pi ikt/M\Delta}$$

converge to f(t) uniformly on each compact set in R as $\Delta \to 0$.

9.14   The Xu-Chamzas Algorithm

The modification of the Papoulis-Gerchberg algorithm to signals corrupted by noise was considered by Xu and Chamzas. They treated the following two problems:

(p1) given that the energy of f is less than or equal to R^2, i.e., $\|f\|^2 \leq R^2$, then approximate f with f_* in the class F_1 of Ω–band-limited functions satisfying the energy constraint and

$$I_1 = \|f_* - g\|_T^2 = min_{F_1}\|f - g\|_T^2 \qquad (9.89)$$

(p2) given that g = f+n and the energy of the noise n in (-T,T) is less than or equal to ϵ^2, $\|n\|_T^2 \leq \epsilon^2$, find an estimate F_* in F_2 such that

$$\|f_* - g\|_T^2 \leq \epsilon^2 \qquad (9.90)$$

$$I_2 = \|f_*\|^2 = min_{F_2}\|f\|^2$$

where F_2 is the family of Ω–band-limited function satisyfing .

THEOREM 9.25. Let $g(t) = \sum_{-\infty}^{\infty} b(k)\phi_k(t)$ for $|t| < T$ be the prolate spheroidal expansion of g. Then the solution to (p1) and (p2) has the form

$$f_*(t) = \sum_{k=0}^{\infty} b(k)\lambda_k \phi_k(t)/(\lambda_k + \mu) \qquad (9.91)$$

where for (p1) $\mu = 0$ or $\mu = \mu_k$ where μ_k is the solution of

$$R(\mu_k) = \sum b(k)\lambda_k^2/(\lambda_k + \mu_k) = R^2. \qquad (9.92)$$

And for(p2) $\mu = 0$ or $0 \leq \mu < \mu_\epsilon$ where μ_ϵ is the solution of

$$J(\mu_\epsilon) = \sum \lambda_k \mu_k^2 b(k)/(\lambda_k + \mu_\epsilon) = \epsilon^2. \qquad (9.93)$$

Proof. Let $f = \sum a(k)\phi_k$ and set $I = \|f - g\|_T^2 = \sum (b(k) - a(k))^2 \lambda_k$. So (p1) is equivalent to determining the a(k) which minimize I under the constraint $\sum a(k)^2 \leq R^2$. Using Lagrange multipliers we have the unconstrained minimization of

$$H(a(k), \mu) = \sum \lambda_k(b(k) - a(k)) + \mu \sum a(k)^2. \qquad (9.94)$$

Setting $\partial H/\partial a(k) = 0$ we find

$$a(k) = \lambda_k b(k)/(\lambda_k + \mu) \qquad (9.95)$$

and substitution shows that μ must satisfy $R(\mu) = \sum \lambda^2 b(k)^2/(\lambda_k + \mu)^2 \leq R^2$. The remainder of the proof is left to the reader.

The Xu-Chamzas algorithm is given in the following result:

THEOREM (Xu-Chamzas) 9.26. Let

$$f_{n+1}(t) = ((1 - \alpha_k)f_n(t) + \alpha(g(t) - f_n(t)P_T(t))) \star sin(\Omega t)/(\pi t). \quad (9.96)$$

Then for $0 < \alpha < 2/(1 + \mu)$, $lim_{n \to \infty} f_n(t) = f_*(t)$.

Proof. Since f_n is Ω-band-limited

$$f_n(t) = \sum a(k)_n \phi_k(t). \quad (9.97)$$

Substituting this in the iteration gives the recursive equation for $a(k)_n$:

$$a(k)_{n+1} = (1 - \alpha(\mu + \lambda_k))a(k)_n + \alpha\lambda_k b(k) \quad (9.98)$$

$$a(k)_o = 0;$$

thus, $a(k)_n = \lambda_k b(k)(1 - (1 - \alpha(\lambda_k + \mu)^n)/(\lambda_k + \mu)$. For α with $0 < \alpha < 2/(1 + \mu) < 2/(\lambda_k + \mu)$ we have

$$lim_{n \to \infty}\|f_n - f_*\| = lim_{n \to \infty} \sum_{k=0}^{\infty} \lambda_k b(k)/(\lambda_k + \mu))^2(1 - \alpha(\lambda_k + \mu))^{2n} = 0.$$
$$\quad (9.99)$$

Noting that for Ω-band-limited functions we have

$$max|f_n - f_*| \leq \sqrt{\Omega/\pi}\|f_n - f_*\| \quad (9.100)$$

we have the desired result.

COROLLARY 9.27. For $\mu > 0$ and $0 < \alpha < 2/(1 + \mu)$ we have

$$\|f_n\|^2 \leq \|g\|_T^2/\mu^2. \quad (9.101)$$

The case $\mu = 0$ and $\alpha = 1$ is precisely the Gerchberg-Papoulis algorithm, which converges if and only if $\sum b(k)^2$, i.e., n(t) is a segment of a Ω-band-limited function.

Since $f_n(t)$ are Ω-band-limited we can rewrite 9.96 as

$$f_{n+1} = (1 - \alpha_\mu)f_n(t) + \alpha \int_{-T}^{T} (g(s) - f_n(s))\frac{sin\Omega(t - s)}{\pi(t - s)}ds \quad (9.102)$$

which is Cadzow's iteration in case $\alpha = 1, \mu = 0$. Since f_n converges to f_* uniformly, we can take the limit $n \to \infty$ on both sides of the last equation to give

$$\mu f_*(t) + \int_{-T}^{T} f_*(s)\frac{sin\Omega(t - s)}{\pi(t - s)}ds = \int_{-T}^{T} g(s)\frac{sin\Omega(t - s)}{\pi(t - s)}ds \quad (9.103)$$

which is a Fredholm integral equation of the second kind.

9.15   The Discrete Xu-Chamzas Algorithm

Consider the sequences for which

$$\|f(n)\|^2 = \sum_{-\infty}^{\infty} |f(n)|^2 < \infty \qquad (9.104)$$

and assume f(n) is Ω-band-limited, i.e.,

$$F(w) = \sum_{-\infty}^{\infty} f(n)e^{-inw} = 0 \text{ for } \pi > |w| > \Omega. \qquad (9.105)$$

Let $g(n) = f(n) + \eta(n), |n| \leq M$ be the specified data. The problem is to estimate f(n) for all n in terms of the noisy data segment g(n). The discrete analog of theorem 9.26 is given by the following result of Xu-Chamzas and Jain-Ranganath.

THEOREM 9.28. Let h(m) be known for $|m| \leq M$; then the function $f^\dagger(m)$ satisfying the conditions:
$f^\dagger(m)$ is Ω-band-limited
$f^\dagger(m) = h(m), |m| \leq M$

$$\|f^\dagger(m)\|^2 =$$

$$min\{\|f\|^2 | f(m) \text{ is } \Omega-\text{band-limited and } f(m) = h(m) \text{ for } |m| \leq M \}$$

is given by

$$f^\dagger(m) = \sum_{k=0}^{2M} a(k)\phi_k(m) \qquad (9.106)$$

where the ϕ_k are the discrete prolate spheroidal sequences and

$$a(k) = \frac{1}{\lambda_k} \sum_{-M}^{M} h(m)\phi_k(m). \qquad (9.107)$$

The discrete iterative equation is

$$f_{n+1}(m) - ((1 - \alpha_\mu)f_n(m) + \alpha(g(m) - f(m))P_M(m)) \star sin(\Omega m)/(\pi m) \qquad (9.108)$$

$$f_o(m) = 0$$

where

$$P_M(m) = \left\{ \begin{array}{ll} 0 & |m| > M \\ 1 & |m| \leq M \end{array} \right.$$

The analogous equations to 9.102 and 9.103 are

$$f_{n+1}(m) = (1 - \alpha\mu)f_n(m) + \alpha \sum_{l=-M}^{M} (g(l) - f_n(l))sin\Omega(m - l)/(\pi(m - l))$$

$$(9.109)$$

$$\mu f_*(m) + \sum -M^M f_*(l)sin\Omega(m - l)/(\pi(m - l)) =$$

$$\sum f(l)sin\Omega(m - l)/(\pi(m - l)).$$

Similar results can be obtained in the case of K-band-limited N-periodic sequences (i.e. f(n) = f(n+N)) and $F(m) = \frac{1}{N}\sum_{n=0}^{N-1} f(n)e^{-i2\pi mn/N} = 0$ for $k < m < N - K$. The case $K \geq M$ and $K < M$ must be examined separately. For $K \geq M$ we have the equivalent of theorem 9.21 where

$$f_*(m) = \sum_{i=0}^{2M} b(i)\lambda_i\phi_i/(\lambda_i + \mu) \qquad (9.110)$$

where ϕ_i are now the periodic discrete prolate spheroidal functions. The iterative algorithm is

$$f_{n+1}(m) = \sum_{l=l_o}^{l_o+N+1} ((1 - \alpha_\mu)f_n(l)+$$

$$\alpha(g(l) - f_n(l)P_M(l))sin\frac{\pi}{N}(2K + 1)(m - l)/(Nsin\frac{\pi}{N}(m - l). \quad (9.111)$$

For the treatment of the other cases see Xu-Chamzas.

The digital implementation of the analog algorithm provides the following connection. We sample g(t) with a sampling interval $T_s \leq \pi/\Omega$ to get $g(l) = g(lT_s), l = -M, ..., M, M = T/T_s$. Select N so large that for $K = [\frac{(N-1)\Omega T_s}{2\pi}]$, then $N > 2M + 1$ and $K > M$. The iterative equation 9.11 gives an N-periodic sequence $f_*^{N,T_s}(mT_s)$. Letting $N \to \infty$ and $T_s \to 0$ we see that this solution tends toward the analog solution $f_*(t)$.

9.16   Adaptive Extrapolation

Consider the problem of determination of frequency components of a signal

$$f(t) = \sum_{k=1}^{m} c(k)e^{i\omega_k t} \qquad (9.112)$$

where one has only the data

$$w_1(t) = \begin{cases} f(t) + n(t) & |t| < T \\ 0 & |t| > T. \end{cases} \qquad (9.113)$$

Clearly the unknown frequencies ω_k and coefficients c(k) can be estimated by Fourier analysis if the observation interval 2T is large compared with all the periods $T_k = 2\pi/\omega_k$ and their differences. The algorithm presented below addresses the case of short observation intervals T and noisy data.

We assume the function f(t) has Fourier transform $F(\omega)$ where $F(\omega) = 0$ for $|\omega| > \Omega$. If f(t) is the sum of m exponentials, then

$$F(\omega) = 2\pi \sum_{k=1}^{m} c(k)\delta(\omega - \omega_k)$$

and the problem is to determine the locations ω_k and their amplitudes c(k). We take Ω larger than the largest possible ω_k. The algorithm is then the Papoulis-Gerchberg-Saxton method:

(1) $w_1(t) = f(t)$ $|t| < T$
(2) form the Fourier transform $W_1(\omega)$
(3) set $F_1(\omega) = W_1(\omega)$ $|\omega| < \Omega$
(4) form the inverse Fourier transform $f_1(t)$ of $F_1(\omega)$
(5) define

$$w_2(t) = f(t) = w_1(t) \quad |t| < T$$

(6) return to step (1).
However, we replace step (3) by

$$F_n(\omega) = W_n(\omega) \quad \omega_k \in B_n$$

where B_n are bands where $W_n(\omega)$ takes its significant values. Here B_n^c denote the complement of B_n. The subsets B_n are derived in terms of a threshold method as follows:

$$|W_n(\omega)| \geq \epsilon_n \quad \omega \in B_n \tag{9.114}$$

$$|W_n(\omega)| < \epsilon_n \quad \omega \in B_n^c;$$

let $M_{n-1} = min_{\omega \in B_{n-1}}|W_{n-1}(\omega)|$ and set $\epsilon_n = max\{\epsilon_{n-1}, \mu M_{n-1}\}$ where μ is a constant less than one.

The discrete implementation of this algorithm takes $f_n = f(nt_o), F_n = F(n\omega_o)$. If t_o is the sample interval, then $\omega_o = 2\pi/(Nt_o)$ where N is the size of the FFT. The data interval is 2T so the number of data samples is $M = [2T/t_o]$. An appropriate choice, which balances aliasing errors and improper estimates, is m = N/4. This leads to $t_o \sim 8T/N$. examples of this technique in Papoulis-Chamzas include $f(T) = 1.5cos(30\pi t + \pi/3) + 1.25cos(20\pi t + \pi/6)$, taking N = 256, $f_o = 1Hz, t_o = 1/256$ seconds, M = 41,51 and n(t) = 0. Taking uniformly distributed noise over (-c,c), cases were considered where S/N = 15 dB (c = .375) and 11 dB (c = .625). Cases were also tested where the frequencies f_q were not integral multiples of f_o, viz., $f(t) = 1.5cos(4.8\pi t) + 1.5cos(18\pi t + \pi/3) + 1.25cos(29.2\pi t + \pi/6)$. Finally, cases where the data interval is less than one-half the unknown period – e.g., f(t) = $1.25cos(5.4\pi t + \pi/6)$, T = .08 seconds, M = 41, and samples with S/N down to 2 dB were successfully tested.

A similar algorithm was proposed by Papoulis and Chamzas to improve range resolution in ultrasound processing. Here the medium is examined by a narrow beam of sound and the reflected signal y(t) is used to characterize the medium; e.g., the location of surface discontinuities. If x(t) is the transmitted signal, then $y(t) = \int_{-\infty}^{\infty} x(t - s)h(s)ds$. If the medium consists of homogeneous layers, then h(t) is a sum of impulses

$$h(t) = \sum c(k)\delta(t - t_k). \tag{9.115}$$

Then $y(t) = \sum c(k)x(t - t_k)$ and the problem is to determine c(k) and t_k from y(t). As before, the observed data is a(t) = y(t) + n(t) and the object is to determine h(t) in terms of z(t) and x(t). We assume the unknown impulses are in the interval (-T,T). The algorithm is precisely as above: viz.,

(1) starting with $W_1(\omega) = H(\omega)$ $\omega \in B$ set

$$w_1(t) = \frac{1}{2\pi} \int_B W_1(\omega)e^{i\omega t} d\omega$$

(2) $h_1(t) = w_1(t)$ $|t| < T$
(3) $H_1(\omega) = \int_{-T}^{T} h_1(t)e^{-\omega t} dt$
(4) $W_2(\omega) = W_1(\omega) - H(\omega)$ $\omega \in B$
(5) return to (1).

Thus, in general $W_{n+1}(\omega) = P(\omega)H(\omega) + (1 - P(\omega))H(\omega)$. Here

$$P(\omega) = \begin{cases} 1 & \omega \in B \\ 0 & \omega \in B^c \end{cases} \tag{9.116}$$

However, any $P(\omega)$ with $0 \leq P(\omega) < 1$ would work. For adaptive extrapolation we assume h(t) is nonzero only in a subset B of (-T,T); e.g., B may be the finite set of points $\{t_k\}$ above. We take

$$w_n(t) \geq \epsilon_n \text{ t in } B_n \tag{9.117}$$

$$w_n(t) < \epsilon_n \text{ t in } B_n^c$$

where $\epsilon_n = min_{t \in B_{n-1}}\{|w_{n-1}(t)|\}$. The function $h_n(t)$ is defined by:

$$h_n(t) = \begin{cases} w_n(t) & \text{t in } B_n \\ 0 & \text{t in } B_n^c. \end{cases} \tag{9.118}$$

The object is a nested sequence of set $b_n, B_{n+1}, ..$ which converges to the set of point t_k. The function used for $P(\omega)$ in Papoulis-Chamzas is not the pulse mentioned above but rather $P(\omega) = \frac{1}{M}|X(\omega)|$ where $M = max|X(\omega)|$. This function favors the frequencies where $|X(\omega)|$ is large.

9.17   Karhunen-Loeve Expansion

Consider the stochastic process y(t) with autocorrelation

$$R(t_1, t_2) = E(y(t_1)y^*(t_2)). \tag{9.119}$$

Let $Y(\omega)$ denote the Fourier transform of y(t). Then

$$\Gamma(u, v) = E(Y(u)Y^*(v))$$

is related to R by

$$\Gamma(u, v) = \int_{-\infty}^{\infty}\int R(t_1, t_2)e^{-i(ut_1 - vt_2)}dt_1 dt_2. \tag{9.120}$$

The eigenfunctions $\psi(v)$ of the equation

$$\int_{-\Omega}^{\Omega}\Gamma(u, v)\psi(v)dv = \gamma\psi(v) \tag{9.121}$$

provide the Karhunen-Loeve expansion

$$Y(\omega) = \sum_{k=0}^{\infty} z(k)\psi_k(\omega) \quad |\omega| \le \Omega \tag{9.122}$$

where z(k) are orthogonal random variables.

In the case n(t) is a white noise process, $R(t_1, t_2) = q(t_2)\delta(t_1 - t_2)$. And if Q is the Fourier transform of q, then $\Gamma(u, v) = Q(u - v)$. If we now take $y(t) = n(t)P_T(t)$ where n(t) is a stationary white noise process with autocorrelation $S\delta(t_1 - t_2)$, then

$$R_y(t_1, t_2) = SP_T(t_1)\delta(t_1 - t_2). \tag{9.123}$$

So $q(t) = SP_T(t)$ and $Q(\omega) = 2S\sin(\omega)/\omega$ and the eigenvalue equation 9.121 is precisely the equation for prolate spheroidal functions

$$2S\int_{-\Omega}^{\Omega}\frac{\sin T(u - v)}{u - v}\psi(v)dv = \gamma\psi(u). \tag{9.124}$$

9.18   Stopping Algorithms

As we saw in Section 1.13 extrapolation of a band-limited function is an ill-posed problem. Suggestions have been made to use more constraints to deal with noise sensitivity problems or to use a spatial frequency filter. Furthermore, simulations have shown that even moderately corrupted or noisy images still lead to stable reconstruction if additional constraints are used. However, there is a trade-off between noise amplification and signal enhancement, which leads to the discussion of stopping rules.

Using the Karhunen-Loeve expansion a stopping formula is generated in Yeh and Chin (1985). Estimates of the nature of the statistics of the noise in the Fourier and image planes are required.

9.19 Regularization

It is well-known that a Fredholm integral equation of the first kind

$$\int_a^b k(s,t)x(t)dt = g(s) \quad a \le s \le b \tag{9.125}$$

where $k(.,.)$ is a square integrable kernel is ill-posed – i.e., the solution x does not depend continuously (in the L^2- sense) on the data g. In real world situations g results from measurements and thus is often only imprecisely known. The discontinuous nature of the solution operator can naturally lead to serious numerical problems.

We can phrase 9.125 abstractly as $Kx = g$ where $K : H_1 \to H_2$ is a compact linear operator from Hilbert space H_1 to H_2 and g belongs to $R(K) =$ the range of K. Above K is the compact operator on $H = L^2(a,b)$.

Let $\tilde{K} = K^*K$ and let $R_\alpha(t)$ be a continuous function on $[0, \|K\|^2]$ which satisfies:

(1) $R_\alpha(t) \to 1/t$ as $\alpha \to 0$ for $t > 0$

(2) $|tR_\alpha(t)|$ is uniformly bounded.

THEOREM (Groetsch) 9.29. Suppose R_α is as above. Then we have $R_\alpha(\tilde{K})K^*g \to K^\dagger g$ as $\alpha \to 0$ for each g in dom$(K^\dagger)$.

This provides a class of general approximation methods when the data g is known exactly. Suppose now only an approximate version g^δ of g is known which satisfies

$$\|g - g^\delta\| \le \delta \tag{9.126}$$

where δ is the know error level. The approximations $x_\alpha^\delta = R_\alpha(\tilde{K})K^*g^\delta$ are said to be regular if they converge in some sense to the minimal norm solution as $\delta \to 0$. And a regularization method for 9.125 is a family $\{R_\alpha\}$ of continuous operators and a parameter choice $\alpha = \alpha(\delta)$ such that $\alpha(\delta) \to 0$ and $R_{\alpha(\delta)}g^\delta \to x$ as $\delta \to 0$ where x is the minimal norm solution of 9.125 or the minimal norm least squares solution if 9.125 is least squares soluble. Assume now that R_α satisfies

$$|tR_\alpha(t)| \le c^2$$

for t in $\sigma(\tilde{K}) = [0, \|K\|^2]$. Here c is a positive constant, $\alpha > 0$ and $r(\alpha) = max_{t \in \sigma(\tilde{K})}\{|R_\alpha(t)|\}$.

THEOREM (Groetsch) 9.30. If g belongs to $dom(K^\dagger), \alpha(\delta) \to 0$ and

$$\delta^2 r(\alpha(\delta)) \to 0$$

as $\delta \to 0$, then $x^\delta_{\alpha(\delta)} \to K^\dagger g$ as $\delta \to 0$.

9.20   Tikhonov Regularization

Set $R_\alpha(t) = (t + \alpha)^{-1}, \alpha > 0$. Then $x_\alpha = (\tilde{K} + \alpha I)^{-1} K^* g$. Tikhonov took the unique minimizer x^δ_α of the functional $F_\alpha(z) = \|Kz - g^\delta\|^2 + \alpha\|z\|^2$. And for fixed $\alpha > 0$ the term $\alpha\|z\|^2$ has a stabilizing effect, making x^δ_α a continuous function of g^δ.

9.21   Landweber Regularization

Consider the Landweber-Friedman iteration

$$x_o = aK^* g \qquad (9.127)$$

$$x_{n+1} = (I - a\tilde{K})x_n + aK^* g$$

where $0 < a < 2\|K\|^2$. This method is given by the regularization function

$$R_n(t) = a \sum_{k=0}^{n} (1 - at)^k. \qquad (9.128)$$

Set $\alpha(\delta) = 1/\beta(\delta)$ where $[\beta(\delta)] = n(\delta)$, the iteration number. Take c = 1, r(n)=a(n+1); then $\alpha(\delta)R_{\alpha(\delta)}(\alpha(\delta))^2 \sim n(\delta)$. Let $M = limsup_{\delta \to 0}\delta^2 n(\delta)$ and $m = liminf_{\delta \to 0}\delta^2 n(\delta)$. Then the following theorem holds for the Landweber iteration regularization:

THEOREM (Groetsch-Engl) 9.31. Let

$$M = limsup\delta^2/\alpha(\delta)$$

and

$$m = liminf\delta^2/\alpha(\delta);$$

then $\{x^\delta_{\alpha(\delta)}\}$ is a strongly regular, weakly regular, strongly divergent or weakly divergent as $M = m = 0, M < 0, M = m = \infty$ or $m > 0$ respectively.

In particular, if the kernel is continuous, we have the following generalization of Landweber's result:

THEOREM 9.32. If K is continuous and x belongs to $R(K^*)$ then $\|x - x_n\|_\infty \to 0$ as $n \to \infty$. And for imprecise data, if $n = n(\delta)$ and $\delta =$

$\mathcal{O}(1/n)$, then $\|x - x_n^\delta\|_\infty \to 0$ as $\delta \to 0$. Moreover, if x belongs to $R(\tilde{K}^\nu K^*)$ for some $\nu \geq 1$ and $n = [\delta^{-1/(\nu+1)}]$ then $\|x - x_n^\delta\|_\infty = \mathcal{O}(\delta^{\nu/(\nu+1)})$.

Thus, we see that if x is "smooth" enough, i.e., x belongs to $R(\tilde{K}^\mu K^*)$ for ν large enough, then a uniform order of accuracy arbitrarily near to the optimal order $\mathcal{O}(\delta)$ can be attained by 9.127, Landweber's iteration.

9.22 Regularization in Optics

Consider the case of a diffraction-limited coherent image formed by a one-dimensional clear pupil extending over $(-\Omega, \Omega)$. This is described by

$$g(y) = \int_{-X}^{X} \frac{sin(\Omega(y-x))}{\pi(y-x)} f(x)dx. \tag{9.129}$$

Let B_Ω denote the band-limiting operator

$$B_\Omega(h(x)) = h * sin(\Omega x)/\pi x = \int_{-\infty}^{\infty} \frac{sin\Omega(x-y)}{\pi(x-y)} h(y)dy \tag{9.130}$$

and the domain-limiting operator

$$D_X(h(x)) = \begin{cases} h(x) & |x| \leq X \\ 0 & |x| > X \end{cases} \tag{9.131}$$

for h in $L^2(-\infty, \infty)$. Let $\bar{D}_X = I - D_X$. Equation 9.129 becomes

$$g = B_\Omega D_X g. \tag{9.132}$$

We are interested in objects which satisfy $D_X f = f$. We first note the following Fourier transform relationships:

THEOREM 9.32.

$$\widehat{B_\Omega h} = D_\Omega \hat{h}$$
$$\widehat{D_X h} = B_X \hat{h}$$
$$\widehat{B_\Omega D_X h} = D_\Omega B_X \hat{h}$$
$$\widehat{D_X B_\Omega h} = B_X D_\Omega \hat{h}.$$

The Gerchberg method of solving 9.132 is

$$\hat{f}_n = \hat{g} \bar{D}_\Omega \widehat{D_X f_{n-1}} = \hat{g} \bar{D}_\Omega B_X \hat{f}_{n-1} \tag{9.133}$$

where $\hat{f}_o = \hat{g}$. And the nth estimate of the object is therefore

$$f_n = g + D_X f_{n-1} - B_\Omega D_X f_{n-1} = g + (I - B_\Omega)D_X f_{n-1}. \tag{9.134}$$

Following the Tikhonov regularization method Abbiss, de Mod and Dhadwal considered the minimization of a functional

$$F(u) = \|g = B_\Omega D_X u\|^2 + \alpha \|(I - B_\Omega) D_X u\|^2 \qquad (9.135)$$

which gives the equation

$$\tilde{f} = D_X B_\Omega + (1 - \alpha)(D_X - D_X B_\Omega D_X)\tilde{f}_o. \qquad (9.136)$$

This leads to the iterative formula

$$\tilde{\tilde{f}}_n = B_X(D_\Omega \hat{g} + \lambda \bar{D}_\Omega B_X \tilde{\tilde{f}}_{n-1}) \qquad (9.137)$$

$$\lambda = (1 - \alpha).$$

However, B_X is redundant except for the last step; thus, we write

$$\tilde{\tilde{f}}_n = D_\Omega \hat{g} + \lambda \bar{D}_\Omega B_X \tilde{\tilde{F}}_{n-1}. \qquad (9.138)$$

And for $\alpha = 0$ this reduces to the Gerchberg-Papoulis equation.

Since $D_\Omega \hat{g} = D_\Omega \tilde{\tilde{F}}_{n-1}$ for all n we can rewrite 9.138 as

$$\tilde{\tilde{f}}_n = (D_\Omega + \lambda \bar{D}_\Omega B_X)\tilde{\tilde{f}}_{n-1}. \qquad (9.139)$$

Hence,

$$\tilde{\tilde{f}}_n = (d_\Omega + \lambda \bar{D}_\Omega B_X)^n D_\Omega \hat{g} \qquad (9.140)$$

where we take $\tilde{\tilde{f}}_o = D_\Omega \hat{g}$.

EXERCISE 9.33. Let ψ_i be the prolate spheroidal wave functions with eigenvalues λ_i. Write $B_\Omega g = \sum g(i)\psi_i$. Show that

$$\tilde{f}_n = \sum q_i(\alpha, n)/\lambda_i g(I) D_X \psi_i$$

where $q_i(\alpha, n) = \lambda_i(1 - (1 - \alpha)^n(1 - \lambda_i)^n)/(\lambda_i + \alpha(1 - \lambda_i))$. (The case $\alpha = 0$ was derived by DeSantis and Gori.) Since $0 < (1 - \alpha)(1 - \lambda_i) < 1$ we have convergence of $\tilde{f}_n$ to $\tilde{f}$, the solution of the regularized equation.

9.23   Moment Discretization

The Hilbert space $L^2(0, 1)$ is not a reproducing kernel Hilbert space since L^2-convergence does not imply point wise convergence. However, if we form the inner product [,] on R(K) given by

$$[f, g] = (K^\dagger f, K^\dagger g)_{L^2} \qquad (9.141)$$

then R(K) is a Hilbert space. And setting $\rho(s,t) = \int_0^1 k(s,u)k(t,u)du$ and $\rho_s(t) = \rho(s,t)$ we see that $\rho_s(t) = (Kk_s)(t)$ where $k_s(u) = k(s,u)$; and for x in $N(K)^\perp$ and $P : L^2 \to N(K)^\perp$ we have for f = Kx

$$[f, \rho_s] = (K^\dagger f, K^\dagger \rho_s) = (x, K^\dagger \rho_x) = \qquad (9.142)$$

$$(x, K^\dagger K k_s) = (x, P k_s) = (P x, k_s) = (x, k_s) = \int_0^1 k(s,u)x(u)du = f(s).$$

Thus, we have:

THEOREM 9.34. (R(K),[,]) is a RKHS with kernel $\rho(s,t)$.

Since $g(s_i) = [g, \rho_{s_i}]$, i=1,...m, we see that if k is continuous, then so is each g in R(K). Let $\Delta_m = sup inf_{s \in [0,1]}\{|s - s_i| i = 1, ...m\}$ it can be shown that:

THEOREM 9.35. If k(.,.) is continuous and g belongs to R(K); if $\Delta_m \to 0$ as m, then $\|x_m - K^\dagger g\| \to 0$ as $m \to \infty$ where x_m is the unique element x in $L^2(0,1)$ which satisfies $(Kx)(s_i) = [g, \rho_{s_i}], i = 1, ...m$.

Wahba has gone on to study the moment discretization problem

$$\int_0^1 k(s_i, t)x(t)dt = d_i, i = 1, ...m \qquad (9.143)$$

where $d_i = g(s_i) + \epsilon_i$ where ϵ_i represent the errors in the measured data values. One might try to regularize 9.143 by minimizing the expression

$$\frac{1}{m}\sum_{i=1}^m (Kx(s_i) - d_i)^2 + \alpha\|x\|^2. \qquad (9.144)$$

The choice of α appears to require knowledge of the actual errors. Wahba's method of cross validation attempts to incorporate this knowledge. We refer the interested reader to Wahba's papers.

9.24 Biraud's Algorithm

Consider the problem where f(x) is a real function which is known to be nonnegative on a closed interval [a,b] and zero elsewhere. By an appropriate change of variables the interval may be assumed to be $[-\pi, \pi]$. We assume f(x) is square-integrable on this region. Thus, we can represent f(x) as a Fourier series

$$f(x) = \sum \hat{f}(n)e^{inx}. \qquad (9.145)$$

Biraud suggests that by the nonnegativity of f we substitute $f(x) = u(x)^2$. Representing u as a Fourier series on $[-\pi, \pi]$ we have

$$u(x) = \sum \hat{u}(j)e^{ijx} \tag{9.146}$$

and so by the convolution theorem

$$\hat{f}(q) = \sum \hat{u}(j)\hat{u}(q - j) \quad q = 0, \pm 1, \pm 2, .. \tag{9.147}$$

Truncating $\hat{u}(j)$ to $|j| \leq M$ gives the estimates

$$\hat{f}(q)_M = \sum_M^M \hat{u}(j)_M \hat{u}(q - j)_M \quad q = 0, \pm 1, ...; \tag{9.148}$$

i.e., knowledge of $\hat{f}_M(q)$ over $-2M \leq q \leq 2M$. The approach is then:

(1) the Fourier coefficients $\hat{f}(0), \hat{f}(\pm 1), ...\hat{f}(\pm N)$ of a nonnegative function are known

(2) fit the finite autocorrelation 9.148 $(M > N/2)$ to these known coefficients by least squares minimization of

$$\sum_{n=0}^N w(n)(\hat{f}(n) - \sum_{j=n-m}^m \hat{u}(j)\hat{u}_M(n - j));$$

here w(n) are weights

(3) use 9.147 to provide extrapolated Fourier coefficients.

The Biraud method can be generalized as follows: assume there are available noisy estimates $\hat{f}^\epsilon$ of $\hat{f}(w)$ on a frequency interval $|w| < W_o$. Express $f(x) = U(x)^2$ or $\hat{f}(w) = \hat{u} \star \hat{u}(w)$. Since f is real, $\hat{u}$ is hermitian $\hat{u}(-w) = \bar{\hat{u}}(w)$. To stabilize the extrapolation Biraud constructs a sequence of approximations $\{\phi_n(w)\}$ to $\hat{f}(w)$ where each $\hat{\phi}_n$ has the form $\hat{\phi}_n = \hat{\gamma}_n(w) \star \hat{\gamma}(w)$ with each $\hat{\gamma}_n$ hermitian and of finite support $(-w_n, w_n)$ where $w_n \geq W_o/2, w_n$ increasing with n. So $supp\phi_n = (-2w_n, 2w_n) \supseteq (-W_o, W_o)$ and $\phi_n = \gamma_n^2 \geq 0$.

Define the spaces

$$E_W^1 = \{h|h \in L^1(R)\hat{h}(w) = 0|w| \geq W\}$$

$$E_W^2\{h|h \in L^2(R)\hat{h}(w) = 0|w| \geq W\}.$$

Clearly $E_W^1 \subset E_W^2$. Let $E_W^{1,r}$ denote the subset of real-valued functions in E_W^1 and let $E_W^{1,+}$ denote the further closed convex subset of nonnegative functions.

For a given $w_n \geq W_o/2, \gamma_n$ above may be found as a minimizer of the nonlinear optimization problem

$$min_{\gamma \in E_{w_n}^{2,r}} \|\hat{f}^\epsilon - \hat{\gamma} \star \hat{\gamma}\|_H^2 \tag{9.149}$$

where $H = L^2(0, W_o)$.

Let $B_W = \{\phi | \phi \in E_W^{1,+}, \phi = \gamma^2, \gamma \in E_{W/2}^{2,r}\}$. We view the minimization problem now as

$$min_{\phi \in B_{w_n}} \|\hat{f}^\epsilon - \hat{\phi}\|_H^2.$$

Davies has characterized B_W as a non-empty proper subset of $E_W^{1,+}$ consisting of those functions which has zeros of even multiplicities only:

THEOREM (Davies) 9.36. A function ϕ in $E_W^{1,+}$ has the representation $\phi(x) = \gamma(x)^2$ for $\gamma(x)$ in $E_{W/2}^{2,r}$ if and only if the zeros of $\phi(z)$ have even multiplicities. The nonreal zeros of $\gamma(z)$ are reflected in the origin and imaginary axis. When this representation exists, $\gamma(x)$ is unique up to a change of sign.

Biraud's approach considers an increasing sequence of band-limits converging to W. Generally we start with $W/2 \leq w_o \leq W_o/2$. The following result of Boas and Kac (1945) is pertinent:

THEOREM (Boas-Kac) 9.37. For any $W > 0$ if ϕ belongs to $E_W^{1,+}$ then

$$\hat{\phi}(w)| \leq |\phi(0)|cos(\pi/(1 + [W/w])), |w| \leq W$$

where $[\ ]$ denotes the greatest integer function.

So for ϕ_o in B_{2w_o} if $\phi_o(w)$ is to approximate $\hat{f}^\epsilon(w)$ closely on $(-W_o, W_o)$ we require

$$|\hat{f}^\epsilon(w) \leq |\hat{f}^\epsilon(0)|cos(\pi/(1 + [2W_o/w])), |w| \leq W_o$$

which yield $w_o \geq W_{min}$ where

$$W_{min} = \frac{1}{2}sup_{0 \leq w \leq W_o} w[\pi/arcos(\hat{f}^\epsilon(w)/\hat{f}^\epsilon(0)) - 1].$$

Consider now the original problem where $f(x) = \sum_{-\infty}^{\infty} \hat{f}(q)e^{iw_q x}$, $w_q = 2\pi q$, and $\hat{f}(-q) = \bar{\hat{f}}(q)$. We assume we have estimates of N+1 Fourier coefficients $\hat{f}(q)$. Now N_o plays the role of W_o and E_W^2 is replaced by the space T_N of trigonometric functions of period one and degree N. Define similarly

$$T_N^r = \{h | h \in T_N, h(x) \text{ real valued}\}$$
$$T_N^+ = \{h | h \in T_N^r, h \geq 0\}$$
$$S_{2N} = \{\phi | \phi \in T_{2n}^+, \phi = \gamma^2, \gamma \in T_N^r\}.$$

The minimization problem becomes

$$min_{\hat{\gamma}} F(\hat{\gamma}) = \sum_{q=0}^{N_o} |\hat{f}(q) - \sum_{r=-N}^{N} \hat{\gamma}(r)\hat{\gamma}(q - r)|^2$$

where $\hat{\gamma}(-q) = \bar{\hat{\gamma}}(q), 0 \leq q \leq N, \gamma(q) = 0, q > N$. Nonlinear least-squares solution of this problem provides $\hat{\gamma}(N)$ and the non-negative polynomial ϕ_N in S_{2N} where $\phi_N = \gamma_N^2$ with $\gamma_N(x) = \sum_{-N}^{N} \hat{\gamma}(q)e^{iw_q x}$. The Boas-Kac inequalities indicate a degree $N \geq N_{min}$ where

$$N_{min} \geq \frac{1}{P2} max_{0 < q \leq N_o} q(\pi / arccos(\hat{f}^\epsilon / \hat{f}^\epsilon(0) - 2).$$

The relationship of N_o and the optimal degree N_{cv} predicted by cross validation of Wahba is given as follows. If a total of 2M+1 Fourier coefficients $\hat{g}_q^\epsilon$ are available, then N_{cv} is the minimizer $(N < M)$ of the following function of N:

$$v(N) = \frac{1}{2M+1} \sum_{|q|=N+1}^{M} |\hat{g}(q)^\epsilon|^2 / (1 - \frac{2N+1}{2M+1})^2.$$

For further details the reader is referred to Davies' thesis and Davies (1983).

Chapter 10

Two Dimensional Signal Recovery Problems

10.1 Introduction

In the last chapters we have seen that certain properties hold for image reconstruction in one dimension. For example, a certain class of bandpass signals were found to allow unique specification by zero crossings. However, most one-dimensional band-limited signals encountered in practice do not fall into this class. Similarly, for the phase retrieval problem we have seen that the solution is generally nonunique. Viz., if $F(x)$ is the Fourier transform of the finite source $f(t)$,

$$F(x) = |F(x)|e^{i\phi(x)} = \int_a^b f(t)e^{-ixt}dt \ |a| \leq b < \infty \qquad (10.1)$$

then by the Paley-Wiener theorem $F(x + iy)$ is an entire function of exponential type where

$$|F(z)| \leq Me^{|z|\sigma}. \qquad (10.2)$$

Here $\sigma = (b - a)/2$ and the indicator function is $h(\alpha) = ((b - a)/2)|sin\alpha|$. By Hadamard's factorization theorem we have

$$F(x + iy) = F(z) = F(0) \prod_j (1 - z/z_j). \qquad (10.3)$$

The observed intensity function $I(x) = F(x)F^*(x)$ has the zeros of $F(z)$ and in conjugate pairs, those of $F^*(z^*)$. And $F_1(z) = F(z)\frac{1-z/z_n^*}{1-z/z_n}$ has the same modulus as $F(x)$. Thus, if $I(x)$ has $2N$ nonreal zeros, then there are 2^N possible functions having the same intensity $I(x)$, all with the same bandlimits.

135

In two dimensions we will discuss how these results change for reconstruction from zero crossings and phase retrieval. In 2-D the Fourier transform

$$F(x) = \int_{-\infty}^{\infty} f(t)e^{itx}dt \qquad (10.4)$$

where $x = (x_1, x_2)$ and $t = (t_1, t_2)$ in case f(t) has compact support can be extended to C^2 to be an entire function of exponential type in terms of $z_j = x_j + iy_j$. This is given by the Plancherel-Polya theorem. Here we say a function $F(z_1, z_2)$ is holomorphic if it is holomorphic in each variable separately. And if F is holomorphic for all finite values of z_1, z_2, then F is said to be entire. If F has at most an exponential growth rate in any direction in complex space then we say F is of exponential type. And the Polya-Plancherel theorem states that any band-limited function of real variables can be uniquely extended to complex space as an entire function of exponential type (EFET). The indicator function also generalizes where

$$h_F(\alpha, x) = lim_{|z| \to \infty} suplnF(x_1 + i|z|cos\alpha_1, x_2 + i|z|cos\alpha_2)/|z| \quad (10.5)$$

and $h_F(\alpha) = suph_F(\alpha, x)$. Here h_F is called the P-indicator and it is equal to the support function of the smallest convex region outside of which $f(t) = 0$. Thus, the support shape in object space determines the rate of growth of $F(z)$ in z-space, viz.,

$$|F(z) \le Me^{|z|h_F(\alpha)} \qquad (10.6)$$

where M is a constant.

However, Hadamard's theorem is now replaced by Osgood's factorization theorem which states that if $F(z)$ is an EFET in 2-D, then $F(z)$ can be uniquely represented by a product of the form

$$F(z) = \prod_{j=1}^{N}[F_j(z)e^{\gamma_j}]^{L_j}N \le \infty. \qquad (10.7)$$

Here $F(z)$ is a product of either a finite or infinite set of factors $F_j(z)$, each of which is entire and globally irreducible, but not necessarily of exponential type. Here γ_j are polynomial functions and L_j are integers. In contrast to the 1-D case, the number of factors is unknown and the form of the $F_j(z)$ are generally not known. Furthermore, in contrast to the 1-D case if F(z) is irreducible, then N = 1 and arg(F) is uniquely defined by $|F|$, given the support of f(t).

In 1-D each F_j has the simple form $(1 - z/z_j)$ and N is infinite. Each factor F_j corresponds to a point zero in C^1. In contrast in n-D the zeros are not isolated; rather the set of zeros of analytic functions are n-1 dimensional surfaces.

In particular in 2-D the zeros form lines. Let V_F denote the set of zeros of F, $V_F = \{z|F(z) = 0\}$. We say the entire function F is irreducible if it cannot be written as $F = GH$ where G,H are entire functions and V_G, V_H are both nonempty. If two irreducible signals F,G have identical zero crossings contours, then V_F, V_G must intersect. However, as we will see below this implies that V_F and V_G must be identical. So F,G must be equal to within a multiplication by an EFET which never vanishes, i.e., an exponential factor. This leads to the result below stating conditions on real 2-D band-limited signals, whose complex extensions are irreducible as entire functions when they are uniquely specified by their zero crossings.

An analytic set is defined to be the intersection of the zero sets of one or more holomorphic functions. So V_F, V_G and $V_F \cap V_G$ are analytic sets. An irreducible analytic set is one which cannot be expressed as the union of two distinct sets. If F is irreducible as above, then V_F is an irreducible analytic set.

In summary, we see that each F_j in the Osgood product describes an irreducible analytic set and the union of all these sets forms the zero set of $F(z)$.

In the case of discrete signals f(j), j=0,1,...M, then

$$F(z) = \sum f(j)e^{ijz} \qquad (10.8)$$

is a multivariate polynomial of degree M. We will see below that almost all polynomials in 2-D are irreducible. However, the same statement does not apply to reducible EFETs. (In 1-D the only irreducible EFETs are of the form $f(z) = e^{az+b}(z+c)$.) For a large class of of problems we will see that F(x) in 2-D is irreducible and F(x) can be uniquely recovered from F(x) or arg F(x). We turn in later sections to stable algorithms for achieving the reconstruction. We consider now the 2-D phase retrieval problem in more detail.

10.2   Phase Retrieval from Magnitude

Consider the case where $f : A \to C$ where A is a compact set in $R^n, n \geq 2$ and $|\hat{f}(w)|$ w in R^n is known for all frequencies. Then the problem of phase retrieval from magnitude is to reconstruct f(x) for x in A. Clearly this reconstruction is not unique since if f is a solution, then so is (1) kf where $|k| = 1$, (2) $f^*(-y)$ and (3) $f_\alpha(x) = f(x+\alpha)$, if $supp(f_\alpha) \subset A$. These ambiguities are referred to as "trivial associates" of f. Thus, the uniqueness question centers on f up to trivial associates.

In the discrete case where $\{a(j)\}, j = (j_1, ..., j_n)$ is a multi-dimensional sequence and $\hat{a}(w)$ is given by the Fourier series

$$\hat{a}(w) = \sum_k a(k)e^{a\pi wk} \tag{10.9}$$

the problem is: given $|\hat{a}(w)|$ for $0 \leq w_i < 1$, i=1,...n reconstruct a(j). We define the z-transform

$$A(z) = \sum a(j)a^j. \tag{10.10}$$

Then Hayes has shown that:

THEOREM (Hayes) 10.1. If the z-transform A(z) is irreducible $(n \geq 2)$ then the sequence a is uniquely determined up to trivial associates by $|\hat{a}|$.

THEOREM (Sanz-Huang) 10.2. Any real sequence $n \geq 2$ except for at most an algebraic set can be uniquely recovered (up to trivial associates) from the magnitude of its Fourier sequences.

As an example let

$$x(m,n) = a\delta(m,n) + b\delta(m-1,n) + c\delta(m,n-1) + d\delta(m-1,n-1).$$

Then the z-transform $X(z_1, z_2)$ will be reducible if and only if ad = bc. So in the space R^4 defined by the four coefficients, the subset of reducible polynomials forms a hyperquadratic surface, a nontrivial algebraic set.

This last result implies that Hayes' irreducibility criterion is stable in the sense that for any irreducible sequence a(j) there exists a small neighborhood of a(j) whose elements are all irreducible. So Hayes' condition is not sensitive to small random perturbations of the unknown input sequence.

As a trivial discrete 2-D problem consider the Fourier transform

$$F(u,v) = A + Bv + Cu \tag{10.11}$$

where $u = e^{-2\pi ix}$ and $v = e^{-2\pi iy}$. The autocorrelation polynomial is

$$Q(u,v) = au + bv + cuv + du^2v + euv^2 + fu^2 + gv^2 \tag{10.12}$$

where $a = AB^*, b = AC^*, c = |A|^2 + |B|^2 + |C|^2, d = A^*B, e = CA^*, f = CB^*$ and $g = CA^*$. Now each of these complex values a,...g can be read off from the autocorrelation assuming A, B, C are nonzero and the measurements are noise free. Since one phase is arbitrary we take arg(B) = 0. Then we can "peel-off" the answer:

$$B = \sqrt{df/g} \tag{10.13}$$

$$A = a/B$$

and

$$C = g/B.$$

This example illustrates the possibility of uniquely reconstructing 2-D problems. It also illustrates the underlying impact of the nature of the support on reconstruction. We turn later to this discussion in respect to the Eisenstein criterion.

For the continuous case we have

THEOREM (Sanz-Huang) 10.3. Let $\hat{f} : R^N \to C$ be a band-limited function $(n \geq 2)$ and let $\hat{f}(z)$ be its unique entire exponential extension on C^n. If $\hat{f}(z)$ is irreducible on C^n, then f can be uniquely reconstructed up to trivial associates from $|\hat{f}|$.

We turn now to a more detailed discussion of band-limited and entire functions in 2-D.

10.3 Entire Functions in Two Dimensions

In one dimension we found that the finite Fourier transform of an object function f(t) with finite support may be extended into the complex plane to give

$$F(z) = \int_a^b f(t)e^{izt}dt. \tag{10.14}$$

And by the Paley-Wiener theorem F(z) is an entire function of exponential type. By the Hadamard factorization theorem this can be written as an infinite product

$$F(z) = F(0)\prod_{j=1}^{\infty}(1 - z/z_j) \tag{10.15}$$

where $\{z_j\}$ are the zeros of F(z). The asymptotic zero distribution is determined by the object support and the edge values of the function – i.e., the mth asymptotic zero is given by

$$z_m = (2m\pi - iln(f(b)/f(a))/(b - a). \tag{10.16}$$

Thus, if the values of the object at the edge of the support are identical, then the asymptotic zeros of F(z) are equally spaced along the real axis.

In higher dimension the Paley-Wiener theorem is replaced by the theorem due to Plancherel-Polya:

THEOREM (Plancherel-Polya) 10.4. For a function of n complex variables to be entire of exponential type and square integrable when considered for real values of its n arguments, it is necessary and sufficient that it can be represented in the form

$$F(z_1, ...z_n) = \int_{-\infty}^{\infty} ... \int_{-\infty}^{\infty} f(t_1, ...t_n)e^{i(z_1 t_1 + ... z_n t_n)}dt_1...dt_n \tag{10.17}$$

where f(t) is square-integrable for all its n variables and vanishes at every point outside a bounded domain.

The Hadamard factorization theorem is replaced by the Osgood factorization theorem. We need some definitions of terms first.

DEFINITION 10.5. An entire function $F(z)$ is called globally reducible if it can be written as the product of two entire functions for all z in C^n. Otherwise it is called globally irreducible.

THEOREM (Osgood) 10.6. An entire function $F(z)$ which has zeros and does not vanish identically can be uniquely decomposed in a finite or infinite product of globally irreducible factors:

$$F(z) = \prod_{m=1}^{N} [F_m(z)e^{\gamma_m}]^{l_m} \quad N \leq \infty. \tag{10.18}$$

Here $F_m(z)$ are globally irreducible entire functions and e^{γ_m} are convergence factors, where γ_m are polynomials in z and l_m are integers.

To see how Osgood's theorem arises, let $M(F)$ be the zero set of some entire function F. Any analytic set can be uniquely decomposed into locally finite countable subset (M_j) such that $\cup M_j = M$. Let $M_j = M_j(f_j)$. By Cousin's second theorem the functions defining each M_j can be connected to form functions F_j. Let $G(z) = \prod_{j=1}^{N}[F_j(z)e^{\gamma_j}]^{l_j}$ where l_j exist because components M_j may be equal. Since F/G is an entire function without zeros we have $F/G = e^H$ where H is entire. This implies the result.

Each F_m in Osgood's product defines an irreducible set. In the neighborhood of every point of this set the Weierstrass preparation theorem can be used to write an unique decomposition of the form $F(z_1, z_2) = (z_1 - a_1)^\mu G(z_1, z_2)\Omega(z_1, z_2)$ where G has the form

$$G(z_1, z_2) = (z_2, a_2)^k + A_1(z_1)(z_2 - a_2)^{k-1} + ... + A_k(z_1) \tag{10.19}$$

and $\Omega(z_1, a_2)$ is holomorphic at (a_1, a_2) with $\Omega(a_1, a_2) \neq 0$. For each pseudopolynomial G it can be expressed as a product $G(z_1, z_2) = (z_2 - f_1(z_1))...(z_2 - f_k(z_1))$ where f_k are holomorphic functions for which $f_i(z_1) \neq f_j(z_1)$ for $i \neq j$. This product now locally describes the zero lines of F_m.

So if F_m is irreducible at a point and $k \geq 2$ zero lines coincide there we have a branch point of multiplicity k. If F_m is reducible there, then we have a pseudobranch point of order k.

We leave it to the reader to check that if a factor from a term F_m is removed then the object wave is no longer band-limited. Thus one must invert each F_m as a whole and not individual zero lines; i.e., each F_m describes an unique and interrelated set of lines. The ambiguity of signal reconstruction is directly related to the number of branch points; viz., if there are m branch points of multiplictiy $k_1, ...k_m$, then the ambiguity is reduced from 2^N to $2^{N-\sum_{i=1}^m k_i}$.

THEOREM 10.7. If two solutions of the phase problem have the same set of zeros, then their Fourier transforms differ by at most a translation.

Proof. If $F_1(u,v), F_2(u,v)$ are two solutions, their quotient $r(u,v) = F_2/F_1$ will be zero free on C^2 and has unit modulus or u,v real. It is, therefore, an entire function of the form $e^{iP(u,v)}$, P real, entire. It can be shown that P has to be of the form $P = \gamma u + \delta v$ where γ, δ are real, which completes the proof.

In the discrete signal case where $f(t) = \sum_{j,k} f(j,k)\delta_j \delta_k$, the band-limited function F(z) is a polynomial

$$F(\zeta_1, \zeta_2) = \sum_{j=0}^{M} \sum_{k=0}^{M} f(j,k) e^{ij\zeta_1 + ik\zeta_2}. \tag{10.20}$$

If we set $z = e^{i\zeta}$, we have

$$F(z_1, z_2) = \sum_{j,k=0}^{M} f(j,k) z_1^j z_2^k. \tag{10.21}$$

THEOREM (Hayes-McClellan-Lawton) 10.8. The set of reducible polynomials is of measure zero in $R^N, N = (M+1)^2$; i.e., in any neighborhood of a given reducible polynomial there exists irreducible polynomials.

To show that in general polynomials in C[z], $z = (z_1, ...z_n)$ for $n \geq 2$ are irreducible we follow the original proof given to the author by W. Lawton. We need the following version of Sard's theorem:

THEOREM (Sard) 10.9. Let $f : R^m \rightarrow R^n$ be infinitely differentiable where $n > m$. Then the image $f(R^m) \subset R^n$ has Lebesgue measure zero.

For any (a+1)x(a+1) matrix M over C, define the polynomial $M[z_1, z_2]$ by

$$M[z_1, z_2] = [1, z_2, ...z_2^a] M [1, z_1, ...z_1^a]. \tag{10.22}$$

Every polynomial of degree a in both z_1 and z_2 can be expressed as above. And the set of all such matrices M can be parametrized by R^b where $b = 2(a+1)^2$.

THEOREM 10.10. If $a = 1$, then $M[z_1, z_2] = M_{11} + M_{21} z_2 + M_{12} z_1 + M_{22} z_1 z_2$ is irreducible if and only if $M_{11}M_{22} - M_{12}M_{21} = 0$ and this occurs on a set of Lebesgue measure zero.

Proof. If $M[z_1, z_2]$ is reducible, then there exist complex constants $\alpha_1, \alpha_2, \beta_1, \beta_2$, such that $M[z_1, z_2] = (\alpha_1 + \alpha_2 z_1)(\beta_1 + \beta_2 z_2)$ and, therefore, $M_{11} = \alpha_1\beta_1, M_{22} = \alpha_2\beta_2, M_{12} = \alpha_2\beta_1$, and $M_{21} = \alpha_1\beta_2$. The second statement follows by Sard's theorem stated above.

THEOREM (Lawton) 10.11. For any $a \geq 1$ the set of polynomials $M[z_1, z_2]$ which are reducible forms a set of Lebesgue measure zero.

Proof. By the above result we need consider only $a \geq 2$. If $M[z_1, z_2]$ can be factored in PQ where P has degree p_1 in z_1 and p_2 in z_2 and Q has degree a_1 in z_1 and q_2 in z_2, and $p_1 + q_1 = p_2 + q_2 = a$, then the $2a^2 + 4a + 2$ real numbers which parameterize $M[z_1, z_2]$ are infinitely differentiable functions of the $2(p_1 + 1)(p_2 + 1) + 2(q_1 + 1)(q_2 + 1)$ real numbers which parametrize P and Q. Since the latter integer is less than the former, the result follows from Sard's theorem.

In summary, in the neighborhood of a given reducible polynomial there exist irreducible polynomials – i.e., reducibility is an unstable property. Stefanescu has developed a slightly weaker case of this result for entire functions:

THEOREM (Stefanescu) 10.12. Let (u_o, v_o) be a point of the zero set z of an entire function $F_o(u, v)$ where z is reducible with respect to a sufficiently small neighborhood of (u_o, v_o); then, in any neighborhood of $F_o(u, v)$ there exits an entire function F(u,v) such that the set of zeros of F is irreducible with respect to a sufficiently small neighborhood of (u_o, v_o).

In one dimension if M(u) is of exponential type τ, positive and integrable, then there exists a function (in fact an infinity of them) such that $M(u) = F(u)F^*(u)$ for u in R.

However, in two dimensions there are polynomials which are positive over the Re u-Re v plane and are not reducible. Similarly, there are entire intensity distributions (i.e., entire functions of exponential type, positive and integrable over the Re u -Re v plane) which do not correspond to any amplitude.

Finally, in one dimension if $M(x) = |F(x)|^2 = F(x)F^*(x)$ and $F(z) = Cz^p \prod(1 - z/z_n)$, then

$$M(x) = Cx^{2p} \prod(1 - x/z_n)(1 - x/z_n^*). \qquad (10.23)$$

Thus given an intensity distribution with 2N complex zeros there are 2^N possibilities for zero locations of F(z). In particular we have the Riesz-Fejer theorem for positive definite trigonometric polynomials.

In two dimensions, the discrete problem is far different. Let P(u,v) be a polynomial of degree (M,N) with respect to (u,v). Assume P decomposes into Q prime factors $P_i(u, v)$. Then the modulus squared of P(u,v) in the

Re u-Re v plane

$$M(u,v) = P(u,v)P^*(u^*,v^*) = \prod_{i=1}^{Q} P_i(u,v)P_i^*(u^*,v^*) \qquad (10.24)$$

contains 2Q factors. Again, the ambiguity of M(u,v) is 2^Q (if the Q factors are distinct). However, the number Q depends on the specific polynomial.

As noted above in contrast to 1-D, there are 2-D functions which are positive in the Re u -Re v plane and entire which do not contain pairs of irreducible manifolds which are conjugate to each other. For example, $M(u,v) = e^u + e^v + 1$; here the zeros consist of one irreducible manifold. However, in the general case Stefanescu has shown that M(u,v) is an admissible intensity distribution if it has a finite number of irreducible zero sets of two classes where to each class M_i there corresponds a class M_i^* and each class separately can support an EFET. As a corollary if Z consists of N irreducible zeros sets, then the ambiguity of the phase retrieval problem is at most 2^N. More generally under the impact of sufficiently small errors will M(u,v) still stay admissible? The answer is yes and the ambiguity is still less than or equal to 2^{N_Z} where N_Z is the number of irreducible zero sets of F_o. We refer the reader to Stefanescu's paper for details.

In two dimensions Barakat and Newsam have shown that

THEOREM 10.13. Let g,G be a solution pair to the 2-D phase retrieval problem. Then any other solution pairs h,H must have the form

$$h(z_1, z_2) = e^{i(\alpha_1 + \alpha_2 z)} g_1(z_1, z_2) g_2^*(z_1^*, z_2^*) \qquad (10.25)$$

where $g_1 g_2$ is a factorization of g.

They have also presented evidence that nonuniqueness in 2-D depends on two conditions: (1) that the zero space of g is decomposable into a union of several submanifolds and (2) that the support of g possesses a suitable combination of convexity and symmetry.

10.4 Radial Geometry

Radial geometry provides specific examples of nonuniqueness for the phase retrieval problem. Let $g(x,y) = \theta(x^2 + y^2)e^{-x}$ where $\theta(r) = 1$ for $r \le 1$ and $\theta(r) = 0$ for $r > 1$. Then the Fourier transform of g is $h(u,v) = J_1(R)/R$ where $J_1(R)$ is the first-order Bessel function and $R = \sqrt{(u+i)^2 + v^2}$. This function can be written in the product form as

$$h(u,v) = \prod_{n=1}^{\infty} (1 - R^2/z_n^2) \qquad (10.26)$$

where z_n are the zeros of $J_1(z)$ arranged in order of ascending magnitude.

THEOREM (Huiser-van Toorn) 10.14. Let

$$h(u,v) = \int_S \int dx dy g(x,y) e^{iux+ivy} \qquad (10.27)$$

where S is a closed convex set in R^2 and g vanishes outside of S and is of bounded variation in S. Suppose we can write h(u,v) = p(u,v)f(u,v) where f is entire and p is a polynomial. Let $\tilde{h}(u,v) = q(u,v)f(u,v)$ where q is a polynomial such that q/p is bounded for $|u|^2 + |v|^2 \to \infty$. Then the support of $\tilde{h}$ is S.

For example, the functions

$$\tilde{h}(u,v) = \{(z_n^2 - (u-i) - v^2)/(z_n^2 - R)\}h(u,v) \qquad (10.28)$$

have $|\tilde{h}| = |h|$ for u,v real. And so $\tilde{h}$ and h have the same support. In Figure 10.1, 10.2 and 10.3 we see the inverse Fourier transforms of functions obtained from h by zero flipping:

$$\tilde{h}_n(u,v) = \{[z_n^2 - (u-i)^2 - v^2]/[z_n^2 - R]\}h(u,v)$$

for n = 1,5,6, respectively. Clearly $|\tilde{h}_n(u,v)| = |h(u,v)|$. Figure 10.4 shows the transform of

$$h(u,v) = p(u,v)(z_6^2 - R^2)^{-1}J_1(R)/R$$

where $p(u,v) = z_6^2 - [(u+i)^2 - v^2]$.

DEFINITION 10.15. Let F(x) be a function on $H = (R^N, (,))$. We say F is radial if F(x) = F(y) whenever (x,x) = (y,y) for x,y in H. Let Rad(H) denote the set of radial functions on H.

Lawton has developed the following representation theorem for band-limited functions on H whose moduli are radial. Let B(H) denote the set of band-limited functions on H.

THEOREM (Lawton) 10.16. If $N > 1$ and F is in B(H) and F belongs to Rad(H), then F can be extended to an analytic function $F : C^N \to C$ which has the representation

$$F(z) = P(z)e^{2\pi i(x_o,z)} \prod(1 - (z,z)/\lambda_k^2) \qquad (10.29)$$

(1) x_o belongs to H
(2) $\{\lambda_k\}$ is an infinite sequence of nonzero complex numbers
(3) P(z) is a polynomial of the form

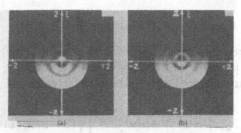

(a) The original real and positive object: e^ξ. (b) The object obtained by flipping zero ν_1. The resulting object is real.

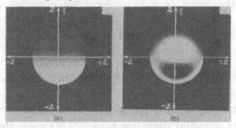

(a) The object obtained by flipping zero ν_5. The result is positive. (b) As in (a) with ν_6.

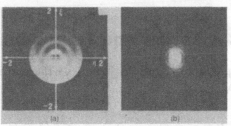

(a) Both zero ν_5 and ν_6 have been flipped. (b) The modulus of the Fourier transforms of the above objects.

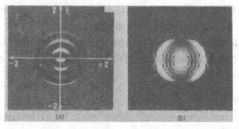

(a) The result obtained by replacing a zero (ν_6) by a different polynomial of degree 2. (b) The modulus of the Fourier transform of (a).

Figure 10.1: Examples with circular symmetry.

(a) $P(z_1, z_2) = A(z_1 + iz_2)^{m_1}(z_1 - iz_2)^{m_2}$ if N = 2, m_1, m_2 nonzero integers

(b) if $N \geq 3$, then $P(z) = A(z, z)^m, m \geq 0$, A in C.

The key observation for the proof is to examine $F_v(t) = F(tv)$ for v in $S^{N-1} = \{x \in H | (x, x) = 1\}$. Noting that F_v is in B(H), we extend $F_v(z)$ to an analytic function with the Hadamard factorization

$$F_v(z) = z^{L(v)} A(v) e^{ia(v)z} \prod_{k=1}^{\infty} (1 - z/\lambda_k(v)) \qquad (10.30)$$

where $\lambda_k(v)$ are the zeros of $F_v(z)$. Next we need to note that (z,z) is reducible when N = 2, i.e., $(z, z) = z_1^2 + z_2^2 = (z_1 + iz_2)(z_1 - iz_2)$ while (z,z) is irreducible for $N \geq 3$. For the rest of the details we refer the reader to Lawton's paper.

COROLLARY 10.17. For $N > 1$ the following two conditions are equivalent:

(1) F, G are in B(H) with $|F| = |G|, |F|$ radial and $\hat{F}$ and $\hat{G}$ are real;

(2) there exists a radial function h in $L^2(H)$ such that h vanishes off some bounded subset of H and both $\hat{F}$ and $\hat{G}$ are translates of h.

10.5   Factorization and Exponential Type

Let P(z,w) be a polynomial and let T(u,v) be the corresponding trigonometric polynomial,

$$T(u, v) = P(e^{2\pi iu}, e^{2\pi iv}). \qquad (10.31)$$

By the result of Hayes we know that T, considered as a trigonometric polynomial, is uniquely determined by its modulus (up to conjugacy and trivial factors) if and only if P is irreducible (up to monomial factors). As we know irreducibility is generic for multidimensional polynomials. However, the example P(x,y) = 1-x and $T(u, v) = 1 - e^{2\pi iu} = (1 + e^{\pi iu})(1 - e^{\pi iu})$ shows that irreducibility of P does not imply the irreducibility (and unique reconstruction from its modulus) of T considered as an entire function of exponential type.

DEFINITION 10.18.   A 2-D trigonometric polynomial is said to be collinear if it is the Fourier transform of a distribution having support contained in some line.

The polynomial T just cited is collinear. Let Z(T) denote the zero set of T and RZ(T) the set of regular points. Then RZ(T) consists of an infinite number of disconnected surfaces.

Consider the irreducible polynomial P(x,y) = 1+x+y. In this case RZ(T) is connected and the complex gradient of T does not vanish on RZ(T). Thus, T is noncollinear and is irreducibile as an entire function. One might hope that if P is irreducible, and if T is non-collinear, then RZ(T) is connected. However, the case $P(x, y) = (1 - x)^2 - 2(1 + x)y + y^2$ is a counter-example; viz., in this case RZ(T) consists of four connected surfaces.

Lawton and Morrison have completely characterized the irreducible factors of non-collinear trigonometric polynomials T when P is irreducible.

THEOREM (Lawton-Morrison) 10.19. Let T be the trigonometric polynomial corresponding to the irreducible polynomial P. Assume T is non-collinear. Then the number M of connected components of RZ(T) is finite. The condition M = 1 is generic (i.e., it holds except for a subset of coefficients of P having measure zero and contained in a nowhere dense closed subset).

Proof. Let $Z(P) = \{(x, y) \in C^2 | P(x, y) = 0\}$ and let S denote the closure of Z(P). So S is compact orientable Riemann surface and the coordinate functions x,y are meromorphic maps from S onto C. Let Z denote the group of integers and for any nonzero meromorphic function w on S define the divisor of w by [w](p) = -(k) if w has a pole (zero) of order $k > 0$; otherwise, [w](p) = 0. Here p is a point in S. Let E(u,v) be defined by $E(u, v) = (e^{2\pi i u}, e^{2\pi i v})$.

Let $D = \{p \in Z(P) | [x](p) \neq 0 \, or \, [y](p) \neq 0\}$ and let B denote the branch point of either x or y.

LEMMA 10.20. $RZ(T) = E^{-1}(S - B - D)$.

The proof is straight forward and will be omitted.

Let $Lat = Z^2$ denote the lattice group which defines a group of transformations g(q) = g+q for all g in Lat and q in RZ(T). Let G denote the subgroup of Lat consisting of all transformations which map the connected component $RZ(T)_o$ into itself. Since Lat acts transitively on each fiber $E^{-1}(p)$, it follows that the set of connected components of RZ(T) is in one-one correspondence with the quotient group Lat/G.

Let Div denote the subgroup of Lat generated by the set $\{[x](p), [y](p)\}$ where $p \in S$.

LEMMA 10.21. Div is a subgroup of G with finite index M; and M equals one generically.

We refer the reader to Lawton and Morrison for the proof. The theorem follows from this Lemma.

Next a trigonometric polynomial whose zero set of $RZ(T)_o$ is constructed. Viz., let (a,b) and (c,d) be generators in G such that M = ad-bc

= index of G in Lat = number of connected components of RZ(T). Let $w = y^a/x^b$ be the meromorphic function on S. One finds that w can be expressed as $w = z^M$ where z is a single-valued, meromorphic function on S. Let $z_1(x), ...z_N(x)$ denote the components of z and for any (x,w) in S-B-D we define the trigonometric polynomial $S_x(s)$ in terms of $z = e^{2\pi i s/M}$ by

$$S_x(s) = \prod (1 - e^{2\pi i s/M}/z_j(x)). \qquad (10.32)$$

THEOREM (Lawton-Morrison) 10.22. There exists a polynomial Q(x) such that the function $F(u, s) = e^{-2\pi i s}Q(e^{2\pi i u})S_x(s)$ where $x = e^{2\pi i u}$ is a trigonometric polynomial in terms of $e^{2\pi i u}$ and $e^{2\pi i s/M}$ which is irreducible as an entire function. And $RZ(F) = RZ(T)_o$ and up to a trivial factor T has a complete factorization into irreducible factors

$$T(u, v) = \prod_j F(u + p_j, s + q_j) \qquad (10.33)$$

where for j = 1,...M (p_j, q_j) in Lat belong to distinct cosets of G.

Proof. By construction $S_x(s)$ has zero set such that $RZ(F) = RZ(T)_o$, which is connected and the gradient of F is not identically equal to zero on $RZ(T)_o$. So F is irreducible as an entire function. Since the elements (p_j, q_j) map RZ(T) onto the distinct connected components of RZ(T) and, therefore, transform G into the irreducible entire functions corresponding to these components, the factorization is a consequence of Osgood's theorem up to a nonvanishing entire factor. Since the only nonvanishing entire function that is a ratio of two trigonometric polynomials is a trigonometric polynomial, the proof is complete.

In the context of phase retrieval from the Fourier modulus these results imply that although the reconstruction of discrete images may be ambiguous when modelled as continuous images and unambiguous when modelled as discrete images, this situation is exceptional.

10.6   Eisenstein Criterion for Irreducibility

In two dimensions given an autocorrelation function Q(z,w) one can construct no more than two solutions p(z,w) and $p(z^{-1}, w^{-1})z^m w^n$ differing by 180 degree rotation if p(z,w) is nonfactorizable. One case which ensures nonfactorizable polynomials is to take $p(x, w) = q(z) + w^k$ which is nonfactorizable for any $k \geq 1$.

A second case is that provided by the Eisenstein criterion:

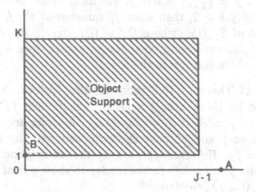

Figure 10.2: Sufficient conditions for a 2-D array to be irreducible

THEOREM (Eisenstein) 10.23. Consider $F(z,w)$ as a polynomial in z, i.e.,

$$F(z,w) = a_o(w) + a_1(w)z + ... + a_{N-1}(w)z^{N-1}$$

where $a_j(w)$ are polynomials in w. If there exists a prime (irreducible) factor $p(w)$ which divides $a_0, ..., a_{N-2}$ but not a_{N-1}, and if $p^2(w)$ does not divide a_0, then $F(z,w)$ is irreducible.

EXAMPLE (Off Axis Holography) 10.24. Let

$$F(z,w) = \sum_0^J \sum_0^K f(j,k)z^j w^k.$$

Assume the object support is a rectangle $0 \le j \le J-1, 1 \le k \le K$. A single reference point at $(J,0)$ ensures irreducibility provided that the point $(0,1)$ is nonzero. This follows since the simplest prime w divides all coefficients except that of the z^J term and w^2 does not divide the z term. Clearly both A and B in the figure must be nonzero. The similarity to off axis holography is clear.

10.7 The Result of Sanz-Huang

In this section we outline the proof of Theorem 10.2. Let Z_f, Z_h denote the zero sets of the analytic functions $f, h : C^n \to C$.

LEMMA (Sanz-Huang) 10.25 If $Z_h = Z_f$ and if f is irreducible, then there exists an entire function $g : C^n \to C$ such that $h = gf$.

The proof of the lemma follows by local analysis. Assume $Z_f = Z_h$ and z^0 is such that $f(z^0) = 0$. Now any prime factor of f around z^0 is a factor of h. Let $f = \prod f_i \gamma$ where f_i are local prime factors $\gamma(z^0) \neq 0$. If f_i has multiplicity $k \geq 2$, then since f_i vanishes at z^0, f_i should vanish at a regular point of Z_f (i.e. where $\nabla f = (\partial f/\partial z_1, ..., \partial f/\partial z_n)$ is not zero). However, ∇f_i^k also vanishes at that point; and so does ∇f giving rise to a contradiction, which completes the proof.

The proof of Theorem 10.2 relies on this lemma. Viz., assume f is irreducible and let $|h(x_1, ...x_n)| = |f(x_1, ...x_n)|$; then $ff^* = hh^*$ and one can check that $Z_f = Z_h \cap Z_f = Z_{h^*} \cap Z_f$. A short argument shows that either f and h or f and h^* have the same zero set. By the lemma either $h = fg$ or $h^* = fg$. However, $gg^* = 1$ by the assumption on f and h. The remainder of the proof is left to the reader to show g is of exponential type and $|g(s_1, ...s_n)| = e^{\sum z_i Im(s_i)}$.

Chapter 11

Reconstruction Algorithms in Two Dimensions

11.1 Canterakis Algorithm

The algorithm of Canterakis is a method for reconstructing a two-dimensional, complex-valued finite sequence from an adequate set of samples of its Fourier transform. The algorithm is based on mapping 2-D finite sequences x(m,n) into 1-D sequences, $\xi(\nu)$.

Let $C = C(N_1, N_2) = \{\xi(\nu)|\xi(\nu) = 0\}$ for $\nu < N_1$ or $\nu < N_2$ and let $CC = C(M_1, M_2, N_1, N_2) = \{x(m,n)|x(m,n) = 0\}$ if $m < M_1$ or $m \geq M_2$ or $n < N_1$ or $n \geq N_2$.

DEFINITION 11.1. Two sequences ξ_1, ξ_2 in C are called equivalent $\xi_1 \sim \xi_2$ if $\xi(\nu) = k\xi_1(\nu + \nu_o)$ for some complex k, $|k| = 1$ and for integer ν_o, for all $N_1 \leq \nu < N_2$. Similarly, x_1, x_2 in CC are called equivalent if $x_2(m,n) = kx_1(m + m_o, n + n_o)$ for integers n_o, m_o for all $M_1 \leq m < M_2$ and $N_1 \leq n < N_2$. Let $<\xi>$ or $<x>$ denote the equivalence class.

Consider the 2N-1 point DFT of ξ in C(0,N)

$$\Xi(k) = \sum_{n=0}^{N-1} \xi(\nu)e^{-i2\pi\nu k/2N-1} \qquad (11.1)$$

k = -(N-1),...,0,...N-1, and the (2M-1)(2N-1) point DFT of x in C(0,M,0,N)

$$X(p,q) = \sum_{m=0}^{M-1}\sum_{n=0}^{N-1} x(m,n)e^{-2\pi i(pm/M-1+qn/N-1)}. \qquad (11.2)$$

151

Clearly if $\xi_1 \sim \xi_2$, then $|\Xi_1| = |\Xi_2|$ and $x_1 \sim x_2$ then $|X_1| = |X_2|$. Let C_s denote the sequences which are symmetric, i.e., $\rho(-\lambda) = \rho^*(\lambda)$ or $r(-m - n) = r^*(m, n)$.

Searching for all ξ_w in $C(0,N)$ with the same Fourier magnitude $|\Xi_w| = |\Xi|$ is equivalent to searching for all ξ_w with the same autocorrelation

$$\rho(\lambda) = R_1(\xi_w)(\lambda) = \sum_{\nu=0}^{N-1} \xi_w^*(\nu)\xi_w(\nu + \lambda) \qquad (11.3)$$

since

$$\rho(\lambda) = \frac{1}{2N-1} \sum |\Xi(k)|^2 e^{-2\pi i k\lambda/2N-1}. \qquad (11.4)$$

Hence, ρ belongs to $C_s(N)$. As we know there are 2^{N-1} different classes $<\xi_w>$ with the same modulus. Thus, $R_1 : C(0, N) \to C_s(N)$ is a many to one onto mapping. The zeros of the z-transform of $\rho(\lambda)$

$$Z(z) = \sum_{\lambda} \rho(\lambda)z^{-\lambda}$$

allow us to construct all possible classes $<\xi_w>$. Let R_1^{-1} denote this map.

Let x(i,j) belong to $C(0,I,0,J)$ and y(k,l) belong to $C(0,K,0,L)$ and let r(m,n) in $C(-(I-1),K,-(J-1),L) = C(r)$ denote the correlation sequence

$$r(m, n) = \sum_{i-0}^{I-1}\sum_{j=0}^{J-1} x^*(i, j)y(i + m, j + n). \qquad (11.5)$$

Let $\phi : C(r) \to C(-(I-1)(L+J-1)-(J-1), (K-1)(L+J-1)+L)$ be the map given by concatenating the rows of r to get ρ:

$$r(m, n)\ \rho(m(L + J - 1) + n). \qquad (11.6)$$

Now form the sequences ξ in $C(0,(I-1)(L+J-1)+J)$ from x and η in $C(0,(K-1)(L+J-1)+L)$ from y by appending L-1 zeros to each row of x (resp J-1 zeros to each row of y) and concatenate the rows.

DEFINITION 11.2. Let $C_o(I, J, L) = \{\xi \in C(0, (I - 1)(L + J - 1) + J)|\xi(\alpha) = 0\}$ if $\alpha mod(L + J - 1) \geq J$.

THEOREM (Canterakis) 11.3. $\xi \times \eta = \rho$.
Take K = I, L = J, y=x and $\eta = \xi$ then we have the maps
$R_2 : C(0, I, 0, J) \to C_s(i, J)$ many to one onto
$\phi : C_s(I, J) \to C_s((I - 1)(2J - 1) + J)$ one-one onto
$f : C(0, I, 0, J) \to C_s(I, J, J)(x \to \xi)$ one-one onto
$R_1 : C(0, (I - 1)(2J - 1) + J) \to C_s((I - 1)(2J - 1) + J)$ many to one,
onto.

The last theorem states that $\phi R_2 = R_1 f$ if and only if $R_2 = \phi^{-1} R_1 f$. The object is to form the inverse $R_2^{-1} = f^{-1} R_1^{-1} \phi$, which will compute all sequences x in $C(0,I,0,J)$ with the same autocorrelation r in $C_s(I, J)$.

The algorithm is:

(1) for r in $C_s(I, J)$ form $\rho = \phi(r)$ in $C_s((I-1)(2J-1)+J)$

(2) compute the set $R_1^{-1}\phi(r)$ of all ξ in $C(0,(I-1)(2J-1)+J)$ with autocorrelation $\rho = \phi(r)$

(3) reject all ξ not in $C_o(I, J, J)$ (i.e., $\xi(\alpha) = 0$ if $\alpha \bmod (2J-1) \geq J$)

(4) apply f^{-1} to obtain ξ in $C_o(I, J, J)$.

THEOREM 11.4. All 2-D sequences x in $C(0,I,0,J)$ obtained by R_2^{-1} have the same autocorrelation r. And there are no other sequences with this property.

EXAMPLE 11.5. Construct all 2-D sequences x of the form

$$\begin{array}{ccc} 0 & 0 & 0 \\ 0 & . & . \\ 0 & . & . \end{array}$$

with the power spectrum

$$X = \begin{array}{ccc} 91 & 79 & 4 \\ 73 & 289 & 73 \\ 4 & 79 & 91 \end{array}$$

The inverse DFT gives the autocorrelation

$$r = \begin{array}{ccc} 6 & 29 & 35 \\ 31 & 87 & 31 \\ 35 & 29 & 6 \end{array}$$

of x. The function $\rho = \phi(r)$ is $\rho = (6, 29, 35, 31, 87, 31, 35, 29, 6)$. If the zeros u of $Z_\rho(z)$ are written as $u(\mu, \nu)$ in a 2x4 matrix, we can arrange so that $u(2, \nu) = u^{*-1}(1, \nu)$. Let ξ_ν belong to $C(1,2)$. Then accroding to (3) we must check each of the 2^4 combinations to see whether

$$\sum_{n_1=1}^{3} \sum_{n_2=n_1+1}^{4} u(\xi_{n_1}, n_1) u(\xi_{n_2}, n_2) = \xi(2) \tag{11.7}$$

vanishes. Two symmetrical cases are found by examination:

$$< \xi_1 >= k(2, 5, 0, 7, 3) \tag{11.8}$$

$$< \xi_2 >= k(3, 7, 0, 5, 2).$$

They correspond to the two classes of solutions of the original problem:

$$< x_1 >= k \begin{array}{ccc} 0 & 0 & 0 \\ 0 & 2 & 5 \\ 0 & 7 & 3 \end{array} \qquad (11.9)$$

$$< x_2 >= k \begin{array}{ccc} 0 & 0 & 0 \\ 0 & 3 & 7 \\ 0 & 5 & 2 \end{array}$$

$$|k| = 1.$$

As Canterakis points out this algorithm is largely a conceptual approach to phase retrieval since it requires, in general, to determine the roots of a polynomial of order 2IJ-(I+J) and the number of cases that have to be checked in step (3) increases rapidly with I and J.

11.2   Berenyi-Deighton-Fiddy Algorithm

The Berenyi-Deighton-Fiddy (BDF) phase retrieval algorithm applies to the case

$$F(u,v) = \sum_{j,k} f(j,k)e^{i(ju+kv)} \qquad (11.10)$$

where f(j,k) are integers. Let $I = |F|^2$. The algorithm is to recover F from the intensity data by applying a 2-D factorization algorithm of Berlekamp. The first step is to assume I(u,v) is monic. Any non-monic polynomial $I(u,v) = P_m(v)u^m + P_{m-1}(v)z^{m-1} + ...P_o(v)$ can be converted to a monic polynomial

$$\tilde{I}(u,v) = P_m^{m-1}I(u/P_m,v). \qquad (11.11)$$

A low valued integer j (usually one) is inserted to form a univariate polynomial I(u,j). Then a prime number q is selected (preferably as small as possible) such that $I_q(u) = I(u,j)mod q$ is square free. Berlekamp's algorithm is then applied to find a set of factors of $I_q(u)$.

Berlekamp's algorithm is based on Fermat's theorem $c^q = cmod q$ and the Chinese remainder theorem. The extended Hensel-Zassenhaus algorithm is applied to generate factors in multivariate form. This algorithm is used on say two factors from Berlekamp's algorithm, one of which is a prime polynomial factor and the other is a product of all the remaining factors. Each of the Berlekamp prime polynomial factors is taken in turn until all the multivariate prime polynomials have been found. For details of the algorithm we refer the reader to the BDF paper.

EXAMPLE 11.6. Consider the 3x4 autocorrelation function

$$
\begin{array}{ccc}
216 & 474 & 238 \\
456 & 953 & 456 \\
238 & 474 & 216
\end{array}
$$

Then $I(u, v) = (216v^2 + 456v + 238)u^2 + (474v^2 + 953v + 474)u + (138v^2 + 456v + 216)$. The solution in this case is $F = (12v+14)u+(17v+18)$ and $F^* = uv(12/uv + 14/u + 17/v + 18) = (18v + 17)u + (14v + 12)$ or in matrix form

$$
\begin{array}{cc}
18 & 14 \\
17 & 12
\end{array}
$$

$$
\begin{array}{cc}
12 & 17 \\
14 & 18
\end{array}
$$

11.3   Logarithmic Hilbert Transforms

Consider the function

$$F(z) = \int_a^b f(t)e^{izt}\,dt \tag{11.12}$$

where f(t) is integrable on (a,b). As we know $F(z)$ is an entire function of order one and type b. When $z \geq 0$, then $F(z)$ is a causal transform and we have the following theorem covering dispersion relations:

THEOREM (Titchmarsh) 11.7. If $F(z)$ fulfils one of the following, it fulfills all of them.

(1) $F(x)$ is for almost all x the limit as $y \to 0+$ of an analytic function $F(z)$ regular for $y > 0$ and L^2 over any line in the upper half plane, parallel to the real axis

$$\int_{-\infty}^{\infty} |F(z)|dx < c \quad y > 0 \tag{11.13}$$

(2) the inverse Fourier transform f(t) of $F(x)$ vanishes for $t < 0$

(3) $ReF(x') = \frac{P}{\pi} \int_{-\infty}^{\infty} \frac{ImF(x)dx}{x-x'}$

(4) $ImF(x') = \frac{-P}{\pi} \int_{-\infty}^{\infty} \frac{F(x)dx}{x-x'}$.

Writing $F(z) = |F(z)|e^{i\phi(z)}$ and taking the logarithm we have

$$lnF(z) = ln|f(z)| + i(\phi(z) \pm 2n\pi) \tag{11.14}$$

where n specifies the Riemann surface in which the log function is defined.

Thus, the real part of ln $F(z)$ is determined solely by the modulus and if the dispersion relations could be used we could reconstruct the function.

However, the problem is that although $\ln F(z)$ has the same region of analyticity as $F(z)$ (except at points where $F(z) = 0$), the function $\ln F(z)$ diverges to $-\infty$ as $|x| \to \infty$; so it is not square integrable.

To achieve a function having a logarithm which is square integrable, one way is to add a finite constant A, so that $G(x) = A + F(x)$ is such that $\ln G(x)$ is square integrable.

Consider now the case that $a = 0$. Examining the real and imaginary parts of the function

$$\int \frac{\ln F(z) dz}{z(x - x')}$$

then, if there are no zeros in the upper half plane, one finds that the disperison relations give:

THEOREM (Burge et al.) 11.8. Under the conditions above

$$ln|F(x')| = \frac{x'P}{\pi} \int_{-\infty}^{\infty} \frac{\phi(x) dx}{x(x - x')} + ln|F(0)| \qquad (11.15)$$

$$\phi(x') = -\frac{x'}{\pi} P \int_{-\infty}^{\infty} \frac{ln|F(x)| dx}{x(x - x')} + \phi(0).$$

However, many functions have zeros in the upper half plane. To deal with these let

$$F_h(x) = |F(x)|e^{i\phi_h(x)} \qquad (11.16)$$

where ϕ_h is the Hilbert phase calculated by the last theorem and all the zeros in the upper half plane have been reflected onto the lower half plane; viz.,

$$F(x) = F(x) \prod_{j=1}^{N} \frac{x - z_j^*}{x - z_j} \qquad (11.17)$$

where the product is taken over the zeros in the upper half plane. Thus, we have

$$|F(x)|e^{i\phi_h(x)} = |F(x)|e^{i\phi(x)}e^{-2i \sum_{j=1}^{N} arg(x - z_j)}. \qquad (11.18)$$

The Hilbert phase is given by

$$\phi_h = \phi(x) - 2 \sum_{j=1}^{N} arg(x - z_j). \qquad (11.19)$$

Putting this in equation 11.15 gives

$$\phi(x) - 2arg(x - z_j) = \frac{xP}{\pi} \int_{-\infty}^{\infty} \frac{ln|F(x')| dx'}{x'(x - x')} + \phi(0) - 2 \sum_{j=1}^{N} arg(z_j). \quad (11.20)$$

In summary, we have

THEOREM (Burge et al.) 11.9. Under the conditions above

$$\phi(x) = \frac{xP}{\pi} \int_{-\infty}^{\infty} \frac{\ln F(x')dx'}{x'(x-x')}dx' - 2\sum_{j=1}^{N} arg(z_j) + \sum_{j=1}^{N} arg(x-z_j) + \phi(0).$$

(11.21)

To use this result, of course, the zeros in the upper half plane must be known. For the case $a \neq 0$ see Burge et al.

The next problem with this approach is that knowledge of the zeros of F(z) are required. However, only information about $|F(z)|^2 = F(z)F^*(z^*)$ is known, which has the zeros of both F(z) and $F^*(z^*)$. The two sets of zeros here are situated in complex conjugate positions. One approach around this situation is to use a nonlinear least-squares method of Nakajima and Asakura, if the object intensity $|f(t)|^2$ is known. We rewrite 11.19 as

$$\phi(x) = \phi_h(x) + 2\sum_{j=1}^{N} tan^{-1}(\frac{x-x_j}{y_j}) + c.$$

(11.22)

Let

$$f_r(t) = \int_{-\infty}^{\infty} |F(x)|e^{i\phi(x)-2\pi ixt}dt.$$

(11.23)

Then the following function is minimized

$$S = \int_{a}^{b} [|f(t)|^2 - |f_r(t)|^2]dt.$$

(11.24)

To evaluate this minimization problem the Powell technique is used in the above cited reference. Without knowledge of the number of upper half plane zeros this has to be done for each possible number until S assumes its least value.

11.4 Algorithm of Nakajima and Asakura

The approach of Nakajima and Asakura to two dimensional phase retrieval is an extension of their 1-D algorithm. Consider the 2-D Fourier transform

$$F(x_1, x_2) = \int\int f(u_1, u_2)e^{-2\pi i(x_1u_1+x_2u_2)}du_1du_2.$$

(11.25)

Let $F'(x_1, x_2)$ be a modified version of F which is bounded in the direction of the x axis

$$F'(x_1, x_2) = rect(x_2/L)F(x_1, x_2)$$

(11.26)

where

$$rect(x/L) = \begin{cases} 1 & |x| \le L/2 \\ 0 & \text{otherwise.} \end{cases}$$

By the sampling theorem we can write

$$F'(x, c) = F(x_1, c) = \int_{-\infty}^{\infty} \frac{1}{L} \sum f'(u_1, n/L) e^{-2\pi i c n} e^{-2\pi i x_1 u_1} du_1 \quad (11.27)$$

where $-L/2 \le c \le L/2$. Thus, the 2-D problem has been reduced to a 1-D problem for which the log Hilbert transform can be applied.

The bandwidth L of F' is selected to be large enough to reconstruct the approximate object function. Here data is for each modulus $|F(x_1, mc)|$ on the 2M+1 lines with equal interval c, i.e., m = -M,...-1,0,1,...M, c=L/2M. The interval c between these data lines is taken to be small enough for sampling the change of the modulus $|F(x_1, x_2)|$ in the direction of the x_2 axis. By using the log Hilbert transform, the phase $\phi_m(x_1, mc)$ of $F(x_1, mc)$ on each data line is evaluated from $|F(x_1, mc)|$ and the zero information of $F(x_1, mc)$. Using the nonlinear least-squares estimation technique one obtains $\phi_m(nd, mc)$, n = -N,...N. Here d is the sampling interval.

The phase distributions evaluated by this procedure have an ambiguity of constant phase differences existing on different sample lines. To resolve this we solve for $\phi(0, mc)$ from the magnitude data $|F(0, x_2)|$, at the point $x_2 = mc$. The sampled 2-D phase distribution is then

$$\phi(nd, mc) = \phi_m(nd, mc) + \phi(0, mc) \quad (11.28)$$

n = -N,...N,m = -M,...M, which completes the reconstruction.

11.5   Zero Crossings in 2-D

Let $p(w, z) = \sum_{m=0}^{M} \sum_{n=0}^{N} p(m, n) w^m z^n = \sum_{m=0}^{M} p_m(z) w^m$ denote a bivariate polynomial. For each complex number z the equation p(w,z) = 0 defines an algebraic function w(z) by p(w(z),z) = 0. Here x(z) is multivalued where $w(z_o)$ will nominally consist of the M roots of $P(w, z_o)$, say $w_1(z), ...w_M(z)$. These are called the branches of the algebraic function.

DEFINITION 11.10. A point z = a in the complex z plane where there are M distinct finite zeros of p(w,a) (i.e., where all the branches of w(z) are finite and do not intersect) is called an ordinary point of w(z).

A value of z where one of the branches becomes infinite or two branches coalesce into a single root is called a singular point.

Near an ordinary point the functions $w_i(z)$ possess a Taylor series expansion: $w = w_{io} + w_{i1}(z - a) + w_{i2}(z - a)^2 + ..., i = 1, ...M$, where w_{io}

are the M distinct roots of p(w,a) = 0. A weak form of Bezout's theorem states:

THEOREM (Bezout) 11.11. If p(w,z) and q(w,z) are bivariate relatively prime polynomials (i.e., no common factors) of degrees r,s respectively, then there are at most rs distinct pairs (w,z) where p(w,z) = 0 and q(w,z) = 0.

COROLLARY 11.12. If two irreducible bivariate polynomials have an infinite number of common zeros, then they are related by a constant.

Let $p_w(w, z)$ denote the partial derivative of p. At a branch point z_o of w(z), we have some w_o for which $p(w_o, z_o) = p_w(w_o, z_o) = 0$. To apply Bezout's theorem to bound the number of zeros of p(w,z) we need to know the circumstances under which p(w,z) and $p_w(w, z)$ are relatively prime.

THEOREM 11.13. Let p(w,z) be expressed as

$$p(w, z) = \alpha \prod_{i=1}^{p} p_i(w, z) \tag{11.29}$$

where P_i are irreducible primitive polynomials and α is a constant. Then is $p_i(w, z) \neq p_k(w, z)$ for $i \neq k$, then p(w,z) and $p_w(w, z)$ are relatively prime.

Let x(m,n) denote a two dimensional discrete signal with z transform $Z(z, w) = \sum \sum x(m, n) z^{-m} w^{-n}$. Let $x_e(m, n) = (x(m, n) + x(-m, -n))/2$ denote the even components of the signal and let $X_e(w, z)$ denote its z transform.

DEFINITION 11.14. A z transform is said to be symmetric if $X(w, z) = X(w^{-1}, z^{-1})$.

DEFINITION 11.15. A signal is said to have a region of support over a nonsymmetric half plane (NSHP) if (m,n) is in the support implies (-m,-n) is not in the region, unless m = n = 0.

Thus, if a signal has a NSHP support then the signal is uniquely specified by its even component since

$$x(m, n) = \begin{cases} x_e(m, n) & m = n = 0 \\ 2x_e(m, n) & \text{otherwise} \end{cases} \tag{11.30}$$

for all m,n in the region of support. Furthermore, in this case the Fourier transform of x_e is $Re\{X(\omega_1, \omega_2)\}$.

DEFINITION 11.16. A sequence x(m,n) is said to be conjugate symmetric if $x(m, n) = x^*(-m, -n)$.

THEOREM 11.17. Let x(m,n) and y(m,n) be two dimensional discrete signals with region of support over a nonsymmetric half-plane contained in $R(N) = \{-N \leq m, n \leq N\}$. If X_e and Y_e are nonfactorizable and $ReX(\omega_1, \omega_2) = ReY(\omega_1, \omega_2)$ at more than $16N^2$ distinct points, then x(m,n) = cy(m,n) for some real constant c.

Proof. Let x,y be two signals with support over a finite NSHP. Then there are integers N_1, N_2 such that $w^{N_1} z^{N_2} X_e(w, z)$ and $w^{N_1} z^{N_2} Y_e(w, z)$ are polynomials in w,z and these two polynomials of degree 4N both vanish at more than $16N^2$ points; by Bezout's theorem they can have at most $16n^2$ common zeros. Thus, $w^N z^N X_e(w, z) = c w^N z^N Y_e(w, z)$.

Note that the number of arbitrarily chosen zero crossing points sufficient to uniquely specify a signal is somewhat greater than the number of points of the signal. For example, if x(m,n) has support over the largest NSHP R(N), then it has $2N^2 + 2n + 1$ distinct points. And by the last result x(m,n) is uniquely specified with m zero crossing points if $m > 16N^2$. If x(m,n) is real, then by symmetries x(m,n) is uniquely specified by $m > 8N^2$ points if these points are chosen over the appropriate range of frequencies.

THEOREM 11.18. Let x,y be real two dimensional sequences with finite NSHP support. Assume $S_x(\omega_1, \omega_2) = S_y(\omega_1, \omega_2)$. If $ReX(\omega_1, \omega_2)$ takes on both positive and negative values and X_e, Y_e are nonfactorizable, then x(m,n) = cy(m,n).

Proof. As in the last theorem $z^{N_1} w^{N_2} X, z^{N_1} w^{N_2} Y$ are polynomials in w,z and they both vanish on the contour in the ω_1, ω_2 plane where $ReX(\omega_1, \omega_2) = 0$. Thus, the theorem follows by Bezout's theorem.

COROLLARY 11.19. If x, y are 2-D conjugate symmetric sequences with finite support with $S_x = S_y$, X,Y are nonfactorizable and X takes on both positive and negative frequencies, then x = cy for some positive constant c.

Let f be a real band-limited continuous time periodic signal with periods T_1, T_2 in the x,y directions. If we set $\omega_1 = 2\pi x/T_1, \omega_2 = 2\pi y/T_2$, then

$$f(x,y) = \sum_m \sum_n F(m, n) e^{2\pi i x m/T_1 - 2\pi i y n/T_2}. \qquad (11.31)$$

By the results above we can view f as a polynomial in $w = e^{-i\omega_1}$ and $z = e^{-i\omega_2}$.

THEOREM 11.20. Let f(x,y) and g(x,y) be real 2-D doubly periodic continuous band-limited signals with sign f = sign g where f takes on both positive and negative values; if f and g are nonfactorizable as polynomials, then f(x,y) = cg(x,y).

Proof. Since f and g have Fourier series coefficients that are complex

conjugate symmetric, have finite support and have nonfactorizable z transforms, the corollary to theorem 11.18 applies.

Curtis, Shitz and Oppenheim have extended this result on periodic
band-limited signals to the case of nonperiodic 2-D signals. In this case
Bezout's theorem is replaced by:

THEOREM 11.21. Two surfaces, analytic over a closed bounded region
D, intersect in at most a finite number of points in D. And two surfaces
analytic over all of C^2 coincide if they have in common some sequence of
points along with their limit point.

Let f,g be real band-limited functions with finite energy. Then by the
Polya-Plancherel theorem these functions can be extended to C^2 as entire
functions of exponential type, f(s,w), g(s,w), which are also EFET in each
variable separately. If f takes on positive and negative values in a closed
bounded region D, then, since f is continuous, there is a contour where f =
0. Similarly, for g. If there exists at least an infinite number of points (x,y)
in D where f and g vanish, then there exists a limit point which is contained
in D. The set of zeros of f, $V_f = \{(s,w)|f(s,w) = 0\}$ is an analytic surface
over a closed bounded domain E where $D \subset E \subset C^2$. Since f is irreducible,
V_f is irreducible. In any bounded, closed set E if two distinct analytic
surfaces have a sequence of points in common along with their limit point,
then by theorem 11.21 $V_f = V_g$ on C^2.

We need the following result of Sanz and Huang.

THEOREM 11.22. If $f, g : C^n \to C$ are entire functions with $V_f = V_g$
and if g is irreducible, then there exists an entire, nonzero function $h :$
$C^n \to C$ where f = hg.

Considering any one dimensional slice of f, g we see that h is an EFET.
But the only nonzero EFET is $e^{\alpha s + \beta w + \gamma}$. Since f,g are real for real s,w,
it follows that α, β, γ are real. And since f,g have polynomial growth in
the real plane, $\alpha = \beta = 0$. Thus, $h = e^{\gamma} = c$ and f(x,y) = cg(x,y) which
completes the proof of theorem 11.22.

If an algorithm were available to reconstruct a function from its zero
crossings it would be nice to know that there are no other signals g with sign
f = sign g without requiring g to be nonfactorable; that is, no other reducible
functions with the same zero crossings. If the exact size of the bandwidth
of f is known, then this information together with the zero crossings of f is
sufficient to uniquely specify f(x,y). Viz.

THEOREM (Curtis) 11.23. Let f,g be real 2-D signals with F(m,n) =
0 and G(m,n) = 0 outside the region $-N_1 \leq m \leq N_1, -N_2 \leq n \leq N_2$ with
sign f = sign g. If $F(N_1, N_2) \neq 0$, f takes on both positive and negative
values and f is nonfactorable, then f = cg for some positive constant c.

Proof. Since sign f = sign g, f and g must have a common factor. Since f is nonfactorable, g = fh for some real periodic bandlimited h. If h is not constant, $H(k,l) \neq 0$ for some k,l, $k > 0$, or $l > 0$. Since $F(N_1, N_2) \neq 0$ we have $G(N_1 + k, N_2 + l) \neq 0$ which violates the assumption. Thus, h is a constant.

A signal of this type is thus uniquely specified to within a scale factor by its zero crossings and the known bandwidth.

By theorem 11.17 a particular number of zero crossing points is sufficient for unique specification. A possible reconstruction algorithm involves solving the set of equations

$$\sum_m \sum_n F(m,n) e^{i2\pi x_j m/T_1} e^{i2\pi y_j n/T_2} = 0 \qquad (11.32)$$

where each equation uses a different pair of points (x_j, y_j) on a zero crossing contour. Assume f is real 2-D, doubly periodic, band-limited nonfactorable, $F(N,N) \neq 0$, with support R(N). If a sufficient number of equations are used (say p equations for $p > 16N^2$), then these equations are guaranteed to have a unique solution.

Curtis has used the real form of the equations

$$\sum_m \sum_n F_R(m,n) cos(2\pi x_j m + 2\pi y_j n) - F_I(m,n) sin(2\pi x_j m + 2\pi y_j n) = 0$$
$$(11.33)$$

where F_R, F_I are the real and imaginary parts of F. The scale factor is determined by setting $F_R(0,0)$ to the mean values of the signal and $F_I(0,0) = 0$ since the image is real. The solution of the equations is obtained by the QR decomposition in Curtis' paper.

11.6   The Israelevitz-Lim Algorithm

Let x(m,n) denote a signal of finite extent and let X(w,z) denote the z-transform, $X(z,w) = \sum_{n_1} \sum_{n_2} x(n_1, n_2) z^{-n_1} w^{-n_2}$.

DEFINITION 11.24. A polynomial $p_x(w,z)$ is said to be associated with x if it is the polynomial of least total degree such that $X(w,z) = p_x(x,w) w^{k_1} z^{k_2}$, for some integers k_1 and k_2.

Let $r(m,n) = \sum_k \sum_l x(k,l) x(k+m, l+n)$ denote the autocorrelation of x(m,n); then select $p_r(w,z)$ and $\tilde{p}_x(w,z)$ to be the polynomials associated with r(w,z) and x(-m,-n). Clearly we have then

$$p_r(w,z) = p_x(w,z) \tilde{p}_x(w,z) \qquad (11.34)$$

since $R(w,z) = X(w,z) \tilde{X}(w,z)$ where $\tilde{X}$ is the z-transform of x(-m,-n).

If $p_x(w, z)$ is irreducible and x(m,n) is not centrosymmetric, then $p_r(w, z)$ and $p_{r,w}(w, z)$ are relatively prime. Thus, by Bezout's theorem, the set of branch points of the algebraic function of $p_r(w, z)$ consists of a finite number of points in the z plane. Therefore, the number of singular points in the z plane is finite so almost all paths in the z plane will consist exclusively of ordinary points.

Since w(z) is analytic at an ordinary point the path w(t), such that $p_r(w(t), a(t))$ is zero for all t, satisfies the differential equation

$$\frac{dw(t)}{dt} = -\frac{p_z \, dz}{p_w \, dt}.$$
(11.35)

First find a complex number c such that $p_r(c, z(t)) = 0$, and then find w(t) which satisfies w(0) = c and equation 11.34. Whether the pair path satisfies $p_x(w(t), z(t)) = 0$ or $\tilde{p}_x = 0$ depends solely on whether the original zero (c,z(0)) is a zero of p_x or $\tilde{p}_x$.

By the result of Curtis, we may sample (w(t),z(t)) at $t_k, k = 1, ...K$ and solve the following equations for the coefficients of $p_x(w, z)$:

$$\sum_{m=0}^{M} \sum_{n=0}^{N} x(M - m, N - n) w_k^m z_k^n = 0$$
(11.36)

where $w_k = w(t_k), z_k = z(t_k)$. By Bezout's theorem if we take $K > (N + M)^2$, then we have a single nontrivial solution to 11.36.

In summary, we have presented the phase reconstruction algorithm of Izraelivitz and Lim for a real signal x(m,n) of finite support:

(1) calculate r(m,n) from $|X(e^{iu}, e^{iv})|$

(2) form $p_r(w, z)$ from r(m,n)

(3) pick z_o such that it is an ordinary point of $p_r(w_o, z_o)$ (almost any value of z_o will do)

(4) find w_o such that $p_r(w_o, z_o) = 0$

(5) pick a path in the z plane consisting of ordinary points such that z(0) = z_o

(6) solve 11.34 with initial condition w_o

(7) pick enough (w,z) pairs so that 11.36 has a unique solution (viz., 2(M-1)(N-1) + 1 for an MxN image)

(8) the solution (if p_x is irreducible) to the equations is a set of coefficients proportional to image values x(m,n) or $\tilde{x}(M - n, N - n)$

(9) normalize the coefficients so that $\sum_{m,n} x(m, n)^2 = r(0, 0)$.

11.7   The Rotem-Zeevi Algorithm

The following one dimensional algorithm was suggested by Voelcker and
Requicha for reconstruction of a bandpass signal from its zero crossings:

$$s_{n+1}(t) = s_n(t) - c(Bsgn(x_n(t) - s_o(t)))) \qquad (11.37)$$

where B denotes the bandpass filter. The Rotem-Zeevi approach extends
this algorithm to two-dimensional signals which satisfy the following conditions:

(1) f(x,y) is real

(2) $\int_{-\infty}^{\infty} \int f^2(x,y)dxdy < \infty$

(3) $\int_{-\infty}^{\infty} f^2(x,y_o)dx < \infty$ for a certain y_o

(4) $\int_{-\infty}^{\infty} f^2(x_k,y)dy < \infty$ for $\{x_k\}$ defined below

(5) F(u,v) vanishes for $u < -b \cup -a < u < a \cup b < u$ and $v < -d \cup -c <
v < c \cup d < v$

(6) for y_o above the function $f(x,y_o)$ and its Hilbert transform do not
have zeros in common except for real zeros of degree one

(7) similarly for $f(x_k,y)$

(8) $\{x_k\}$ constitute a sampling set for these bandpass functions.

By Logan's results in Chapter V we have:

THEOREM (Rotem-Zeevi) 11.25. If there exists another function g(x,y)
which satisfies the above conditions, and if sign f = sign g, then g(x,y) =
A f(x,y).

Their approach is to construct functions along vertical lines x = x_k and
then scale them according to the functions along the horizontal line y_o. For
examples of this algorithm, the reader is referred to Rotem-Zeevi (1986).

Chapter 12

Nonexpansive Maps and Signal Recovery

12.1 Introduction

Let X be a Banach space and U a (possibly) nonlinear mapping of X in X.

DEFINITION 12.1. U is said to be nonexpansive if $\|U(x) - U(y)\| \leq \|x - y\|$ for each pair x, y in X.

The study of the existence of fixed points for nonexpansive mappings is an extension of the classical theory of successive approximations for strict contractions due to Picard, where U is said to be a strict contraction with ratio $k < 1$, if for each pair x, y of X

$$\|U(x) - U(y)\| \leq k\|x - y\|. \tag{12.1}$$

THEOREM (Picard) 12.2. If U is a strict contraction, the Picard sequence $x_n = U(x_{n-1})$ given x_o converges strongly to the unique fixed point of the mapping U.

In the case of nonexpansive mapping U, the Picard sequence need not converge, nor need the fixed point be unique if it does exist.

DEFINITION 12.3. U is said to be strictly nonexpansive if we have $d(U(x), U(y)) < d(x, y)$ whenever $x \neq y$.

Strict nonexpansive maps have at most one fixed point. However, additional constraints are required to guarantee that a fixed point exists.

165

Namely:

THEOREM 12.4. Let $U : A \subset X \to X$ be a strictly nonexpansive mapping which maps a subspace A of a complete metric space X into itself. If the image of A under U is compact, then U has a unique fixed point x^* in A and the sequence $x_k = U^k(x_o)$ converges to x^* for any x_o in A.

A slightly more general concept is a map which is nonexpansive around each of its fixed points.

DEFINITION 12.5. U is said to be quasi-nonexpansive provided that if $Up = p$ then $\|Ux - p\| \leq \|x - p\|$ for all x in X.

Clearly nonexpansive maps are quasi-nonexpansive. However, there are quasi-nonexpansive mappings which are not expansive, e.g., $U(x) = (x/2)sin(1/x), U(0) = 0$ for x in R.

If X is a uniformly convex Banach space, C is a closed convex subset of X and U is a quasi-nonexpansive mapping on C, then if C is bounded, U has at least one fixed point by the Tychonoff fixed point theorem; and if U is nonexpansive, then U has at least one fixed point by the Browder-Kirk fixed point theorem.

THEOREM (Browder-Petryshyn) 12.6. If X is a uniformly convex Banach space for any $\lambda, 0 < \lambda < 1$ the mapping $S_\lambda = \lambda I + (I - \lambda)U$ is nonexpansive, has the same fixed points as U and is asymptotically regular – i.e., $S_\lambda^{n+1} x - S_\lambda^n x$ converges (strongly) to 0 for each x in X.

THEOREM (Dotson) 12.7. If X is a uniformly convex Banach space, and $U : X \to X$ is a linear nonexpansive map, then for $0 < \lambda < 1$ and x_o in X $\{S_\lambda^n x_o\}$ converges strongly to a fixed point of X.

Proof. The proof here is to show that $\{S_\lambda^n\}$ is a system of almost invariant integrals for the semigroup U^m, m=0,1,2... and then apply the mean ergodic theorem of Eberlein. Since U is linear we can write

$$S_\lambda^n = (\lambda I + (1 - \lambda)U)^n = \sum_{j=0}^{n} \binom{n}{j} \lambda^{n-j}(1 - \lambda)^j U^j \qquad (12.2)$$

so S_λ^n is a convex combination of $I, U, U^2, ...U^n$ and $\|S_\lambda^n\| \leq 1$ for all n. The remainder of the proof is Eberlein's mean ergodic theorem.

EXAMPLE 12.8. The requirement of uniform convexity can be seen by taking the not uniformly convex space l_1 and by defining the map $U(x_1, x_2, ...) = (0, x_1, x_2, ...)$. Then, U is nonexpansive, linear and has the unique fixed point $(0,0,...)$. But for $x_o = (1, 0, 0, ...)$ we have $\|S_{1/2}^n x_o\| = 1$ for all n. So $S_{1/2}^n x_o$ does not converge strongly to $(0,0,...)$. And in this example weak convergence is the same as strong convergence.

An example of a uniformly convex Banach space is a Hilbert space. For Hilbert spaces Browder has developed the following extension of the Banach contraction theorem for complete metric spaces.

THEOREM (Browder) 12.9. Let H be a Hilbert space and let U a nonexpansive mapping of H into H. Suppose there exists a bounded, closed, convex subset C of H mapped by U into itself. Let x_o be an arbitrary point of C and for each $0 < k < \lambda$ let $U_k(x) = kU(x) + (1 - k)x_o$. Then U_k is a strict contraction with ratio k, U has an unique fixed point u_k in C, and u_k converges as $k \to 1$ strongly in H to a fixed point u_o of U in C. The fixed point u_o in C is uniquely specified as the fixed point of U in C closest to x_o.

This theorem follows from the next two results.

THEOREM 12.10. Let X be a uniformly convex Banach space (e.g. a Hilbert space), U a nonexpansive mapping of X into X. Suppose there is a bounded, closed, convex subset C of X which is nonempty and is mapped into itself by U. Then, if $F = \{u|u \in C, U(u) = u\}$ is the fixed point set of U in C, we have that F is closed, convex and nonempty.

Proof. The fact that U has a fixed point follows from the Browder-Kirk fixed point theorem. Hence, F is nonempty and obviously closed. The convexity of F follows from the following. If u_o and u_1 are two fixed points of U and if $u_t = (1-t)u_o + tu_1, 0 \le t \le 1$, then,

$$\|U(u_t) - u_o\| + \|U(u_t) - u_1\| \le \|u_i - u_o\| + \|u_t - u_1\| = \|u_o - u_1\|. \quad (12.3)$$

By the strict convexity of X it follows that $U(u_t) = u_t$ and hence, F is convex.

THEOREM 12.11. If H is a Hilbert space, and x_o is in H, then there exists a unique point u_o in F closest to x_o. This point x_o is characterized among the points of F by

$$(x_o - u_o, u_o - x) \ge 0 \quad (12.4)$$

for x in F.

THEOREM 12.12. For Hilbert space H and nonexpansive mapping U on H, if $\{v_j\}$ is a sequence in H and v_j converges weakly to v and $v_j - U(v_j) \to 0$, then v is a fixed point of U.

Proof. Let T = I-U. Then for all u and v in H

$$(Tu - Tv, u - v) = \|u - v\|^2 - (Uu - Uv, u - v) \ge 0. \quad (12.5)$$

(Any operator with this property is called monotone.) By hypothesis $Tv_j \to 0$. For any u in H

$$0 \ge (Tu - Tv_j, u - v_j) \to (Tu - u, v). \quad (12.6)$$

Setting $u = v+tw$ for w in H and $t > 0$, then $t(T(u_t), w) \geq 0$ or $(Tu_t, w) \geq 0$. Letting $t \rightarrow 0+$ we obtain $(Tv, w) \geq 0$. Since this is true for all w in H, Tv $= 0$.

Let B be the unit ball in Hilbert space $H, B = \{x \in H | \|x\| \leq 1\}$.

THEOREM (Browder) 12.13. Let $f : B \rightarrow B$ be a nonexpansive map and let y_k be the unique element of B which satisfies $y_k = kf(y_k)$ for $|k| < 1$. Then $lim_{k \rightarrow 1} y_k = y$ where y is the unique fixed point of f with smallest norm.

DEFINITION 12.14. Let C be a closed convex set in a Banach space X. Then a mapping $T : C \rightarrow C$ is said to be asymptotically regular if for any x in C the sequence $\{T^{n+1}x - T^n x\} = \{(I - T)(T^n x)\}$ tends to zero as $n \rightarrow \infty$.

THEOREM (Opial) 12.15. Let C be a closed convex set in a Hilbert space X and let $T : C \rightarrow C$ be a nonexpansive asymptotically regular mapping for which the set F of fixed points is nonempty. Then for an x in C the sequence of successive approximations $\{T^n x\}$ is weakly convergent to an element of F.

THEOREM (Opial) 12.16. Let C be a closed convex set in a uniformly convex Banach space X having a weakly continuous duality mapping. Assume the nonexpansive asymptotically regular mapping $T : C \rightarrow C$ has in C at least one fixed point. Then for any x in C the sequence $\{T^n x\}$ is weakly convergent to a fixed point of T.

In particular this last theorem of Opial applies to the spaces l_p for $1 < p < \infty$, since these spaces are uniformly convex and have weakly continuous duality mappings.

COROLLARY 12.17. If T is nonexpansive with at least one fixed point, then for any x in C and any λ in $(0,1)$ the sequence of successive approximations $\{T_\lambda^n x\}$ where $T_\lambda = \lambda I + (1 - \lambda)T$ is weakly convergent to a fixed point of T.

EXERCISE (Halperin) 12.18. Let B be a Hilbert space; then show that P is an idempotent contraction of B if and only if P is a projection onto $\{x|Px = x\}$.

EXERCISE (Halperin) 12.19. Let S be a bounded linear operator, with bounded linear inverse. Suppose each ST_iS^{-1}, $i = 1,...r$ is a strict contraction and set $T = T_1...T_r$. Show that STS^{-1} is a strict contraction and Tx $= x$ if and only if $T_ix = x$ for all i.

Let $P_1,...P_r$ denote projection operators onto subspaces $M_1,...M_r$ of a

Hilbert space and let $P_1 \vee ... \vee P_r$ denote the projection onto $\bigcap_{i=1}^{r} M_i$.

THEOREM (von Neumann) **12.20.** Let P denote the product $P = P_1...P_r$. Then P^n converges strongly to $P_1 \vee ... \vee P_r$.

This was originally proven by von Neumann for the case r = 2.

Given two projections P_1 and P_2 in a Hilbert space then the product $T_n...T_2T_1$ converges strongly as $n \to \infty$ where $T_j = P_1$ or $T_j = P_2$ at random.

Let T be a contraction in a Hilbert space, i.e., $\|T\| \leq 1$. Then by the mean ergodic theorem the average $\frac{1}{n} \sum_{j=1}^{n} T^j$ converges strongly as $n \to \infty$ to the projection onto the subspace of all vectors invariant under T – i.e., the null space of I-T. And the orthogonal complement of the null space of I-T coincides with the closure of its range.

Now what about the iterates T^m itself; when do they converge strongly and weakly?

DEFINITION 12.21. We say that a contraction T has property W if $\|f_n\| \leq 1, \|Tf_n\| \leq 1$ implies $(I - T)f_n \to 0$ weakly.

Let r(.) be a mapping from $\{1, 2, ...N\}$ which we call a random selection. For contractions T_j and map r(.) define the sequence of contractions

$$S_n = T_{r(n)}...T_{r(2)}T_{r(1)}. \tag{12.7}$$

By the random ergodic theorem $\frac{1}{n} \sum_{j=1}^{n} S_j$ converges strongly for almost all selections. The problem of strong convergence of random products of contractions is presently unsettled. However, for weak convergence we have:

THEOREM (Amemiya-Ando) **12.22.** If T_j is a contraction with property (W) then for any random selection r(.) the sequence S_n conveges weakly as $n \to \infty$.

Browder, Kirk and others have shown that in a uniformly convex Banach space X every nonexpansive map U of a closed, bounded, convex subset C of X into C must have a fixed point. For Hilbert spaces this was also shown by using results from monotone operators by Browder.

THEOREM (Browder) **12.23.** Let C be a closed, bounded, convex subset of a Hilbert space H and let U be a nonexpansive mapping of C into C. Then U has at least one fixed point in C.

As we noted above C is called asymptotically regular at x if $\|U^n x - U^{n+1}x\| \to 0$ as $n \to \infty$.

DEFINITION 12.24. A mapping U from C into C is called a reasonable wanderer in C if starting at any x_o in C its successive steps $x_n = U^n x_o$ are such that the sum of squares of their lengths is finite, i.e.,

$$\sum_{n=0}^{\infty} \|x_{n+1} - x_n\|^2 < \infty. \qquad (12.8)$$

Clearly every operator which is a reasonable wanderer is asymptotically regular.

THEOREM (Browder-Petryshyn) 12.25. If U is nonexpansive, the set F of fixed points of U in C is not empty and if $U_\lambda = \lambda I + (1 - \lambda)U$ for any given λ with $0 < \lambda < 1$; then U_λ is a reasonable wanderer from C into C with the same fixed points as U. In addition U_λ is asymptotically regular.

For strong convergence we need the notion of demicompact mapping U of C into H which means that whenever $\{u_n\}$ is a bounded sequence in H and $\{Ux_n - x_n\}$ is strongly convergent then there is a subsequence $\{u_{n_i}\}$ of $\{u_n\}$ which is strongly convergent.

THEOREM 12.26. If U is nonexpansive and U maps a bounded closed convex set C into itself and U is demicompact, then the set F of fixed points of U in C is nonempty convex set and for any given x_o in C and any fixed $\lambda > 0, 0 < \lambda < 1$, the sequence $\{x_n\} = \{U_\lambda^n x_o\}$ determined by

$$x_n = \lambda U x_{n-1} + (1 - \lambda)x_{n-1} \qquad (12.9)$$

converges strongly to a fixed point of U in C.

The class of demicompact operators includes the class of compact operators for which versions of this theorem were proven by Krasnoselsik and Schaefer. If we drop the demicompact condition, we have:

THEOREM (Opial) 12.27. If U is nonexpansive, U_λ is as above, and U maps a bounded, closed, convex set C into C. Then for an x_o inC, $U_\lambda^n x_o$ converges weakly to y where y is a fixed point of U in C.

DEFINITION 12.28. Let C be a closed convex set in a Hilbert space H and let $R_C x$ denote the closest point to x in C, i.e., R_C is the projection of H on C.

EXAMPLE 12.29. If $C = B_r(x_o)$ then

$$R_C x = \begin{cases} x & \text{if } \|x - x_o\| \le r \\ \frac{r(x - x_o)}{\|x - x_o\|} & \text{if } \|x - x_o\| \ge r. \end{cases} \qquad (12.10)$$

THEOREM 12.30. Let C be a closed convex subset of the Hilbert space H and for each point x of H let $R_C(x)$ denote the point of C closest to x.

Then R_C is a well-defined mapping of H into C and R_C is a nonexpansive mapping of H which leaves all of the points of C fixed.

Proof. Let x and y be two points of H and let $x_1 = R_C(x), y_1 = R_C(y)$. For z in C, t real $0 < t \leq 1$, since x_1 is the nearest point in C to x and since C is convex we have

$$\|x - x_1\|^2 \leq \|x - (1-t)x_1 - tz\|^2 = \|x - x_1\|^2 + 2t(x - x_1, x_1 - z) + t^2 \|x_1 - z\|^2.$$
$$(12.11)$$

Thus, for $t > 0$

$$2(x - x_1, x_1 - z) \geq -t\|x_1 - z\|^2. \qquad (12.12)$$

Letting t go to zero we obtain

$$(x - x_1, x_1 - z) \geq 0. \qquad (12.13)$$

Setting z to y_1 we have

$$(x - x_1, x_1 - y_1) \geq 0; \qquad (12.14)$$

interchanging x, x_1, y_1 with y, y_1, x_1 gives

$$(x - y_1, x_1 - y_1) \geq 0. \qquad (12.15)$$

Adding these two inequalities together gives

$$(x - y, x_1 - y_1) \geq \|x_1 - y_1\|^2 \qquad (12.16)$$

or

$$(R_C(x) - R_C(y), x - y) \geq \|RC(x) - R_C(y)\|^2. \qquad (12.17)$$

THEOREM 12.31. For Hilbert space H and closed convex set C in H which contains 0, if U is a nonexpansive mapping of C into H with $U(0) = 0$, and if we set for real λ

$$U_\lambda = \lambda U(x) + (1 - \lambda)x \qquad (12.18)$$

for $0 < \lambda < 1$, then U_λ is a nonexpansive mapping of C into H which has the same fixed points as U and for all x in C

$$\|U_\lambda(x)\|^2 + \frac{(1-\lambda)}{\lambda}\|(I - U_\lambda(x)\|^2 \leq \|x\|^2. \qquad (12.19)$$

Proof. Since U and I are both nonexpansive and U_λ is a convex linear combination of U and I, then U_λ is nonexpansive. Clearly every fixed point of U is also a fixed pooint of U_λ; while if $U_\lambda(x) = x$, then $\lambda U(x) = \lambda x$ − i.e., x is also a fixed point of U.

Taking the norms on both sides of $U_\lambda(x)$ gives

$$\|U_\lambda(x)\|^2 = \lambda^2\|U(x)\|^2 + (1-\lambda)^2\|x\|^2 + 2\lambda(1-\lambda)(U(x), x). \quad (12.20)$$

If we take the norm of c(x-U(x)), set $c = \lambda(1-\lambda)$ and add this to the last equality, the inequality follows.

DEFINITION 12.32. If j is a function from Z^+ into a subset A of Z^+, then j is said to be an admissible sequence of integers in A, if for each integer k in A there exists a positive integer m(k) such that the image of each block of m(k) successive integers n,n+1,...n+m(k) must contain the integer k.

Based on this concept Browder has generalized the balayage process to give the following convergence theorem.

THEOREM (Browder) 12.33. Let H be a Hilbert space, let C be a closed convex subset of H, and let $\{C_j | j \in A\}$ be a sequence, infinite or finite of closed convex subsets of H. Here A is a subset of Z^+. Suppose $C = \bigcap_{j \in A} C_j$. Let j(r) be an admissible sequence of integers in A and set $V_k = R_{C_{j(k)}}$. Then if $S_n(x) = V_n V_{n-1}...V_1(x)$ for a given x in H, it follows that $S_n(x)$ converges weakly to an element y_o of C where y_o depends on the initial choice x.

12.2 Time and Frequency Mappings

Let h(n) be a sequence of length N which has an N-point discrete Fourier transform H(k). In polar form we set

$$H(k) = |H(k)|e^{i\theta(k)}. \quad (12.21)$$

DEFINITION 12.34. The mapping which incorporates known values of an N-point sequence h(n) into an arbitrary sequence x(n) will be defined by

$$Tx(n) = \begin{cases} x(n) & n \text{ not in } I_T \\ h(n) & n \in I_T \end{cases} \quad (12.22)$$

where I_T is a subset of [0,N-1] over which h(n) is known.

DEFINITION 12.35. The frequency inclusion map is defined by

$$BX(k) = \begin{cases} X(k) & k \text{ not in } I_B \\ H(k) & k \in I_B \end{cases} \quad (12.23)$$

where I_B is a subset of [0,N-1] where H(k) is known and X(k) and H(k) are the DFTs of x and h respectively.

DEFINITION 12.36. The frequency inclusion mapping which replaces the phase of X(k) with the known phase of H(k) is

$$\Phi[X(k)] = |X(k)|e^{i\theta_h(k)}. \tag{12.24}$$

If W denotes the discrete Fourier transform, we define the three mappings on R^N by $T, F_g = W^{-1}BW$ and $F_p = W^{-1}\Phi W$.

THEOREM 12.37. T, F_g and F_p are nonexpansive.

Proof. Let $d(x, y) = \{\sum_{n=0}^{N-1}[x(n) - y(n)]^2\}^{1/2}$ denote the metric on the metric space (H,d), $H = R^N$. For any x,y in H we set

$$d^2(x, y) = \sum_{n \in I_T}(x(n) - y(n))^2 + \sum_{n \in I_T^c}(x(n) - y(n))^2 \tag{12.25}$$

where I_T^c is the complement of I_T. Thus, $d^2(x, y) \geq \sum_{n \in I_T^c}(x(n) - y(n))^2 = d^2(Tx, Ty)$. And equality holds if and only if x(n) = y(n) for all n in I_T. The nonexpansiveness of B follows similarly and by Parseval's theorem F_g is also nonexpansive.

To see that F_p is nonexpansive, set

$$d^2(X, Y) = \sum_{k=0}^{N-1}|X(k) - Y(k)|^2. \tag{12.26}$$

Then by the triangle inequality we have

$$d^2(X, Y) \geq \sum \||X(k)| - |Y(k)|\|^2 = \sum \||X(k)|e^{i\theta_h(k)} - |Y(k)|e^{i\theta_h(k)}|^2 \tag{12.27}$$
$$= d^2(\Phi X, \Phi Y).$$

12.3 Time-Limited Extrapolation

Assume that h(n) is the sequence which is zero outside the interval [0,M-1]. For $I_T = [M, N - 1]$ let T be the map

$$Tx(n) = \begin{cases} x(n) & 0 \leq n < M \\ 0 & M \leq n < N. \end{cases} \tag{12.28}$$

Let F_g be the map, $F_g = W^{-1}BW$, where $I_B \subset [0, N-1]$ contains at least M points. The time-limited extrapolation problem requires the alternate imposition of time and frequency domain constraints, $G = TF_g$. Consider now the iteration $x_{k+1} = Gx_k = (TF_g)x_k$. Let $S_r(h)$ denote the closed sphere of radius r about h: $S_r = \{x \in H | d(x,h) \leq r\}$. We note that by composition G is nonexpansive, $Gh = h$ and for x in $S_r(h)$ we have

$$d(Gx, h) = d(Gx, Gh) \leq d(x, h) \leq r; \qquad (12.29)$$

so G maps $S_r(h)$ into itself.

THEOREM 12.38. B is strictly nonexpansive on $S_r(h)$ – i.e., we have $d(Gx, Gy) < d(x, y)$ for all x,y in $S_r(h)$ when $x \neq y$.

Since $S_r(h)$ is compact we have by theorem 12.4:

COROLLARY 12.39. The sequence $x_{k+1} = Gx_k$ converges to h(n) for any x_o in $S_r(h)$. Since the radius r is arbitrary this sequence converges for any x_o in H.

12.4 Phase-Only Reconstruction

In Section 6.1 we have developed conditions under which a sequence is uniquely specified to within a scale factor by the phase of its Fourier transform. In particular from theorem 6.2 we have the following corollary:

THEOREM 12.40. Let h(n) be a real sequence which is zero outside [0,M-1] with $h(0) \neq 0$. Assume the z-transform of h(n) has no zeros in reciprocal pairs or on the unit circle. Let $H(k) = |H(k)|e^{i\theta_h(k)}$ denote the N-point discrete Fourier transform of h(n). Here $N \geq 2M$. If x(n) is any sequence which is zero outside [0,M-1] with an N-point DFT of the form (aa) $X(k) = |X(k)|e^{i\theta_h(k)+i\pi\alpha(k)}$ where $\alpha(k) = 0, \pm 1$ for each k, then $x(n) = \beta h(n)$ for all n and some scalar β.

This result motivates the following iterative technique to reconstruct h(n) from $\theta_h(k)$:

$$x_{k+1} = Px_k = (TF_p)x_k, \qquad (12.30)$$

where

$$Tx(n) = \begin{cases} h(0) & n=0 \\ x(n) & 0 < n < M \\ 0 & M \leq n < N \end{cases} \qquad (12.31)$$

i.e., $I_T = \{0\} \cup [M, N-1]$.

THEOREM 12.41. The sequence $x_{k+1} = Px_k$ converges to h(n) for any x_o in H.

Proof. On $S_r(h)$ we see P is nonexpansive, $Ph = h$ and

$$d(Px, Py) = d(TF_px, TF_py) \leq d(F_px, F_py) \leq d(x, y). \qquad (12.32)$$

Assume that equality hold and $x \neq y$. Then we have

$$d(TF_p x, TF_p y) = d(F_p x, F_p y) = d(x, y) \qquad (12.33)$$

where the left hand equality holds if and only if $F_p x = F_p y$ for points in I_T. Define $z(n)$ by $z = F_p x - F_p y$. Note that $z(n) = 0$ for n in I_T and $z(n)$ has an N-point DFT of the form (aa) in Theorem 12.40. Thus $z(n) = \beta h(n)$ for all n and some scalar β. Since $z(0) = 0$ and $h(0) \neq 0$ we have $\beta = 0$ and $z(n)$ for all n. Therefore, $d(F_p x, F_p y) = 0$, so d(x,y) = 0 and x = y, which is a contradiction. Thus, P is strictly nonexpansive for all x,y in $S_r(h)$ when $x \neq y$. Since $P(S_r(h))$ is a compact subset of S_r we have convergence of the iteration to h(n).

12.5   Minimal Phase Signals

Consider the z-transform H(z) of a sequence h(n). Assume H(z) is a rational function of the form

$$H(z) = \frac{A z^{n_0} \prod_{k=1}^{M_a}(1 - a_k z^{-1}) \prod_{k=1}^{M_b}(1 - b_k z)}{\prod_{k=1}^{P_c}(1 - c_k z^{-1}) \prod_{k=1}^{P_d}(1 - d_k z)} \qquad (12.34)$$

where $|a_k|, |b_k|, |c_k|$, and $|d_k|$ are less than or equal to one. If H(n) is stable, i.e., $\sum h(n) < \infty$, then $|c_k|$ and $|d_k|$ are strictly less than one.

H(z) is of minimum phase if it and its reciprocal $H^{-1}(z)$ are both analytic for $|z| \geq 1$.

DEFINITION 12.42. Sequence h(n) is said to be of minimal phase if its z-transform is of minimum phase.

For H(z) rational as above, the minimum phase condition excludes poles or zeros on or outside the unit circle or at infinity. Thus, the factors $(1 - b_k z)$ and $(1 - d_k z)$ will not be present. Also, $n_0 = 0$ to exclude poles or zeros at infinity.

EXERCISE 12.43. Show that if h(n) is stable and H(z) is rational in the form 12.34 with no zeros on the unit circle, then h(n) is of minimum phase if and only if h(n) is causal, i.e. h(n) = 0, $n < 0$ and $n_0 = 0$.

EXERCISE 12.44. Show that if h(n) and H(z) are as in the last exercise, then h(n) is of minimum phase if and only if h(n) is causal and h(0) = A, for A in 12.34.

It follows from above, that if in addition h(n) is of finite duration (i.e., H(z) has no poles), then the iterative technique, given by imposing finite duration plus causality in the time domain and the specified phase in the Fourier domain, converges to a scalar multiple of the minimum phase signal.

12.6   Min and Max Constraints

In many engineering and physics problems there are additional constraints that possibly can be imposed on the solution. For example, in the case in which h(n) is known to be positive for n not in I_T, let T_o be defined by

$$T_o(x(n)) = \begin{cases} |x(n)| & \text{n not in } I_T \\ h(n) & n \in I_T. \end{cases} \tag{12.35}$$

By the triangle inequality we have

$$d^2(x,y) \geq \sum_{n \in I_T} [\|x(n) - y(n)\|]^2 = d^2(T_o x, T_o y); \tag{12.36}$$

so T_o is nonexpansive.

Another option is to use

$$T_o^*(x(n)) = \begin{cases} max(0, x(n)) & \text{n not in } I_T \\ h(n) & n \in I_T \end{cases} \tag{12.37}$$

which can also be checked to be nonexpansive.

Minimum constraints can be imposed as follows. Assume h(n) is known to be bounded below by m for all n not in I_T. We set

$$T_m(x(n)) = \begin{cases} m + |x(n) - m| & \text{m not in } I_T \\ h(n) & n \in I_T. \end{cases} \tag{12.38}$$

If we set $S_m(x(n)) = x(n) - m$ and $S_m^{-1}(x(n)) = x(n) + m$, then since S_m and S_m^{-1} are nonexpansive, and since $T_m = S_m^{-1} T_o S_m$, we see that T_m is nonexpansive.

If h(n) satisfies $M_1 \leq h(n) \leq M_2$ for n not in I_T, then a combination of minimum and maximum value constraints can be imposed by

$$T_{M_1, M_2} x(n) = \begin{cases} min(M_2, max(M_1, x(n)) & \text{n not in } I_T \\ h(n) & n \in I_T. \end{cases} \tag{12.39}$$

Again, this map can be checked to be nonexpansive.

12.7   Masks for Microlithography

The problem in mask generation in microlithography is to generate a pre-scribed binary image at the output of a diffration-limited imaging system with high-contrast recording. The task is to find a 2-D band-limited function with specified zero crossings.

The imaging system is a band-limited system followed by hard-limiting clipping. Let f denote the input image and g = Gf denote the desired output image. The algorithm of Sayegh et al. is:

(1) select $g_o(x)$ and form $G_o(w) = \int_{-\infty}^{\infty} g_o(x)e^{-iwx}dx$

(2) $F_n(w) = G_{n-1}(w)P_{w_o}(w)$ where

$$P_{w_o} = \begin{cases} 1 & |w| < w_o \\ 0 & \text{otherwise} \end{cases}$$

(3) $f_n(x) = \int_{-w_o}^{w_o} F(w)e^{iwx}dw/2\pi$

(4) $g_n(x) = \lambda g_{n-1}(x) + (1-\lambda)\hat{f}_n(x)g_{n-1}(x)$ where

$$\hat{f}_n(x) = \begin{cases} |f_n(x)| & \text{for } |f_n(x)| > \epsilon \\ \epsilon & \text{otherwise} \end{cases}$$

$$0 < \lambda < 1$$

(5) $G_n(w) = \int_{-\infty}^{\infty} g_n(x)e^{-iwx}dx$

(6) return to (2).

Consider now the discrete implementation of this algorithm. Let $g(m,n)$ $0 \leq m, n \leq M-1$ denote the desired output. Define the maps

$$\phi(f(m,n)) = |f(m,n)|g(m,n) \tag{12.40}$$

$$T_\epsilon(f(m,n)) = max(\epsilon, f)$$

$$T_o(f(m,n)) = |f|$$

$$B_T F(m,n) = F(m,n)H(m,n)$$

$$H(m,n) = \begin{cases} 1 & m, n < M \\ 0 & \text{otherwise} \end{cases}$$

$$F_B = \mathcal{F}^{-1} B_T \mathcal{F}.$$

The algorithm can be expressed as $f_{n+1} = F_B g_n$, $\hat{f}_{n=1} = T_\epsilon T_o f_{n+1}$. Thus, combining we have

$$g_n = \lambda g_{n-1} + (1-\lambda)P g_{n-1} \tag{12.41}$$

where $P = \phi T_\epsilon T_o F_B$.

Let H be the vector space $R^m \times R^m$. By theorem 12.27 we have

THEOREM 12.45. Let D_o in H be a closed, convex set. Then P is nonexpansive on D_o and hence, 12.41 converges to a fixed point of P in D_o.

The value w_o i.e., the number of nonzero frequency points of the ideal low pass filter in each direction must be selected with some care. Sayegh et al. have noted that the algorithm works well when w_o is such that the ratio r of the energy of $F_B(g)$ to the total energy of g satisfies $.6 < r < .8$. For smaller r's the value of w_o must be increased, which decreases r.

12.8   Energy Reduction Algorithms

Certain algorithms have been proposed for which the convergence to a fixed point cannot be shown. However, energy reduction at each step can be shown. Consider the case of a finite length sequence where x(n) is zero outside $0 \leq n \leq N - 1$. Take $x(0) \neq 0$. The M-point DFT of x(n) will be denote $X(k) = |X(k)|e^{i\theta_x(k)}$, where we assume $M \geq 2N$. The object is to reconstruct the sequence x(n) from the M samples of its phase $\theta_x(k)$, k=0,1,...M-1. The algorithm starts with an initial guess of the unknown DFT magnitude $|X_o(k)|$. The initial estimate $X_1(k) = |X_o(k)|e^{i\theta_x(k)}$ is found. The inverse DFT gives the estimate $x_1(n)$. Then the sequence $y_1(n)$ is formed where

$$y_1(n) = \begin{cases} x_1(n) & 0 \leq n \leq N - 1 \\ 0 & N \leq n \leq M - 1. \end{cases} \qquad (12.42)$$

The DFT of y_1 is taken giving $Y_1(k)$, whose magnitude is used to form the next estimate $X_2(k) = |Y_1(k)|e^{i\theta_x(k)}$.

The error after the kth iteration is $E_k = \sum_{n=0}^{M-1}(x(n) - x_k(n))^2$ which by Parseval's theorem is equal to

$$E_k = \frac{1}{M}\sum_{n=0}^{M-1}|X(n) - X_k(n)|^2$$

Since $X(n)$ and $X_k(n)$ have the same phase we have

$$E_k = \frac{1}{M}\sum_{n=0}^{M-1}[|X(n)| - |X_k(n)|]^2 = \frac{1}{M}\sum_{n=0}^{M-1}[|X(n)| - |Y_{k-1}(n)|]^2. \quad (12.43)$$

By the triangle inequality we have

$$E_k \leq \frac{1}{M}\sum_{n=0}^{M-1}|X(n) - Y_{k-1}(n)|^2 = \sum_{n=0}^{M-1}(x(n) - y_{k-1}(n))^2. \qquad (12.44)$$

However, $y_{k-1}(n) = x_{k-1}(n)$ for $0 \leq n \leq N - 1$ and $y_{k-1} = x(n) = 0$ for $N \leq n \leq M - 1$. Thus

$$E_k \leq \sum_{n=0}^{M-1}[x(n) - y_{k-1}(n)]^2 \leq \sum_{n=0}^{M-1}(x(n) - x_{k-1}(n))^2 = E_{k-1}. \quad (12.45)$$

In other words the algorithm produces energy reduction at each step; however, convergence can not be guaranteed unless x(n) is known to satisfy the conditions of theorem 12.40.

Chapter 13

Projections on Convex Sets in Signal Recovery

13.1 Introduction

Consider the Landweber iteration applied to solving the system of linear equations:

$$Ax = b \qquad (13.1)$$

where A is a real MxN matrix, x and b are real Nx1 and Mx1 vectors, respectively. The general iteration is given by

$$x_{k+1} = x_k + DA^t(b - Ax_k). \qquad (13.2)$$

Here D is a diagonal matrix chosen to aid the convergence rate. In the overdetermined case ($M > N$) the iteration will converge to the solution if the system of equations is consistent, and it will converge to the minimum norm least squares solution otherwise. In the underdetermined case ($M < N$) the iteration scheme will converge to the pseudoinverse solution plus a vector (determined by x_o) in the null space of A.

The Kaczmarz method (or K-method), which was investigated by Tanabe, was outlined in Section 9.1; viz., let a_i denote the ith row of A and b_i denote the ith element of b. The ith equation of the system is then $< a_i, x >= b_i$. The K-method is based on the projections P_i where

$$P_i(x) = x - \frac{< a_i, x > -b_i}{< a_i, a_i >}a_i. \qquad (13.3)$$

The K-method iteration scheme is

$$x_{k+1} = P_1...P_M(x_k). \qquad (13.4)$$

179

The relationship between Landweber scheme (or L-method) and the K-method is seen by looking at the iterations for a single element of the solution vector estimate. For the L-method

$$x_{k+1}(j) = x_k(j) + d_j \sum_{i=1}^{M} (b_i - (a_i, x_k)a_{i,j}) \qquad (13.5)$$

where $x_k(j)$ denotes the jth element of the estimate and d_j is the jth element of the diagonal matrix D. The K-method gives

$$x_{k+1}(j) = x_k(j) + \frac{1}{M} \sum_{i=1}^{M} (b_i - \frac{(a_i, x_k)}{(a_i, a_i)} a_{i,j}). \qquad (13.6)$$

If the rows of the matrix A all have the same norm $(a_i, a_i) = a$ for all i (e.g., if A is a circulant matrix), then the K-method and the L-method are equivalent if $d_j = 1/Ma$ for all j. If the system of equations are normalized so that $(a_i, a_i) = 1$ for all i, then the methods are also equivalent.

The generalization of the K-method to arbitrary closed convex sets is as follows. Assume an unknown signal f is known to lie in the intersection of m closed convex sets C_i :

$$C_o = \bigcap_{i=1}^{m} C_i. \qquad (13.7)$$

Let P_i i=1,...m denote the projections onto C_i i=1,...m. The generalized method of projection on convex sets is to solve for f by the iterative method

$$x_{k+1} = Gx_k \qquad (13.8)$$

where $G = P_m P_{m-1}...P_1$. We present now details of the results in this area.

Let H be a (complex) Hilbert space.

DEFINITION 13.1. A subset C of H is said to be convex if together with any x_1, x_2 it also contains $x_1 + (1 - \mu)x_2$ for all $\mu, 0 \leq \mu \leq 1$.

THEOREM 13.2. Let C be a closed convex subset of H and let f belong to H. Then there exists a unique x_o in C such that

$$inf_{x \in C} \|f - x\| = \|f - x_o\| \qquad (13.9)$$

i.e., x_o is the element in C closest to f in norm.

Proof. First we demonstrate uniqueness. Suppose that

$$inf_{x \in C} \|f - x\| = \delta = \|f - x_o\| = \|f - y_o\| \qquad (13.10)$$

where x_o, y_o belong to C. Then by the identity

$$\|\frac{x+y}{2}\|^2 + \|\frac{x-y}{2}\|^2 = \frac{\|x\|^2 + \|y\|^2}{2}, \quad (13.11)$$

using $f - x_o, f - y_o$ for x,y we obtain

$$\|f - \frac{x_o + y_o}{2}\|^2 = \delta^2 - \|(x_o - y_o)/2\|^2 \leq \delta^2. \quad (13.12)$$

But because C is convex $(x_o+y_o)/2$ belongs to C so $\|f-(x_o+y_o)/2\|^2 \geq \delta^2$. Hence, $\|x_o - y_o\| = 0$ and $x_o = y_o$.

By definition of infinum there exists a sequence x_n of elements contained in C such that $lim\|f - x_n\| = \delta$. Again, using the identity 13.11 with $f - x_n, f - x_m$ for x,y we have

$$\|x_n - x_m\|^2 = 2(\|f - x_n\|^2 + \|f - x_m\|) - 4(\|f - (x_n + x_m)/2\|^2). \quad (13.13)$$

It follows that $lim_{n,m\to\infty}\|x_n - x_m\| = 0$. The sequence x_n is, therefore, Cauchy and converges to a limit x_o in C because C is closed.

DEFINITION 13.3. The element in C closest to x for any x in H identified in theorem 13.2 is called the projection $P_C x$ of x onto C.

By theorem 13.2 $P_C x$ exists and is uniquely determined by x and C and $P_C x$ satisfies

$$\|x - P_C x\| = min_{y \in C}\|x - y\|. \quad (13.14)$$

Let C be a closed convex set in H and examine $x_2 = x - P_C x$ for x in H and y in C. Then we have

$$\|x - P_C x\|^2 \leq \|x - \mu y - (1 - \mu)P_C x\|^2. \quad (13.15)$$

Expanding the right-hand side gives

$$\|x-P_C x\|^2 \leq \|x-P_C x\|^2 + \mu^2\|y-P_C x\|^2 - 2\mu Re(x-P_C x, y-P_C x). \quad (13.16)$$

Hence,

$$2Re(x - P_C x, y - P_C x) \leq \mu\|y - P_C y\|^2. \quad (13.17)$$

Letting μ go to zero gives

$$Re(x - P_C x, y - P_C x) \leq 0 \quad (13.18)$$

for all y in C. Thus, we have the first part of the following result:

THEOREM 13.4. Let C be a closed convex set. Then for any x in H and for every y in C

$$Re(x - P_C x, y - P_C x) \leq 0. \quad (13.19)$$

Conversely, if for some z in C we have the property

$$Re(x - z, y - z) \leq 0 \qquad (13.20)$$

for all y in C then $z = P_C x$.

Proof. From 13.20 we have

$$\|x-y\|^2 = \|x-z+z-y\|^2 = \|x-z\|^2 - 2Re(x-z, y-z) + \|z-y\|^2 \geq \|x-z\|^2$$

for all y in C. Thus, by theorem 13.2 $z = P_C x$.

DEFINITION 13.5. A sequence x_n in H is said to converge weakly to f if $lim(g, x_n) = (g, f)$ for all g in H.

THEOREM 13.6. Any closed, bounded, convex set is weakly compact, i.e., every sequence x_n in C has a subsequence which converges weakly to a limit f and all such weak limits belong to C.

Proof. Since C is bounded, every sequence x_n is contained in C. Since H is weakly compact, there is a subsequence $x_{n'}$ of x_n such that $x_{n'}$ converges weakly to f in H. By 13.4 we have

$$Re(f - P_C f, x_{n'} - P_C f) \leq 0 \qquad (13.21)$$

for all n'. Since $x_{n'}$ converges weakly to f, we have

$$0 \geq Re(f - P_C f, f - P_C f) = \|f - P_C f\|^2 \geq 0.$$

Thus, $\|f - P_C f\| = 0$ or $f = P_C f$ in C.

THEOREM 13.7. Let C be a closed convex set. Then for every pair x,y in H we have

$$\|P_c x - P_C y\|^2 \leq Re(x - y, P_C x - P_C y). \qquad (13.22)$$

Proof. Since $P_C x$ and $P_C y$ belong to C, one applies the identity (13.19) to these vectors, so

$$Re(x - P_C x, P_C y - P_C x) \leq 0 \qquad (13.23)$$

$$Re(y - P_C y, P_C x - P_C y) \leq 0.$$

The result follows by addition.

COROLLARY 13.8. Projection operators onto closed convex sets C are nonexpansive and, therefore, continuous.

Proof. Schwartz's inequality applied to 13.21 yields for every x,y in H

$$\|P_C x - P_C y\| \leq \|x - y\|. \qquad (13.24)$$

DEFINITION 13.9. A convex set C is said to be strictly convex if x,y in C with $x \neq y$ implies that $(x+y)/2$ is an interior point of C.

DEFINITION 13.10. A convex set C is said to be uniformly convex if there exists a function $\delta(\tau)$ positive for $\tau > 0$ zero only for $\tau = 0$ such that for x,y in C with $\|z - (x+y)/2\| \leq \delta(\|x - y\|)$ implies z belongs to C. If in addition $\delta(\tau) = \mu\tau^2$ for μ positive, C is said to be strongly convex.

THEOREM (Gubin et al.) 13.11. If C is strictly convex, y in C, x not in C and $y \neq P_C x$, then

$$Re(x - P_C x, y - P_C x) < 0. \tag{13.25}$$

COROLLARY 13.12. If C is strongly convex, then

$$\|P_C x - y\| \leq \rho\|x - y\| \tag{13.26}$$

where $\rho = 1/(1 + 2\mu\|x - P_C x\|)$.

Associate to any projection P_C on a closed convex set C is the operator

$$T = I + \lambda(P_C - I). \tag{13.27}$$

Like P_C, T can be shown to be nonexpansive. This leads to the following result:

THEOREM 13.13. Let $C_o = \cap_{i=1}^m C_i$ where C_i are closed, convex sets. If C_o is nonempty, then for every x in H and every choice of constants $\lambda_1, \lambda_2, ...\lambda_m, 0 < \lambda_i < 2$, the sequence $T_i^n x$ converges weakly to a point of C_o. Here $T_i = 1 + \lambda_i(P_i - 1)$ with P_i the projection on C_i.

Proof. T_i is clearly nonexpansive for $0 \leq \lambda_i \leq 1$. And we have

$$\|T_i x - T_i y\| =$$

$$\|(1-\lambda_i)(x-y)+\lambda_i(P_i x - P_i y)\|^2 \leq (1-\lambda_i)^2\|x-y\|^2 + \lambda_i(2-\lambda_i)\|P_i x - P_i y\|^2 \leq \tag{13.28}$$

$$\lambda_i(2-\lambda_i) + (1-\lambda_i)^2\|x-y\|^2 = \|x-y\|^2$$

so it is nonexpansive.

To show that T is a reasonable wanderer we have for $T = T_1$ and $C_o = C_1$ that

$$\|x - Tx\|^2 = \lambda_1^2\|x - P_1 x\|^2. \tag{13.29}$$

And for any y in C, $Ty = P_1 y = y$ with

$$\|Tx-y\|^2 = \|x-y+\lambda_1(Px-x)\|^2 \leq \|x-y\|^2 - \lambda_1(2-\lambda_1)\|x-P_1\|^2. \tag{13.30}$$

Combining this with 13.29 we have

$$\|x - Tx\|^2 \leq \frac{\lambda_1}{2 - \lambda_1}(\|x - y\|^2 - \|Tx - y\|^2) \qquad (13.31)$$

for $0 < \lambda_1 < 2$.

For arbitrary $m \geq 1$ an induction argument yields the inequality

$$\|x - Tx\|^2 \leq b_m 2^{m-1}(\|x - y\|^2 - \|Tx - y\|^2) \qquad (13.32)$$

where $b_m = sup_{1 \leq i \leq m}(\lambda_i/(2 - \lambda_i))$. We leave the details to the reader.

From this last inequality it follows that

$$\sum_{n=0}^{\infty} \|T^n x - T^{n+1}x\|^2 \leq b_m 2^{m-1}\|x - y\|^2 < \infty \qquad (13.33)$$

and hence, T is a regular wanderer. By theorem 12.15 the sequence $\{T^n x\}$ converges weakly to a fixed point of T.

Note that x inC_o implies $x = Tx$ since x is in C_i for i=1 to m. Conversely if $x = Tx$ and y belongs to C_o, then $\|x - y\|\|Tx - Ty\| \leq \|T_1 x - T_1 y\| = \|T_1 x - y\| \leq \|x - y\|$. Hence, $\|x - y\| = \|T_1 x - y\|$. In view of 13.29 this is only possible if $x = T_1 x$, so $x = T_m T_{m-1}...T_2 x$. A repetition of this argument leads to x belonging to C_i for all i, i.e., x belongs to C_o.

For strong convergence we need a further restriction. Let T be a non-expansive asymptotically regular operator on closed convex cone C in H where the set of fixed points F of T is nonempty. From above we know the sequence $\{T^n x\}$ converges strongly to a member of F if and only if at least one of its subsequences converges strongly. This is assured if all iterates $T^n x$ lie eventually in some compact or finite dimensional subset of H. If each of the C_i is a closed linear manifold (CLM), we have the following generalization of von Neumann's alternating projection theorem. First, we state the original von Neumann result.

THEOREM (von Neumann) 13.14. Let C_a, C_b be two CLMs in a Hilbert space H. Let P_a, Q_a, P_b, Q_b denote the projections onto $C_a, C_a^\perp, C_b, C_b^\perp$. Then for every f in H

$$lim(Q_a P_b)^k f = f_c \qquad (13.34)$$

where f_c is the projection of f onto the CLM $C_b \cap C_a^\perp$.

THEOREM (Halperin) 13.15. If every C_i is a CLM, then T_n converges strongly to P_o, the orthogonal projection operator on C_o.

Proof. Since P_i are bounded linear self-adjoint operators in this case, then $T_i^* = 1 + \lambda_i(P_i^* - 1) = T_i$. Similarly, T^m is linear for all $n \geq 0$. Since T is nonexpansive and T0 = 0, we have $\|Tx\| = \|Tx - T0\| \leq \|x\|$ for all x in H. So $\|T^m\| \leq 1$ for $n \geq 0$.

Since $T^n x - T^{n+1} x = T^n(1-T)x$ converges to 0, we see $T^n y$ converges to zero for all y in the range of I-T. We know from theorem 8.1 $\bar{R}(I-T) = (N(I-T)^*)^{\perp}$. But $(I-T)^* = I - (T_m T_{m-1}...T_1)^* = I - T_1 T_2...T_m$. So x is in the null space if and only if $x = T_1 T_2...T_m x$. This null space is identical with C_o; so by the projection theorem $H = C_o + R(I-T)$.

By the orthogonal projection theorem x in H admits a unique decomposition x = z+y where z is in C_o and y is in R(I-T). Since $T^n y \to 0$ we have $T^n x = T^n z + T^n y \to z = P_o x$ as $n \to \infty$.

DEFINITION 13.16. A linear variety C is the set of all vectors x of the form x = g+y where g belongs to H is fixed and y ranges throughout some CLM S.

A linear variety is thus a closed convex set. The following result of Amemiya and Ando holds:

THEOREM 13.17. If each C_i is a linear variety and if their intersection C_o is nonempty, then the sequence $\{T^n x\}$ converges strongly to $P_o x$ for every x in H.

Gubin, Polyak and Raik have examined the case of strong convergence as follows.

THEOREM (Gubin et al.) 13.18. Either of the two conditions below suffices to guarantee the strong convergence of the sequence $\{T^n x\}$ of theorem 13.13 to its weak limit x^*:

(1) at least one of the C_i is uniformly convex and does not contain x^* in its interior;

(2) for some $\alpha, 1 \le \alpha \le m$, the set intersection

$$C_\alpha \cap (\cap_{i \neq \alpha}^m C_i)^o$$

is nonempty;

and under this condition the convergence of $\{T^n x\}$ to x^* is at a geometric rate.

13.2 Alternating Projections

Let H be a Hilbert space and let C be a closed linear manifold in H. By the projection theorem $H = C \oplus C^{\perp}$. Let P and Q denote the projections on C and $C^{\perp}$ respectively.

The reconstruction problem is as follows: let f in H be known to belong to a CLM C_b but we are given only its projection $g = P_a F$ onto a known CLM C_a; can we reconstruct f from g?

Since f belongs to C_b we have $f = P_b f$ and $g = P_a f = P_a P_b f = (1 - Q_a) P_b f = P_b f - Q_a P_b f = f - Q_a P_b f$. Thus f satisfies the linear equation $Af = g$ where A is the bounded linear operator $A = 1 - Q_a P_b$.

DEFINITION 13.19. The problem of reconstructing f from $P_a f$ given that f belongs to C_b is said to be completely posed if the operator A has a bounded inverse. Otherwise, it is said to be incompletely posed.

By the Schwartz inequality for two elements f,g of H we have

$$0 \le \frac{|(f,g)|}{\|f\|\|g\|} \le 1. \tag{13.35}$$

Thus, we have the angle $\psi(f,g)$ between f ad g which is defined by

$$cos(\psi(f,g)) = |(f,g)|/\|f\|\|g\|. \tag{13.36}$$

DEFINITION 13.20. The angle between two linear manifolds C_a, C_b is defined by $\psi(C_a, C_b) = inf_{f \in C_a} \psi(f, g)$.

THEOREM 13.21. If P_a, P_b are the projections onto CLMs C_a, C_b, then

$$\|P_a P_b\| = cos\psi(C_a, C_b) = \|P_b P_a\|. \tag{13.37}$$

THEOREM 13.22. If C_a, C_b are CLMs with $C_a \cap C_b = \{0\}$, then $C = C_a + C_b$ is closed if and only if $\psi(C_a, C_b) > 0$.

Proof. This follows by the last result since $\psi(C_a, C_b) > 0$ implies $\|P_b P_a\| < 1$.

THEOREM (Youla) 13.23. Let C_a, C_b be any two CLMs in H. Suppose f belongs to C_b. Then:

(1) f is uniquely determined by its projection $P_a f$ if and only if $C_b \cap C_a^{\perp} = 0$.

(2) the problem of reconstruction of f from $P_a f$ is completely posed if and only if $\psi(C_b, C_a^{\perp}) > 0$; (i.e., if and only if $\rho = \|Q_a P_b\| < 1$).

(3) in both cases (1) and (2) the sequence $\{f_k\}$ where

$$f_{k+1} = g + Q_a P_b f_k \quad k = 1,...$$

$$f_1 = g = P_a f$$

converges in norm to f; the convergence is strictly monotone increasing.

Proof. For (1) $Af = g$ is uniquely determined by g if and only if A^{-1} exists, i.e., if and only if $Af_o = (1 - Q_a P_b)f_o = 0$ has the unique solution

$F_o = 0$. However, $f_a = Q_a P_b f_o$ implies $\|f_o\| = \|Q_a P_b f_o\|$ which is possible if and only if f_o belongs to $C_b \cap C_a^\perp$. Conversely any f_o so defined satisfies $Af = 0$; hence, A^{-1} exists if and only if $C_b \cap C_a^\perp = \{0\}$.

For (2) we note that if $\psi(C_b, C_a^\perp) > 0$, then $C_b \cap C_a^\perp = \{0\}$ by theorem 13.22 and the explicit solution of

$$g = (1 - Q_a P_b)f \tag{13.38}$$

is given by

$$f = \sum_{r=0}^{\infty} (Q_a P_b)^r g$$

which converges like a geometric series since $\|Q_a P_b\| = \rho < 1$. Thus, we have

$$\|f\| \leq \frac{1}{1 - \|Q_a P_b\|} \|g\|$$

or

$$\|A^{-1}\| \leq 1/(1 - cos\psi(C_b, C_a^\perp)).$$

The rest of (2) follows straight forwardly.

For (3) if g is in the range of A we have $g = f - Q_a P_b f$ or iteratively

$$f_k = \sum_{r=0}^{k-1} (Q_a P_b)^r g = f - (Q_a P_b)^k f. \tag{13.39}$$

Convergence follows from von Neumann's alternating projection theorem 13.14.

Thus, we see if $\psi(C_b, C_a^\perp) = 0$, then the range of the operator A is dense in H but it is not all of H. This possibility cannot arise in the finite dimensional setting.

THEOREM (Youla) 13.24. If 0 is a limit point of the set of nonzero eigen values of $P_a P_b$, then $\psi(C_b, C_a^\perp) = 0$.

13.3   Reconstruction by Projections

Let H denote the space of square integrable functions $f(t)$ $-\infty < t < \infty$ with Fourier transform $F(w) = \int_{-\infty}^{\infty} f(t) e^{-itw} dt$.

Let C_a denote the subset of H of functions which vanish almost everywhere for $|t| > a$. Let C_b denote the subset of all functions whose Fourier transforms vanish a.e. for $|w| > b$. It is clear that C_a and C_b are CLMs. Moreover, C_b is devoid of interior points, since given any $\varepsilon > 0$ and f in C_b there exists an h not in C_b such that $\|f - h\| \leq \varepsilon^2$; in words, any neighborhood of a band-limited function contains signals that are not band-limited.

Let
$$\chi_a(t) = \begin{cases} 1 & |t| \le a \\ 0 & \text{otherwise.} \end{cases} \qquad (13.40)$$

Then the projections are explicitly

$$P_a f = \chi_a(t) f(t) \qquad (13.41)$$

$$P_b f = \frac{1}{2\pi} \int_{-b}^{b} F(w) e^{iwt} dw = \int_{-\infty}^{\infty} \frac{sinb(t-s)f(s)ds}{\pi(t-s)}. \qquad (13.42)$$

The Gerchberg-Papoulis problem is a special case of Youla's theorem: assume a function is bandlimited to $-b \le w \le b$ but only its segment over $-a \le t \le a$ is known; reconstruct f(t) for $|t| > a$. Since any f in C_b is automatically an entire function, then it will vanish in $|t| \le a$ if and only if it is the zero function. Thus, $C_b \cap C_a^\perp = \{0\}$ and Youla's theorem states that

$$f_{k+1} = g + Q_a P_b f_k \qquad (13.43)$$

$$f_1 = g = P_a f$$

converges to f.

Since $P_a P_b x_i = \lambda_i x_i$ is equivalent to

$$\int_{-a}^{a} \frac{sinb(t-s)x_i(s)ds}{\pi(t-s)} = \lambda_i x_i \quad -a \le t \le a \qquad (13.44)$$

with eigenvalues $\lambda_i, 1 > \lambda_i > 0$, which converge to zero as $i \to \infty$, we have by Youla's theorem the following result due originally to Viano:

THEOREM (Viano) 13.25. The reconstruction problem as defined by Gerchberg-Papoulis is incompletely posed.

13.4   Regularization

Youla has generalized Tikhonov's regularization scheme as follows:

DEFINITION 13.26. $A = 1 - Q_a P_b$ is said to be properly posed if there exists a one parameter family of transformation $T_\mu, \mu > 0$ such that
(1) for every fixed μ, T_μ has domain H
(2) for every x in $C_b, \lim T_\mu A x = x$
(3) $T_\mu g$ is jointly continuous in μ and g from above.

Such a family $T\mu$ is called a regularizer.

DEFINITION 13.27. Let C_b^μ be any one parameter family of subsets of C_b such that
(1) each C_b^μ is closed, convex and compact

(2) $C_b^{\mu_1} \subset C_b^{\mu_2}$ if $\mu_1 < \mu_2$

(3) $\cup_\mu C_b^\mu = C_b$

(4) for every $\mu_o \geq 0, \cap_{\mu > \mu_o} C_b^\mu = C_b^{\mu_o}$.

The set $\{C_b^\mu\}$ with these properties is called a refinement.

Since A is bounded, the monotone family of sets $C_\mu = AC_b^\mu$ is compact, closed and convex. We let $P_\mu g$ denote the (nonlinear) projection of g onto C_μ. Then we state:

THEOREM (Youla) 13.28. Let C_b^μ be a refinement of c_b. For a fixed g in H and $\mu > 0$ let $m_\mu(g) = inf_{x \in C_b^\mu} \|Ax - g\|$. Then $T_\mu g$ is obtained as the unique solution of $Ax = P_\mu g$, i.e., $T_\mu = A^{-1} P_\mu$. $T_\mu g$ is jointly continuous in μ and g from above; and for every x in C_b, $lim_{\mu \to \infty} T_\mu Ax = x$.

If a regularizer exists then the impact of noise on the reconstruction problem is not serious. Say $Af_o = g_o$ where f_o belongs to C_b and suppose $g = g_o + \Delta g$. Then $T_\mu g_o - f_o = T_\mu g_o - f_o + T_\mu g - T_\mu g_o = (T_\mu Af_o - f_o) + (T_\mu g - T_\mu g_o)$. By definition given $\varepsilon > 0$ there is a $\mu(\varepsilon, f_o)$ such that $\|T_\mu Af_o - f_o\| < \varepsilon/2$ for all $\mu > \mu_o$ and by continuity there is a δ_o such that $\|T_\mu g - T_\mu g_o\| < \varepsilon/2$ for $\mu_o < \mu < \mu_o + \delta$ for $\|g - g_o\| < \delta$. Taking the norm of both sides above shows that $\|T_\mu g - f_o\| < \varepsilon$ for $\mu_o < \mu < \mu_o + \delta$ and $\|g - g_o\| < \delta$. That is, the original f_o can be reconstructed to any desired degree of accuracy with the aid of the regularizer T_μ provided $\|\delta g\|$ is sufficiently small.

We note that $x = T_\mu g$ satisfies $Ax = P_\mu g$, so $T_\mu g$ can be found by the iterative technique 13.40 with $f_1 = P_\mu g$.

13.5   Convex Projection in Noise

Let C_a denote the closed linear manifold of functions that are zero outside [-a,a] and let C_b denote the set of functions bandlimited to the interval [-b,b]. The Papoulis algorithm is then

$$f_{k+1} = P_g P_b f_k \tag{13.45}$$

where P_b is the projection operator onto C_b and P_g is the projection operator onto C_g, the closed convex set of all functions whose value over [-a,a] is the prescribed function g(t). This iteration will converge since the intersection of C_g and C_b is nonempty. However, if the recorded segment g is corrupted by noise this intersection may be empty and the iteration will diverge.

To stabilize this process we can imbed C_g in a closed convex set $C_g(\epsilon)$. Viz., consider the set of functions whose distance from the recorded data over [-a,a] is smaller than some positive constant ϵ:

$$C_g(\epsilon) = \{f | \|P_a f - g\| < \epsilon\}. \tag{13.46}$$

This set is closed, convex and has an interior. To see this let $f_n \in C_g(\epsilon)$ be a sequence which converges to f_o. Since P_a the projection onto C_a is a continuous operator

$$P_a f_n - g \to P_a f_o - g \tag{13.47}$$

so $\|P_a f_o - g\| < \epsilon$ i.e., f_o belongs to $C_g(\epsilon)$. And if f_1, f_2 belong to $C_g(\epsilon)$ and $0 \le t \le 1$, then $t f_1 + (1-t)f_2$ belongs to $C_g(\epsilon)$ because

$$\|P_a(t f_1 + (t-t)f_2) - g\| \le t\|P_a f_1 - g\| + (1-t)\|P_a f_2 - g\| \le \epsilon. \tag{13.48}$$

The reader can check that $C_g(\epsilon)$ has interior points.

The projection of f onto $C_g(\epsilon)$ is given by

$$P_g(\epsilon)f = \begin{cases} f & \text{if } \|P_a f - g\| \le \epsilon \\ f - (1-\epsilon)(P_a f - g)/\|P_a f - g\| & \text{otherwise.} \end{cases}$$
$$\tag{13.49}$$

Assume the recorded segment of g has been corrupted by noise – i.e.,

$$g = P_a f + P_a n. \tag{13.50}$$

If n(t) can be modeled by some stochastic process so that $\|P_a n\|$ can be estimated from the statistics, then taking $\epsilon > \|P_a n\|$ we use the iteration formula

$$f^{k+1} = P_b P_g(\epsilon)f_k. \tag{13.51}$$

Since $C_b \cap C_g(\epsilon)$ has a nonempty interior (which contains f), f_{k+1} converges to some fixed point f_o at a geometric rate. Unfortunately, it is most often the case that $\|f_o - f\| \sim \epsilon$, which is a serious limitation.

13.6   Inconsistent Constraints

Let two incompatible properties Π_1, Π_2 define two disjoint, closed, convex subsets C_1, C_2 of a Hilbert space and suppose C_1 is bounded. Let P_1, P_2 be the respective projection operators on C_1, C_2. Then the composition operator $T = P_1 P_2$ possesses a fixed point f and every such fixed pont is a point in C_1 whose distance from C_2 is a minimum. Such a minimizer f may be characterized as a signal which possesses property Π_1 and best approximate Π_2.

THEOREM (Goldburg-Marks) 13.29. Let C_1 and C_2 be two disjoint, closed, convex subsets of a real Hilbert space H and suppose C_1 is bounded. Then any fixed point f of $T = P_1 P_2$ is a point in C_1 whose distance from C_2 is as small as possible and equals the distance from C_1 to C_2.

For any x,y in a complex Hilbert space H let

$$d(x, C) = \inf_{y \in C} \|x - y\| \tag{13.52}$$

and

$$d(C_1, C_2) = inf_{x \in C_1, y \in C_2} \|x - y\|.$$

THEOREM (Youla-Velasco) 13.30. Let C_1 and C_2 be two disjoint, closed, convex subsets of a complex Hilbert space H and suppose C_1 is bounded. Then (1) $d(C_1, C_2) > 0$ and there exists an f in C_1 such that $d(f, C_2) = d(C_1, C_2)$; such an f is said to be minimizing. And (2) f is minimizing if and only if it is a fixed point of $T = P_1 P_2$.

Proof. Since C_2 is convex $\psi(x) = d(x, C_2)$ is a continuous convex function of x defined for all x in H. Such a function assumes its infinum on any bounded, closed, convex set. Hence, for some f in C,

$$d(f, C_2) = inf_{x \in C_1} d(x, C_2) = d(C_1, C_2). \qquad (13.53)$$

But C_2 is closed and convex so that $d(f, C_2) = d(f, P_2 f) = d(C_1, C_2)$. Clearly, $d(C_1, C_2) = 0$ implies $f = P_2 f$ which means f belongs to $C_1 \cap C_2$, a contradiction, which completes the proof of (1). The reader is left to prove (2).

The computation of minimizers f can also be carried out with T replaced by its relaxed version $T_\lambda = 1 + \lambda(T - 1)$ where $0 < \lambda < 1$.

A unique generalization to the case of more than two constraints does not seem possible. For details the reader is referred to the paper of Youla and Velasco.

13.7 Algebraic Reconstruction Techniques

The Algebraic Reconstruction Technique or ART was realized by G. T. Herman et al. to be an example of the Kaczmarz algorithm or more generally the following result on inequalities. Let H be a Hilbert space and suppose we are given a set of linear inequalities

$$(c_\alpha, x) \le a_\alpha \qquad (13.54)$$

for α in A, where c_α belongs to H. The problem of solving this set of inequalities is equivalent to finding a point of the set $C = \cap_{\alpha \in A} C_\alpha$ where $C_\alpha = \{x | (c_\alpha, x) \le a_\alpha\}$ is a half-space. The projection onto C_α is given by

$$P_{C_\alpha}(x) = \begin{cases} x & \text{if } (c_\alpha, x) \le a_\alpha \\ x - \lambda c_\alpha & \text{if } (c_\alpha, x) > a_\alpha \end{cases} \qquad (13.55)$$

where $\lambda = ((c_\alpha, x) - a_\alpha)/\|c_\alpha\|^2$ which provides the natural generalization of the Kaczmarz iteration.

THEOREM (Gubin et al.) 13.31. Let the system of inequalities have a solution and let any of the following conditions hold:
(1) C^o is nonempty
(2) H is finite dimensional
(3) A is finite.
Then the iterative method

$$x_{n+1} = \begin{cases} x_n - ((c_{\alpha(n)}, x_n) - a_{\alpha(n)})c_{\alpha(n)}/\|c_{\alpha(n)}\|^2 & \text{if } (c_{\alpha(n)}, x_n) > a_{\alpha(n)} \\ x_n & \text{otherwise,} \end{cases}$$

(13.56)

where $\alpha(n)$ is given by

$$\rho(x, Q_{\alpha(n)}) = \|x - P_{Q_{\alpha(n)}}(x)\| = sup_{\alpha \in A} \frac{(c_\alpha, x_n) - a_\alpha}{\|c_\alpha\|}, \quad (13.57)$$

converges to a solution.

Many examples of the application of ART have been discussed by Herman and coworkers. Huang et al. have also proposed the use of the Kaczmarz algorithm to restore degraded images as we noted earlier.

EXAMPLE 13.32. Given g(t) in L^2, with g(t) = 0 for $|t| > a$, let $C_a(g)$ denote the set of all f whose projection onto T_a is equal to g(t). Then $C_a(g) = g \bigoplus T_a^\perp$ is a linear variety. The projection onto $C_a(g)$ is given by $P_{C_a(g)}f = g(t) + (1 - P_{a(t)})f(t)$.

DEFINITION 13.33. A convex set C is a cone with vertex 0 if f belongs to C implies μf belongs to C for all $\mu \geq 0$.

EXAMPLE 13.34. Let B_b^+ denote the subset of B_b whose Fourier transforms are nonnegative. Then B_b^+ is a close convex cone with vertix 0. To see this, let f(t), $f_1(t)$ and $f_2(t)$ belong to B_b^+. Then $\mu f(t)$ is clearly in B_b^+. Similarly, the Fourier transform of $\mu f_1 + (1 - \mu)f_2$ is nonnegative. So B_b^+ is a convex cone. We leave it to the reader to check that B_b^+ is closed. The projection onto B_b^+ is $Pf = \mathcal{F}^{-1}(P_b(w)Re(F)^+)$ where $F^+ = F$ for $F \geq 0$ and 0 otherwise.

EXAMPLE 13.35. Let $B_b^+(r)$ denote the subset of B_b^+ composed of all functions f for which $\frac{1}{2\pi} \int_{-b}^b F(w)dw = r$. Since $F(w) \geq 0$, it follows that $r \geq 0$. Clearly, $B_b^+(r)$ is convex. To check closure, let $\{f_n\}$ belong to $B_b^+(r)$ with $f_n \to f$. By the last example f belongs to B_b^+. If F is the Fourier transform of f, then by Schwarz's inequality we have

$$|r - \frac{1}{2\pi} \int_{-b}^b F(w)dw|^2 = \frac{1}{4\pi}| \int_{-b}^b (F_n(w) - F(w))dw|^2 \leq \quad (13.58)$$

$$\frac{1}{4\pi}(\int_{-b}^b |F_n(w) - F(w)|dw)^2 \leq \frac{b}{2\pi^2} \int_{-b}^b |F_n(w) - F(w)|^2 dw \to 0.$$

Thus, $\frac{1}{2\pi}\int_{-b}^{b}F(w)dw = R$ and $B_b^+(r)$ is closed. The projection onto $B_b^+(r)$ is given by $Pf = \mathcal{F}^{-1}(P_b(w)(Re(F)+c)^+)$ where c is a real constant chosen so that $\frac{1}{2\pi}\int_{-b}^{b}(Re(F(w)) + c)dw = r$.

EXAMPLE 13.36. Let $A_b^+(r^2)$ denote the subset of all functions in B_b^+ which satisfy $\int_{-\infty}^{\infty}|f(t)|^2dt = \frac{1}{2\pi}\int_{-\pi}^{\pi}|F(w)|^2dw \leq r^2$. Since $\|\mu f_1 + (1-\mu)f_2\| \leq \mu\|f_1\| + (1-\mu)\|f_2\| \leq \mu r + (1-\mu)r = r$ for any f_1, f_2 in $A_b^+(r^2)$ and $\mu, 0 \leq \mu \leq 1$, we see that $A_b^+(r^2)$ is convex. If $f_n \to f$, then $r \geq \|f_n\| \to \|f\|$ and closure follows.

EXERCISE 13.37. Find the projection onto $A_b^+(r^2)$.

EXERCISE 13.38. Check that $B_b^+, B_b^+(r)$ and $A_b^+(r^2)$ as subsets of B_b are all with interiors.

EXAMPLE (Reconstruction from Phase) 13.39. Let $\phi(w_1, w_2)$ be a prescribed, real valued function on a set D in the w_1, w_2 plane. Let $C_\phi(D)$ denote the set of all L^2-functions whose Fourier transforms are of the form

$$F(w_1, w_2) = A(w_1, w_2)e^{i\phi(w_1,w_2)} \qquad (13.59)$$

a.e. on D, where $A = |F|$. The projection onto $C_\phi(D)$ is

$$PF =$$

$$\begin{cases} \mathcal{F}^{-1}(e^{i\phi(w_1,w_2)}A(w_1,w_2)cos^+(\theta(w_1,w_2) - \phi(w_1,w_2)) & w_1, w_2 \in D \\ \mathcal{F}^{-1}(F(w_1,w_2)) & w_1, w_2 \in D^c. \end{cases}$$

$$(13.60)$$

Here f is assumed to have the Fourier transform $F = |F|e^{i\theta}$.

13.8 Nonnegative Spatial Constraints

In section 9.24 we discussed Biraud's algorithm which dealt with nonnegative images. We present the view of this problem for projection onto convex subsets (POCS). We introduce the following sets in $L^2(R)$ of functions and projections:

(1) P(a) = set of (-a,a) space-limited functions

$$g = P_af = rect(x/2a)f$$

$$h = Q_af = (1 - rect(x/2a))f$$

(2) P(b) the set of b band-limited function

$$g = P_bf = \frac{sinbx}{\pi x} * f(x)$$

$$h = Q_b f = (\delta(x) - \frac{sinbx}{\pi x}) \ast f(x)$$

(3) P(p) = the set of positive functions
if $f = f_r + if_i$, then

$$g = Rf = f_r$$

for $f_r \geq 0$ and

$$g = 0$$

otherwise

(4) P(s) = the set of functions f for which the spectra F(w) is nonnegative

(5) $P^+(b) = P(b) \cap P(s)$

$$g = P_b^+ f = \mathcal{F}^{-1}(RF(w)rect(w/2b))$$

(6) $Q_a^+ = (\perp P(a)) \cap P(p)$ = the set of nonnegative functions that are zero over (-a,a)

$$g = Q_a^+ f = RQ_a f = A_a Rf.$$

The reader can verify that R, P_b^+, and Q_a^+ are projections onto closed convex sets. The following results come directly from Browder's or Youla's work.

THEOREM (Stark et al.) 13.40. Let f belong to $P^+(b)$ and suppose we are given $g = P_a f$. Then the algorithm

$$f_1 = g$$

$$f_{k+1} = g + Q_a P_b^+ f_k \qquad\qquad (13.61)$$

converges weakly to f.

THEOREM (Stark et al.) 13.41. Let f belong to $P(b) \cap P(p)$ and suppose we are given $g = P_a f$. Then the algorithm

$$f_1 = g$$

$$f_{k+1} = g + Q_a^+ P_b f_k \qquad\qquad (13.62)$$

converges weakly to f.

13.9   Reconstruction from Zero Crossings

The results on uniqueness for reconstruction from zero crossings were presented in Section 5.6. Here we present an iterative algorithm to reconstruct the signal from zero crossings by projection on convex sets. Let h(x,y) be the signal we are trying to recover from sign h(x,y). Let W^* denote the set of sequence which satisfy the frequency domain constraints

$$F(n_1, n_2) = 0 \text{ for } n_1 \text{ or } n_2 \text{ not in } [-N, N] \qquad (13.63)$$

$$F(N, N) = H(N, N).$$

Let T^* denote the set of sequences which satisfy the spatial domain constraints

$$f(x, y) \geq 0 \text{ if sign h(x,y)} = 1 \qquad (13.64)$$

$$f(x, y) \leq 0 \text{ if sign h(x,y)} = -1.$$

Let T and W denote the projections onto the sets T^* snd W^*:

$$W(F(n_1, n_2)) = F(n_1, n_2)\, 0 \leq n_1, n_2 \leq N, (n_1, n_2) \neq (N, N) \qquad (13.65)$$

$$W(F(n_1, n_2)) = H(N, N)\ (n_1, n_2) = (N, N)$$

$$W(F(n_1, n_2)) = 0 \text{ otherwise}$$

$$Tf(x, y) = f(x, y) \text{ if sign f(x,y)} = \text{sign h(x,y)} \qquad (13.66)$$

$$Tf(x, y) = 0 \text{ otherwise.}$$

Consider the iterative algorithm

$$x_{k+1} = Gx_k \qquad (13.67)$$

where $G = TW$. By theorem 13.13 this sequence will converge to a point on $G^* = W^* \cap T^*$. If h(x,y) satisfies the constraints of theorem 13.13, then G^* contains exactly one sequence and the iteration converges to that sequence. If sign h(x,y) is sampled, then the iteration will converge to a sequence which satisfies the time and frequency constraints; however, this solution is not unique and the solution obtained will depend on the initial estimate.

13.10 A Phase Retrieval Example

Let $M(w_1, w_2)$ be a prescribed, nonnegative function which is square integrable over a set D in the w_1, w_2 plane. Let $C_M(D)$ denote the set of all square integrable functions f whose Fourier transforms subordinate to M over D – i.e., if f in $C_M(D)$ has Fourier transform F, then $|F(w_1, w_2)| \leq M(w_1, w_2)$ a.e. on D. To see convexity, if f_1, f_2 belong to $C_M(D)$ and have Fourier transforms F_1, F_2, then

$$|\mu F_1 + (1 - \mu)F_2| \leq \mu|F_1| + (1 - \mu)|F_2| \leq \mu M + (1 - \mu)M = M \quad (13.68)$$

for $0 \leq \mu \leq 1$, a.e. on D. The reader can check that $C_M(D)$ is closed.

If f has Fourier transform $F = Ae^{i\theta}$, then the projection onto $C_M(D)$ is

$$Pf = \mathcal{F}(A'e^{i\theta}) \qquad (13.69)$$

where

$$A'(w_1, w_2) = min(A, M) \quad w_1, w_2 \in D$$

$$A'(w_1, w_2) = A(w_1, w_2) \quad w_1, w_2 \text{ not in D.}$$

13.11 Lent-Tuy Algorithm

Let A, B be two subsets of R^m, set $H = L^2(R^m)$ and let f be a function where (1) supp f is in A, (2) the knowledge of $\hat{f}$ is restricted to B, (3) f is real and (4) $l \leq f \leq u$. Thus, f is a real bounded function with compact support contained in A. Let LT(A,B) denote this set of functions. Let P_A, P_B denote the projection operators onto A,B.

Define the closed, convex subsets L,U, Z of H by

$$L = \{g \in H | l \leq g\} \qquad (13.70)$$

$$U = \{g \in H | g \leq u\}$$

$$Z = \{g \in H | Img = 0\}.$$

The projections on these subsets are given by

$$P_L(h) = sup(l, h) \qquad (13.71)$$

$$P_U(h) = inf(u, h)$$

$$P_Z(h) = Re(h).$$

We first note that if f belongs to LT(A,B) and $g = f$, then g belongs to $Ker(I - P_A) \cap (g_o + ker(F^{-1}PF))$ where $g_o = F^{-1}P_B\hat{f}$.

The Lent-Tuy extrapolation algorithm to extrapolate $P_B\hat{f}$ for f in the space LT(A,B) is:

(1) $g_o = F^{-1}P_B\hat{f}$ for n = 0

(2) $g_n^{(1)} = g_n + \lambda_n^{(1)}(Re(g_n) - g_n)$

(3) $g_n^{(2)} = g_n^{(1)} + \lambda_n^{(2)}(P_A g_n^{(1)} - g_n^{(1)})$

(4) $g_n^{(3)} = g_n^{(2)} + \lambda_n^{(3)}(sup(l, g_n^{(2)}) - g_n^{(2)})$

(5) $g_n^{(4)} = g_n^{(3)} + \lambda_n^{(4)}(inf(u, g_n^{(3)}) - g_n^{(3)})$

(6) $g_{n+1} = F^{-1}(\lambda_n^{(5)}P_B f + (1 - \lambda_n^{(5)})P_B F g_n^{(4)} + (I - P_B)F g_n^{(4)})$

(7) let n go to n+1 and return to (2).

By theorem 13.13, for $0 < \lambda_n^{(i)} < 2$, i = 1,...5, then g_n converges weakly to g where g is an extrapolation of $\hat{f}$.

13.12   Example of Lent-Tuy

Let

$$f(x) = 1 \ |x| \le 1/2$$

and 0 otherwise, so $\hat{f} = sin(\pi w)/(\pi w) = sinc(w)$. Let B = $[0,\infty)$. If H(w)
is the Heaviside function (i.e., H(w) = 1 if $|w| \ge 0$ and 0 otherwise), then
$P_B \hat{f}(w) = H(w) sinc(w)$. Since $\hat{H}(x) = \frac{1}{2}\delta(x) - \frac{i}{2\pi x}$ we have by convolution

$$g_o(x) = \int_{-\infty}^{\infty} f(y) H(\hat{x} - y) dy = \frac{1}{2} f(x) - \frac{i}{2\pi} ln \left| \frac{x + \frac{1}{2}}{x - \frac{1}{2}} \right|.$$

Let A = (-1/2,1/2) and let l = 3/4, u = 5/4 and $\lambda_n^{(k)} = 1$ for all k and n.
Then the Lent-Tuy algorithm gives

$$g_o^{(1)} = g_o^{(2)} = \frac{1}{2} f(x)$$

$$g_o^{(3)} = g_o^{(4)} = \frac{3}{4} f(x)$$

$$y_1(x) = F^{-1}(\frac{3}{4} sincw + \frac{1}{4} H(w) sincw) = \frac{3}{4} f(x) + \frac{1}{4}(\frac{1}{2} f(x) - \frac{i}{2\pi} ln \left| \frac{x + 1/2}{x - 1/2} \right| =$$

$$\frac{7}{8} f(x) - \frac{i}{8\pi} ln \left| \frac{x + 1/2}{x - 1/2} \right|.$$

Continuing one finds that

$$g_{n+1} = \frac{2^{n+2} - 1}{2^{n+2}} f(x) - \frac{i}{2^{n+2}\pi} ln \left| \frac{x + .5}{x - .5} \right|$$

which converges to f.

13.13   Divergence-Free Vector Fields

Simard and Mailloux have applied POCS to the problem of restoration of
a velocity field or optical flow computed from an image sequence when the
underlying velocity field is known to be divergence free. The application is
in echocardiography when a 3-D velocity field of the heart is restored from
2-D velocity field computed from cross-section images.

The problem space is $H = L^2(\Omega)$ where Ω is an open subset of R^n with
boundary Γ. The important convex set here is $C_3 = \{g \in H | div(g) = 0\}$.
The projection operator onto C_3 is given by $Pf = f - gradp$ for f in H
where p is the solution of Poisson's equation $\Delta p = div(f) \in H^{-1}(\Omega)$ and

$p|\Gamma = 0$. Here H^{-1} is the dual of the Sobolev space $H_0^1(\Omega)$. Viz., $P_3(f)$ is the element g which satisfies

$$min_{g \in C_3} \|f - g\|.$$

This is equivalent to using a Lagrange multiplier p such that

$$min_{g \in H} max_{p \in H_0^1(\Omega)} \frac{1}{2}\|f - g\| + \int_\Omega pdiv(g).$$

The approach is to view any vector field as locally the linear combination of a divergence free field D, a rotational field R, and two hyperbolic fields H_1, H_2. And we have $P_3(D) = 0$.

Let C_1 be the convex set of all f in H whose Fourier transform is zero outside a region A in the Fourier plane. Let C_2 be the convex set of all f in H whose components (f_x, f_y) projected onto arbitrary subsets in Ω assume prescribed values $(g_x(x, y), h_y(x, y))$. Let P_i denote the natural projections onto C_i. The reconstruction algorithm is then

$$f_{k+1} = S_i g + (1 - S_i)[\mathcal{F}^{-1}(p_b \mathcal{F}(f_k))].$$

Here $p_b(u, v) = 1$ if $u \leq b$ or $v \leq b$. S_i are certain masks where $S_1(x, y) = (1, 0)$ if x mod 4 = 0 and (0,0) otherwise and $S_2(x, y) = (1, 1)$ if x mod 8 = 0 or y mod 8 = 0 and (0,0) otherwise.

Chapter 14

Method of Generalized Projections and Steepest Descent

14.1 Introduction

Levi and Stark have studied restoration by generalized projections where for any closed set C they call $g = P_C h$ the projection of h onto C if g belongs to C and

$$\|g - h\| = min_{y \in C} \|y - h\|. \qquad (14.1)$$

A generalized projection for a nonconvex set is no longer necessarily unique and may not exist.

Assuming P_1, P_2 are generalized projections onto closed sets C_1, C_2, then we define the summed distance error $J(f_n)$ by

$$J(f_n) = \|P_1 f_n - f_n\| + \|P_2 f_n - f_n\| \qquad (14.2)$$

where f_n is the estimate of f which is assumed to lie in $C_o = C_1 \cap C_2$. Let $T_i = 1 + \lambda_i(P_i - 1)$, i = 1,2 and let $f_{n+1} = T_1 T_2 f_n$. Note that $J(f_n) \geq 0$ and $J(f_n) = 0$ if, and only if, f_n belongs to $C_1 \cap C_2$. Let $J(T_2 f_n) = \|P_1 T_2 f_n - T_2 f_n\| + \|P_2 T_2 f_n - T_2 f_n\|$.

THEOREM (Levi-Stark) 14.1. For a restricted set of λ_i

$$J(f_{n+1}) \leq J(T_2 f_n) \leq J(f_n). \qquad (14.3)$$

For a proof see Levi-Stark. The restricted set includes $\lambda_i = 1$, so we have:

COROLLARY 14.2. The iterative algorithm

$$f_{n+1} = P_1 P_2 f_n \qquad (14.4)$$

is always error reducing: $J(f_{n+1}) \leq J(f_n)$.

14.2   Phase Retrieval

The classical restoration from magnitude problem does not fall under the category of problems covered by projection on convex sets for although the set C_1 of functions g where $g = 0$ for $|x| > a$ is convex, as we noted above, the set $C_2 = \{g \| |F(g)|(w) = M(w)$ for all $w\}$ is not convex. However, we still have generalized projections in the sense of Levi-Stark

$$P_1 g = \begin{cases} g(x) & |x| \leq a \\ 0 & |x| > a \end{cases} \qquad (14.5)$$

and

$$P_2 g = F^{-1}(M(w) e^{i\phi(w)})$$

where ϕ is the phase of the Fourier transform of g. The reconstruction algorithm $f_{n+1} = P_2 P_1 f_n$ is the Gerchberg-Saxton algorithm and the algorithm $f_{n+1} = P_2 T_1 f_n$ for a selected λ_1 is Fienup's output-output algorithm. Fienup and Gerchberg-Saxton all noted that these algorithms have the non-increasing error (or "energy reduction") property.

14.3   Deconvolution by POCS

Consider the linear problem of reconstructing a signal f when one measures the data $g = Hf + n$ where n is the signal independent noise and H is the impulse response function; in addition it is assumed that other constraints apply. In this section we view g,f as vectors of length N and H is an NxN matrix. For an estimate $\hat{f}$ of the signal vector, the vector of residuals is given by $r = g - H\hat{f}$.

Using the method of POCS we examine possible convex constraint sets. The following are examples.

Residual Variance Set. Let $C_v = \{f \| \|g - Hf\| \leq \delta_v\}$. This set is clearly convex since if $f_3 = \alpha f_1 + (1 - \alpha)f_2$ for f_1, f_2 in C_v, then by Schwartz inequality $\|g - Hf_3\| \leq \sqrt{\delta_v}$. The selection of δ_v is based on the variance of the actual noise.

Mean of the Residual Convex Set. Let $C_m = \{f \| \sum(g_i - [Hf]_i| \leq \delta_m\}$ where g_i and $[Hf]_i$ are the ith elements of the respective vectors. The bound δ_m represents the confidence limits on the sample mean. C_m captures the constraint that the mean of the residual be zero.

Outliers of the Residual Convex Set. Given knowledge of the probability distribution of the noise, the bound on the deviation from zero may be specified. This is expressed by the convex set $C_o = \{f \| |g_i - [Hf]_i| \leq \delta_i\}$.

Power Spectral Bounds Convex Set. Let G and F represent the DFT of g,f and let H(k) denote the kth frequency component of the DFT of the first row of H. Here H is assumed to be circulant. Since the periodgram has a chi-squared distribution, its confidence limits allow the definition of the convex set

$$C_p = \{f \| |G(k) - H(k)F(k)|^2 \leq \delta_p, 1 \leq k \leq N/2 - 1\}. \quad (14.6)$$

The projection operators for these convex sets are:

On C_v, $f_p = f + (H^tH + \frac{1}{\lambda}I)^{-1}H^t(g - Hf)$ where λ is a Lagrange multiplier (v. Trussell and Civanlar).

On C_m,

$$f_p = f + \sum r_i - \delta_m h_c \sum r_i > \delta_m \quad (14.7)$$

where $h = (\sum[H]_{i1}, \ldots \sum[H]_{iN})$.

On C_o, to find the projection one must find the $min\|f_p - f\|^2$ where $|g_i - [Hf_p]_i| \leq \delta_o$. The Kuhn-Tucker condition gives $f_p = \lambda h_i$ and λ is found to be

$$\lambda = \begin{cases} \frac{1}{\|h_i\|^2}(r_i - \delta_o) & r_i > \delta_o \\ \frac{1}{\|h_i\|^2}(r_i + \delta_o) & r_i < -\delta_o \end{cases} \quad (14.8)$$

Thus, we have

$$f_p = \begin{cases} f + \frac{h_i}{\|h_i\|^2}(r_i - \delta_o) & r_i > \delta_o \\ f & |r_i| \leq \delta_o \\ f + \frac{h_i}{\|h_i\|^2}(r_i + \delta_o) & r_i < -\delta_o \end{cases} \quad (14.9)$$

where h_i is the column vector containing the ith row of H.

On C_p,

$$F_p(k) = \begin{cases} \frac{1}{H(k)}(G(k) - \sqrt{\delta_p}\frac{G(k)-H(k)F(k)}{|G(k)-H(k)F(k)|} & \text{if } |G(k) - H(k)F(k)| > \delta_p \\ F(k) & \text{otherwise.} \end{cases}$$

$$(14.10)$$

Trussel and Civanlar have applied these projections and the iteration $f_{k+1} = P_1...P_mf_k$ to various one and two dimensional signal restoration problems including an example of x-ray fluorescence signal in the presence of noise. These signals are characterized by isolated peaks and large regions near zero. The example shown in this paper indicates the POCS restoration was better than constrained least squares restoration. The addition of the nonnegativity constraint to those above showed even better performance.

14.4   Fuzzy Sets

The combination of ideas from POCS and fuzzy sets has been presented
by Cinvanlar and Trussel, to incorporate imprecise information in signal
restoration. We recall some concepts of fuzzy set theory here. For a set A,
let $\mu_A : X \rightarrow [0,1]$. Here $\mu_A(x)$ is the grade of membership of x in fuzzy
set A. So if $\mu_A(x) = 1$, then it is certain that x is in A. Two fuzzy sets A,B
are said to be equal if and only if $\mu_A(x) = \mu_B(x)$ for all x in X. And A is
a subset of B if and only if $\mu_A(x) \leq \mu_B(x)$ for all x in X. The set of x for
which $\mu_A(x) > 0$ is called the support of fuzzy set A. The height of A is
defined by $hgt(A) = sup_{x \in X} \mu_A(x)$. A measure of energy on a fuzzy set can
then be defined by $e(A) = \frac{1}{hgt(A)} \int \mu_A(x)dx$.

Membership functions can be defined in terms of probability density
functions given by

$$\mu(x) = \begin{cases} \lambda p(x) & \text{if } \lambda p(x) < 1 \\ 1 & \text{otherwise} \end{cases} \qquad (14.11)$$

where λ is determined by the probability distribution function and the confidence level c. For example, for signal restoration, membership functions
might include:

Mean Value of Residual Signal. If the feasible residual signal is a nor-
mal random variable with zero mean and variance σ^2/N, then let $p_m = \frac{1}{\sigma\sqrt{N}} \sum [r]_i$ where $[r]_i$ is the ith sample of the residual signal and define

$$\mu_m(p_m) = \begin{cases} 1 & |p_m| \leq a(c) \\ \lambda(c)e^{-p_m^2/2}/(\sigma\sqrt{2}\pi) & \text{otherwise.} \end{cases} \qquad (14.12)$$

Variance of the Residual Signal. If the noise variance has a chi-squared
distribution with N degrees of freedom, we might set

$$p_v = \frac{\sqrt{2\sum [r]_i}}{\sigma^2} - \sqrt{2N-1}, \qquad (14.13)$$

and we replace p_v by p_m in 14.12 above to get μ_v.

Power Spectrum. The magnitude squared DFT coefficients of a feasible
residual signal has a chi-squared distribution with two degrees of freedom.
Let R(k) denote the kth discrete frequency component of the residual signal
and let P_{pm} denote the maximum of $p(k) = 2|R(k)|/N\sigma^2$. Let $N' = N/2-1$
and set $f_{pm}(x) = n'(1 - e^{-x/2})^{N'-1}e^{-x/2}$ then a membership function is

$$\mu_{pm}(x) = \begin{cases} 1 & \text{if } a < x < b \\ \lambda f_{pm}(x) & \text{otherwise.} \end{cases} \qquad (14.14)$$

To try to use the method of POCS we say a fuzzy set A is convex if and
only if its α−level sets are convex where the α−level sets are given by

$$A_\alpha = \{x | \mu(x) \geq \alpha\}. \qquad (14.15)$$

If $P_1, ... P_m$ are projections onto convex fuzzy sets, we would apply the iteration

$$f_{k+1} = P_1 ... P_m f_k. \tag{14.16}$$

The problem is that the fuzzy sets constructed from the residual signal statistics (with the exception of the one using the mean value) are not convex. If one can assume that the statistics of the residual signal are more likely to be larger than the upper bounds and stay in the same region during the iteration, then the convex hulls of the nonconvex α-level sets may be used. The convex approximations to the α-level fuzzy sets are: C_m, C_v, C_p, C_o as defined in section 14.3. Here α and δ parameters are related. Viz., $\delta_m = c_1 \sigma_n \sqrt{N}, c_1 = \sqrt{a^2(c) - 2ln(\alpha)}, \delta_p^2 = \sigma_n^2 N(b(c) - ln(\alpha))$, etc.

Under these approximations, Civanlar and Trussel have applied this technique of digital signal restoration using fuzzy sets to the x-ray fluorescence spectra problem mentioned in Section 14.2. The reader should refer to the paper for a discussion of the results.

14.5 Method of Steepest Descent

Consider the case of a real function $g(z)$ of a complex variable z. For example, let $g(z) = |a - Az|^2$ where A is an mxn complex matrix and a is a complex m-vector. To find the vector z which minimizes $g(z)$ we form $grad(g) = \partial g / \partial z = -A^T(\bar{a} - \bar{A}\bar{z})$. Setting this equal to zero and taking complex conjugates we have

$$\bar{A}^T A z = \bar{A}^T a. \tag{14.17}$$

Thus, if $\bar{A}^T A$ is nonsingular we have

$$z = (\bar{A}^T A)^{-1} \bar{A}^T z. \tag{14.18}$$

14.6 Least Squares Solutions

Let X and Y be Hilbert spaces and T a bounded linear operator of X in Y. Associate to the linear operator equation Tx = y the quadratic functional

$$J(x) = \frac{1}{2} \|Tx - y\|^2. \tag{14.19}$$

DEFINITION 14.3. The method of steepest descent for minimizing J(x) is defined by the sequence

$$x_{n+1} = x_n - \alpha_n grad J(x_n) \tag{14.20}$$

where α_n is selected to minimize $J(x_{n+1})$.

The reader can verify that:

THEOREM 14.4. $gradJ(x) = T^*Tx - T^*g$.

Thus, the steepest descent sequence is

$$x_{n+1} = x_n - \alpha_n r_n \qquad (14.21)$$

where $r_n = T^*(Tx_n - y)$ and $\alpha_n = \|r_n\|^2/\|Tr_n\|^2$. Here x_o is any initial approximation. The result on convergence of the method of steepest descent is contained in the following:

THEOREM (Nashed) 14.5. Let $T : X \to Y$ be a bounded linear operator with closed range. Then the sequence x_n defined above converges for y in Y and initial approximation x_o in X to the vector $T^\dagger y + (I - P)x_o$, a least squares approximation of $Tx = y$; here $T^\dagger$ is the generalized inverse and $P = P_{N(T^*)^\perp}$. Also

$$\|T^\dagger y + (I - P)x_o - x_n\| \leq c[\frac{M - m}{M + m}] \qquad (14.22)$$

where c is a constant and

$$m\|x\|^2 \leq (T^*Tx, x) \leq M\|x\|^2 \qquad (14.23)$$

for x in $R(T^*)$ with $0 < m \leq M < \infty$.

THEOREM (Kammerer-Nashed) 14.6. Let $T : X \to Y$ be a bounded linear operator and let Q be the orthogonal projection of Y onto $N(T^*)^\perp$. If Qy belongs to $R(TT^*)$, then the steepest descent sequence starting from $x_o = 0$ converges monotonically to the least squares solution of minimal norm of $\hat{x} = T^\dagger y$.

The conjugate gradient method of minimizing J(x) is given by:
(1) given x_o in X
(2) $r_o = p_o = T^*(Tx_o - y)$
(3) if $p_o \neq 0$, let $x_1 = x_o - \alpha_o p_o$ where $\alpha_o = \|r_o\|^2/\|Tp_o\|^2$ and in general
(4) $r_i = T^*(Tx_i - y) = r_{i-1} - \alpha_{i-1}T^*Tp_{i-1}$ where

$$\alpha_{i-1} = (r_{i-1}, p_{i-1})/\|Tp_{i-1}\|^2$$

and if $r_i \neq 0$, we set

$$p_i = r_i + \beta_{i-1}p_{i-1}$$

$$\beta_{i-1} = (r_i, T^*Tp_{i-1})/\|Tp_{i-1}\|^2$$

(5) $x_{i+1} = x_i - \alpha_i p_i$.

The convergence of the conjugate gradient technique is given by:

THEOREM (Nashed) 14.7. If $T : X \to Y$ is a bounded linear operator with closed range, then the conjugate gradient method converges monotonically to $u = T^\dagger y + (I - P)x_o$, a least squares solution of Tx = y. If m,M are as in theorem 14.5, then

$$\|x_n - u\| \leq g(x_o)(\frac{M - m}{M + m})^{2n}, n = 1, 2, \ldots \qquad (14.24)$$

where $g(x) = \|Tx - Qy\|$.

COROLLARY 14.8. If rank T = dim R(T) = r, then the conjugate gradient method converges in at most r steps to the least squares solution $u = T^\dagger y + (I - P)x_o$ for any x_o in X.

EXERCISE (Jain) 14.9. A discrete signal y(k), $k = 0, \pm 1, \ldots$ is called band-limited if its Fourier transform

$$Y(f) = \sum y(k)e^{-i2\pi k f}, -1/2 \leq f \leq 1/2 \qquad (14.25)$$

satisfies Y(f) = 0 for $1/2 > |f| > \sigma$. Let

$$z(k) = \begin{cases} y(k) & -M \leq k \leq M \\ 0 & \text{otherwise.} \end{cases}$$

Define $L = (l_{ij})$ where

$$l_{ij} = \frac{sin2\pi(i - i)\sigma}{\pi(i - j)}.$$

and $W = (w_{ij})$ where

$$w_{ij} = \begin{cases} 1 & i= j \; -M \leq i,j \leq M \\ 0 & \text{otherwise} \end{cases}$$

and let y denote the infinite vector y = (...,y(-k),...y(-1),y(0),y(1),...).

Define S by

$$S(i, j) = \begin{cases} 1 & i = j = \pm 1, \pm 2, \ldots \pm M \\ 0 & \text{otherwise} \end{cases}$$

and let $\hat{L} = SLS^T$. Show that $S^T S = W$ and $L^2 = L$. Show that for $M < \infty$ $\hat{L}$ is positive definite and all the eigenvalues of $\hat{L}$ lie in (0,1). The observations z are given by z = Sy and since y is band-limited we have Ly = y so we write Z = SLy. Set H = SL. Find the minimum norm least squares (MNLS) solution

$$y^\dagger = min\{\|y\|^2 | H^T H y = H^T z\}. \qquad (14.26)$$

Show that the gradient method for this problem gives

$$y_{n+1} = y_n + \varepsilon H^T(z - Yy_n) \tag{14.27}$$

$$y_o = 0$$

$$0 < \varepsilon < 2/\lambda_{max}(H^T H).$$

Show this can be rewritten as

$$y_{n+1} = \varepsilon f_1 + (I - \varepsilon LW)y_n \tag{14.28}$$

where $f_1 = H^T z = LS^T z$. Since f_1 is band-limited (i.e., $Lf_1 = f_1$), each y_n is band-limited; and we can rewrite (14.28) as

$$y_{n+1} = \varepsilon f_1 + L(I_\varepsilon W)y_n \tag{14.29}$$

which is the discrete version of the Papoulis algorithm.

Show that the conjugate gradient method equations in this case are

$$y_{k+1} = y_k + \alpha_k d_k \tag{14.30}$$

$$y_o = 0$$

$$d_{k+1} = -g_{k+1} + \beta_k d_k, d_o = -g_o$$

$$g_k = LWy_k - LS^T z = g_{k-1} + \alpha_{k-1}LWd_{k-1}$$

$$\beta_k = \sum_{n=-M}^{M} g_k(n)d_k(n) / \sum_{n=-M}^{M} d_k(n)^2$$

$$\alpha_k = \sum_{n=-M}^{M} d_k(n)(y_k(n) - z(n)) / \sum_{n=-M}^{M} d_k(n)^2.$$

Show that the generalized inverse of H is

$$H^\dagger = H^T(HH^T)^{-1} = LS^T(SLS^T)LS^T \hat{L}^{-1}$$

which exists since $\hat{L}$ is nonsingular (however, possibly very ill-conditioned). Thus, the MNLS estimate of y is

$$y^\dagger = H^\dagger z. \tag{14.31}$$

Show that the eigenvectors ϕ_k, ψ_k given by

$$LWL\phi_k = \lambda_k \phi_k \tag{14.32}$$

$$SLS^T \psi_k = \lambda_k \psi_k$$

$-M \leq k \leq M$ satisfy

$$\psi_k = \frac{1}{\sqrt{\lambda_k}} S\phi_k \qquad (14.33)$$

$$L\phi_k = \phi_k$$

$$\phi_k = \frac{1}{\sqrt{\lambda_k}} LS^T \psi_k.$$

Show that $H^\dagger = \sum_{-M}^M a(k)\phi_k(m)/\sqrt{\lambda_k}$

$$y^\dagger(m) = \sum_{-M}^M a(k)\phi_k(m)/\sqrt{\lambda_k} \qquad (14.34)$$

where $a(k) = \psi_k a = \sum_{m=-M}^M \psi_k(m)y(m)$. Show that $y^\dagger(m) = y(m)$ for m in $[-M, M]$.

In the presence of noise z =SLy+n. Show that the Wiener filter estimate is

$$\hat{y} = RL^T S^T (SLRL^T S^T + R_n)^{-1} z \qquad (14.35)$$

where R and R_n are the autocorrelation matrices of y and n respectively. Since y is band-limited, $LRL^T = R$. Thus, we have

$$\hat{y} = RLS^T (SRS^T + R_n)^{-1} y. \qquad (14.36)$$

If we do not know R and set R = L, then if $R_n \to 0$ we see that $\hat{y} \to y^\dagger$, the MNLS extrapolated estimate.

14.7 Design of FIR Filters

A two dimensional FIR digital filter is described by an array of filter spatial coefficients $h_{mn}, 0 \leq m \leq M-1, 0 \leq n \leq N-1$. The filter produces output g_{mn} from input d_{mn} according to

$$g_{mn} = \sum\sum h_{kl} d_{m-k,n-l}. \qquad (14.37)$$

Associated to h is the frequency domain array H where

$$H_{kl} = \sum\sum h_{mn} U^{km} V^{ln} \qquad (14.38)$$

where $U = e^{-2\pi i/M}$ and $V = e^{-2\pi i/N}$. Of course,

$$h_{mn} = \frac{1}{MN} \sum\sum H_{kl} U^{-kn} V^{-ln}. \qquad (14.39)$$

DEFINITION 14.10. The magnitude filter design problem is to design a filter with coefficients h_{mn} which are zero except for a small number of specified values m,n and whose corresponding frequency domain coefficient amplitudes will assume (or approximate) prescribed values $A_{kl}, 0 \leq K \leq M-1, 0 \leq l \leq N-1$.

We assume the possible nonzero spatial coefficients H_{mn} are for $0 \leq m \leq M_1 - 1, 0 \leq n \leq N_1 - 1$ where $M_1 \leq M, N_1 \leq N$. The Manry-Aggarwal algorithm to solve the filter design problem is:

(1) given $h^{(i)}$

(2) calculate $H^{(i)}$ where

$$H_{kl}^{(i)} = |H_{kl}^{(i)}|e^{i\theta_{kl}^{(i)}}$$

(3) set $\tilde{H}_{kl}^{(i)} = A_{kl}e^{i\theta_{kl}^{(i)}}$

(4) calculate $\tilde{h}^{(i)}$

(5) truncate $\tilde{h}^{(i)}$ to be zero except for $0 \leq m \leq M_1 - 1, 0 \leq n \leq N_1 - 1$ call it $\tilde{\tilde{h}}_{mn}^{(i)}$

(6) return to step (2) with $h^{(i+1)} = \tilde{\tilde{h}}^{(i)}$.

If we define the error term

$$f(h) = \sum\sum(|H_{kl}| - A_{kl})^2, \tag{14.40}$$

then the gradient of $f(h^{(i)})$ is:

$$gradf(h^{(i)}) = -2MN(\tilde{\tilde{h}}^{(i)} - h^{(i)}). \tag{14.41}$$

Thus, the steepest descent algorithm

$$h^{(i+1)} = h^{(i)} - \alpha_i(gradf(h^{(i)})) \tag{14.42}$$

is equivalent to the Manry-Aggarwal algorithm using the step $\alpha_i = 1/2MN$.

THEOREM 14.11. The Manry-Aggarwal sequence converges to a unique solution or has a continuum of limit points. In either case all limits h satisfy grad f(h) = 0.

In attempts to speed up convergence Manry and Aggarwal experimented with variations on the theme by letting

$$h^{(i+1)} = h^{(i)} + \alpha_i(\tilde{\tilde{h}}^{(i)} - h^{(i)}) \tag{14.43}$$

where α_i is selected to speed up convergence. Daniel went on to propose using the conjugate technique of Fletcher-Reeves.

EXERCISE 14.12. Following the remarks in Section 14.4 extend the Manry-Aggarwal results to the arbitrary complex filter coefficients h_{kl} case.

14.8 Projection onto Nonconvex Sets

Let C_i be a collection of arbitrary sets with projection operators P_i. Since reconstruction problems are ill-conditioned which might imply that $\cap C_i$ is empty, we consider to reconstruct by finding the image x which is closest to all the sets C_i. This of course reduces to a common intersection point if one exists. The steepest descent reconstruction problem from a collection of closed convex sets is the solution x which minimizes

$$F(x) = \sum_{i=1}^{M} \|x - P_i x\|^2. \qquad (14.44)$$

THEOREM (Barakat-Newsam) 14.13. The sequence x_n define by the iteration

$$y_n = \frac{1}{M} \sum_{i=1}^{M} P_i x_n \qquad (14.45)$$

$$x_{n+1} = x_n + \lambda_n(y_n - x_n)$$

for λ_n in (0,2) satisfies $F(x_{n+1}) < F(x_n)$ or $x_{n+1} = x_n$. If the projection operators are single valued, then $x_{n+1} = x_n$ implies $x_{n+k} = x_n$ for all k. If x_n has a limit point at which the projections are continuous and $0 < liminf \lambda_n \leq limsup \lambda_n < 2$, then x is a fixed point.

The proof of this result rests on showing that y is a unique minimum of $G(z,s) = \sum_{i=1}^{M} \|z - b_i\|^2, b_i = P_i x$ and if $y \neq z, G(z + \lambda(y - z), x) < G(z, x)$ for all λ in (0,2). For details the reader should confer Barakat and Newsam.

THEOREM 14.14. If there exists a $\lambda > 1$ such that $P_A(P_A x + \lambda(x - P_A x)) = P_A x$ and $P_A x$ is single-valued, then

$$grad(\|x - P_A x\|^2) = 2(x - P_A x). \qquad (14.46)$$

Proof. By definition of gradient it suffices to show that for some constant c we have

$$\|\|y - P_A y\|^2 - \|x - P_A x\|^2 - 2(x - P_A x, y - x)\| \leq c\|x - y\|^2. \qquad (14.47)$$

Again for details, the reader should see the Barakat-Newsam paper.

COROLLARY 14.15. $grad F(x) = 2 \sum_{i=1}^{M}(x - P_i x) = 2M(x - y)$.

Thus, the directions $y_n - x_n$ used in theorem 14.13 are the steepest descent directions for F(x). And at the limit point $\tilde{y} = \tilde{x}$ where $grad F(\tilde{x}) = 0$; i.e., the limit point is a stationary point of F(x).

In many image reconstruction problems we require, based often on physical principles, that the limit point belongs to a particular set, say C_M. To meet this requirement Barakat and Newsam have proposed the restricted projected algorithm:

THEOREM (Barakat-Newsam) 14.16. If the projections P_i are continuous and unique, then the sequence x_n in C_M defined by

$$y_n = \frac{1}{M-1} \sum_{i=1}^{M} P_i x_n \qquad (14.48)$$

$$x_{n+1} = P_M(x_n + \lambda_n(y_n - x_n)) \quad \lambda_n \in (0,1)$$

satisfies $F(x_{n+1}) < F(x_n)$ or $x_{n+k} = x_n$ for all k. And if $\tilde{x}$ is the limit point of x_n, then $gradF(\tilde{x})$ is normal to C_M.

Proof. Let $z_n = x_n + \lambda_n(y_n - x_n)$. Then

$$\|x_{n+1} - z_n\|^2 = \|x_{n+1} - x_n\|^2 + \|x_n - z_n\|^2 + 2(x_{n+1} - x_n, x_n - z_n) \quad (14.49)$$

and $2(x_{n+1} - x_n, x_n - z_n) =$. Since $P_M z_n = x_{n+1}$ we have

$$2(x_{n+1} - x_n, y_n - x_n) \geq \frac{1}{\lambda_n} \|x_{n+1} - x_n\|^2. \qquad (14.50)$$

Therefore,

$$F(x_{n+1}) \leq \sum_{i=1}^{M} \|P_i x_n - x_{n+1}\|^2 =$$

$$F(x_n) + (M-1)(x_n - x_{n+1}, 2y_n - x_n - x_{n+1}) \leq \qquad (14.51)$$

$$F(x_n) + (M-1)(1 - 1/\lambda_n)\|x_{n+1} - x_n\|^2.$$

Thus, $F(x_{n+1}) < F(x_n)$ since $\lambda_n < 1$. And the inequality is strict unless $x_{n+1} = x_n$. Uniqueness of projections ensures that $x_{n+1} = x_n$ implies that $x_{n+k} = x_n$ for all k. The continuity of P_M implies that $\tilde{x}$ belongs to C_M and

$$gradF(\tilde{x}) = \sum_{i=1}^{M} 2(\tilde{x} - P_i \tilde{x}) = 2(M-1)(\tilde{x} - \tilde{y}). \qquad (14.52)$$

Since $\tilde{x}$ is a fixed point $P_m \tilde{x} = \tilde{x}$; so the sphere

$$S = \{z \mid \|z - \tilde{y}\| < \|\tilde{x} - \tilde{y}\|\} \qquad (14.53)$$

must satisfy $S \cap C_M$. The rest is a geometric check.

COROLLARY 14.17. If C_M is convex, we can have $\lambda \in (0,2)$. And if C_M is linear subspace, then P_M is linear,

$$x_{n+1} = P_M x_n + \lambda P_M(y_n - x_n),$$

and one may select λ to minimize $F(x_{n+1})$ as a function of λ.

Chapter 15

Closed Form Reconstruction of the Support and the Object

15.1 Introduction

Given the modulus of the Fourier transform of an object or its autocorrelation, the phase retrieval problem asks to reconstruct the object. Here we are assuming the object has finite support. To apply the reconstruction algorithms of Fienup or others it is useful to know the support of the object. If this is unknown, the first step would be to determine the support of the object.

In general there may be many solutions for the object's support given the autocorrelation. Techniques to generate all possible solutions have been examined by Fienup, Crimmins and Holsztynski.

Let f be a function with compact support S on $H = R^N$ and let

$$f \star f(x) = \int_H f(y)f(y+x)dy \qquad (15.1)$$

denote the autocorrelation of f. Let A denote the support of the autocorrelation. Then we have:

THEOREM 15.1. (1) $A = \cup_{y \in S}(S - y) = S - S = \{x - y | x, y \in S\}$
(2) A is symmetric – i.e., A = -A where $-A = \{-x | x \in A\}$
(3) 0 is in A if S is nonempty.

DEFINITION 15.2. Two subsets S,T of H are said to be equivalent $S \sim T$, if there is a vector v in H such that $T = v + \beta S = \{v + \beta x | x \in S\}$

211

where $\beta = 1$ or -1.

If S_1 is a solution of S-S $= A$ and if S_2 is equivalent to S_1, then S_2 is also a solution. Thus, we say a solution to S-S $= A$ is unique if S_1 is a solution and all other solutions are equivalent to S_1 and A is said to be unambiguous.

EXAMPLE 15.3. $A = \{-1, 0, 1\}$ is unambiguous having the unique solution $S = \{0, 1\}$.

DEFINITION 15.4. A set L in H is called the locator set for A if for every closed set S of H satisfying $A = $ S-S, then some translate of S is a subset of L – i.e., there exists a vector v such that $v + S \subset L$.

If w is in A, then $L = A \cap (w + A)$ is a locator set for A. This can be generalized to:

THEOREM 15.5. Let W be a set contained in some $S' \sim S$ for every set S satisfying $A = $ S-S (i.e., W is an intersection of all possible solutions), then $L = \cap_{w \in W}(w + A)$ is a locator set for A.
    Proof. If $A = $ S-S and $S' \sim S$ with $W \subset S$; then $s' - x \subset A$ for every x in S'. Hence, $S' \subseteq \cap_{x \in S'}(x + A) \subseteq \cap_{x \in W}(x + A)$.

Not all symmetric sets that contain 0 are necessarily autocorrelation supports as can be seen from the example

$$A = \{(0,0), (1,0), (-1,0), (0,1), (0,-1)\}.$$

The reader can check that there is no solution for $A = $ S-S in this case. However, all convex symmetric sets A have at least one solution, viz., $S = \frac{1}{2}A = \{x/2 | x \in A\}$.

THEOREM 15.6. If A in R^2 is a closed convex symmetric set with nonnull interior, if w_1 belongs to $\partial(A) = $ the boundary of A, and w_2 belongs to $\partial(A) \cap \partial(w_1 + A)$ and if $B = A \cap (w_1 + A) \cap (w_2 + A)$, then B is a solution to $A = $ S-S – i.e., $A = $ B-B.

If all convex solutions are equivalent to $\frac{1}{2}A$ then A is said to be convex unambiguous.

THEOREM 15.7. In two dimensions if A is a parallelogram, then A is convex unambiguous.

However, convex symmetric sets in 2-D which are not parallelograms are reported to have infinitely many nonequivalent solutions to $A = $ S-S. Thus, in general, three translates of the autocorrelation is a solution and these solutions can be combined to form other solutions. However, if the support S is discrete, this simplifies.

Consider the case $f(x) = \sum_{m=1}^{M} f(m)\delta(x - x_m), f(m) > 0$. So f(x) has support $S = \{x_m | m = 1, ...M\}$. The autocorrelation is then

$$f \star f(x) = \sum_{n=1}^{M} \sum_{m=1}^{M} f(n)f(m)\delta(x - x_n - x_m) =$$

$$\sum_{n=1}^{M} f(n)^2\delta(x) + \sum_{n=1}^{M} \sum_{m \neq n} f(n)f(m)\delta(x - x_n - x_m) \qquad (15.2)$$

which has M^2 terms located at $x = x_n - x_m$, M of which are at x = 0.

THEOREM 15.8. In the case above if $x_1, x_2, y_1, y_2, z_1, z_2$ belong to X, $x_1 \neq x_2$ and $x_1 - x_2 + y_1 - y_2 + z_1 - z_2 = 0$, then either $x_1 = y_2$ or z_2 and $x_2 = y_1$ or z_1. Let w_1 be in A, w_2 be in $A \cap (w_1 + A), 0 \neq w_1 \neq w_2 \neq 0$ and let $B = A \cap (w_1 + A) \cap (w_2 + A)$. Then $S \sim B$ - i.e., B is the unique solution to A = S-S.

Thus, for "randomly" situated points, three translates of the autocorrelation yields a solution and it is unique.

There is an algorithm which shows that when the hypothesis of theorem 15.9 is satisfied, then the object function f(x) can be reconstructed exactly up to a translation. Viz., the product of three autocorrelations is formed

$$(f \star f(x))(f \star f(x - w))(f \star f(x - w')) \qquad (15.3)$$

where x is in A and w' is in the support of the product of the first two terms. For the specific formulae the reader is referred to the references.

Fienup has developed a closed form phase retrieval algorithm for reconstructing discrete objects which have a triangular support constraint. Hayes and Quatieri proposed a recursive phase retrieval algorithm for discrete objects which required knowledge of the value of the object along its boundaries. However, Fienup has shown that a priori knowledge of the edges of an object is not sufficient to ensure that it can be uniquely reconstructed from its autocorrelation function.

Crimmins has generalized Fienup's results to a wider class of support constraints.

DEFINITION 15.10. A finite subset of Z^2 is called a mask if it contains at least three noncollinear points and its convex hull (i.e., the smallest convex set containing it) in H has no parallel sides.

Let CH(M) denote the convex hull of a mask M. CH(M) is a convex polygon. A vertex of CH(M) is said to be opposite a side s of CH(M) if the line through v and parallel to s contains no points of CH(M) other than v. A vertex of CH(M) is said to be a reference point if it is opposite some side of CH(M). The set of all reference points of M is denoted R(M).

THEOREM (Crimmins) 15.11. Let f, f_1 be complex valued functions on Z^2 with finite support S(f); if M is a mask and $R(M) \subset S(f) \subset M, S(f_1) \subset M$ with $r = r_1$, then there is a complex number α of modulus one such that $f_1 = \alpha f$.

In other words, if a function is zero outside a given mask and is nonzero at the reference points of the mask, then it is uniquely determined (up to a phase factor) by its autocorrelation function among all other object functions that are zero outside the mask.

Crimmins has developed a reconstruction algorithm for the function. Basically it involves taking the reference points in M in a certain order: (1) $f(q_o)$ is gotten by the quotient of products of autocorrelation factors r(.); (2) $f(q_n) = r(q_n - q_{n-1})/f(q_{n-1})^*$ for the rest of the reference points; (3) $f(q_n) = (r(q_n - q_{m_n}) - \sum f(q_k)f(q_k - q_n - q_{m_n})^*/f(q_{m_n})^*$ for the remaining points.

For details of this algorithm the reader should refer to Crimmins' paper.

Chapter 16

Fienup's Input-Output Algorithms and Variations on this Theme

16.1 Introduction

The Gerchberg-Saxton or error reduction algorithm was developed to solve the problem of reconstructing phase from two intensity measurements. Also, a similar algorithm was proposed by Lesem et al. and Gallagher and Liu for synthesizing phase codes given intensity constraints in the object and Fourier domains. The algorithm can be summarized at the kth iteration as follows:

(1) $G_k(u) = |G_k(u)|e^{i\phi_k(u)} = \mathcal{F}(g_k(u))$

(2) $G'_k(u) = |F(u)|e^{i\phi_k(u)}$ where $|F(u)|$ is the measured intensity (or intensity constraint)

(3) $g'_k(x) = |g'_k(x)|e^{i\theta'_k(x)} = \mathcal{F}^{-1}(G'_k(u))$

(4) $g_{k+1}(x) = |f(x)|e^{i\theta_{k+1}(x)} = |f(x)|e^{i\theta'_k(x)}$

(5) return to (1).

In the case that only a single intensity measurement is available (say the Fourier domain) along with object domain support constraints, the algorithm is identical to the above case with (4) replaced by

(4') $g_{k+1} = 0$ if $g'_k(x)$ violates the constraint (like being negative when the object is nonnegative or exceeds the estimated support) (x in S) and

$g_{k+1}(x) = g'_k(x)$ otherwise (x not in S).

Here S is the set of points where $g'(x)$ violates the object domain constraints. In astronomical reconstruction the object is known to be real and nonnegative. The autocorrelation function is derived from the Fourier modulus and the "diameter" of the object is estimated as half the diameter of the autocorrelation function. We have discussed in Chapter XV that the exact support of the object cannot, in general, be reconstructed from the support of its autocorrelation. Thus, the diameter constraint is only an estimate of the support.

As we described in Section 14.7, if we let the error term be defined by

$$B(k) = N^{-2} \sum_n [|G_k(u)| - |F(u)|]^2, \tag{16.1}$$

then the steepest descent algorithm is

$$g_k''(x) = g_k(x) - h_k \partial B_k / \partial g|_{g_k(x)}. \tag{16.2}$$

Assuming g(x) is real, it is easy to check that

$$\partial B / \partial g = 2N^{-2} \sum [|G(u)| - |F(u)|] \partial |G(u)| / \partial g \tag{16.3}$$

and

$$\partial G(u) / \partial g = a^* \tag{16.4}$$

$$\partial |G(u)| / \partial g = \frac{1}{2|G(u)|} \partial G(u) / \partial g = \frac{G(u)a^* + G^*(u)a^*}{2|G(u)|} \tag{16.5}$$

where $a = e^{2\pi i u x / N}$. A collection of terms using the Fourier constraint $G'(u) = |F(u)|G(u)/|G(u)|$ gives

$$\partial B / \partial g = |_{g_k} = 2(g_k(x) - g_k'(x)). \tag{16.6}$$

Selecting a step size of 1/2 gives

$$g_k''(x) = g_k'(x) \tag{16.7}$$

so the error-reduction algorithm can be viewed as a steepest descent method. To satisfy the object domain constraints the new estimate is taken to be

$$g_{k+1}(x) = \begin{cases} g_k''(x) & x \text{ not in S} \\ 0 & x \in S. \end{cases} \tag{16.8}$$

Similarly, the conjugate gradient method suggests the algorithm

$$g_k''(x) = g_k'(x) + h_k(g_k'(x) - g_{k-1}'(x)) \tag{16.9}$$

where h_k is the step size.

To speed convergence Fienup developed the input-output (IO) algorithm where steps (1)-(3) remain the same and (4') is replaced by

$$g_{k+1}(x) = g_k(x) + \beta \Delta g_k(x) \tag{16.10}$$

where Δg_k are various corrections. The basic IO algorithm is

$$g_{k+1}(x) = \begin{cases} g_k(x) & \text{x not in S} \\ g_k(x) - \beta g_k'(x) & \text{x in S} \end{cases} \qquad (16.11)$$

An example of the IO algorithm is shown in Figure 16.1 which appeared in Fienup's now classic paper. Several variations on the IO algorithm were proposed. The output-output algorithm takes

$$g_{k+1}(x) = \begin{cases} g_k'(x) & \text{x not in S} \\ g_k'(x) - \beta g_k'(x) & \text{x in S.} \end{cases} \qquad (16.12)$$

Of course, if $\beta = 1$, the output-output algorithm is precisely the error reduction algorithm.

The hybrid input-output algorithm uses

$$g_{k+1}(x) = \begin{cases} g_k'(x) & \text{x not in S} \\ g_k(x) - \beta g_k'(x) & \text{x in S.} \end{cases} \qquad (16.13)$$

If the modulus $|f(x)|$ of the object is known, the algorithm becomes

$$g_{k+1} = g_k'(x) + \beta[|f(x)|g_k'(x)/|g_k'(x)| - g_k'(x)]. \qquad (16.14)$$

Another algorithm when $|f(x)|$ is known is to use

$$g_{k+1}(x) = g_k'(x) +$$

$$\beta\{[|f(x)|g'(x)/|g_k'(x)| - g_k'(x)] + (|f(x)|g_k'(x)/|g_k'(x)| - |f(x)|g_k(x)/|g_k(x)|\} \qquad (16.15)$$

where the second term rotates the phase angle of the input toward the phase angle of the output.

The relaxation parameters algorithm was proposed by Oppenheim et al. where

$$g_k''(x) = (1 - \eta_k)g_{k-1}''(x) + \eta_k g_k'(x) \qquad (16.16)$$

and the new estimate $g_{k+1}(x)$ is formed from $g_k''(x)$ by making it satisfy the function domain constraints. Of course the case $\eta_k = 1$ is precisely the error reduction algorithm. If the object domain error is expressed as

$$e_o^2 = \int_S |g_k''(x)|^2 dx \qquad (16.17)$$

where S is the region over which the function is known to be zero, setting the derivative of e_o^2 with respect to η_k equal to zero, one finds the optimal value of η_k:

$$\eta_k = -Re\{\int_S g_{k-1}''(g_k' - g_{k-1}'')^* dx\} / \int_S |g_k''(x) - g_{k-1}''(x)|^2 dx. \qquad (16.18)$$

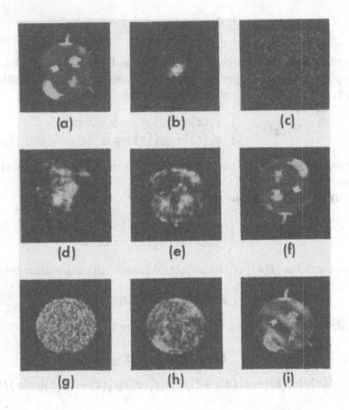

Reconstruction of a nonnegative function from its Fourier modulus. (a) Test object; (b) modulus of its Fourier transform; (c) initial estimate of the object (first test); (d)-(f) reconstruction results – number of iterations; (d) 20 ,(e) 230, (f) 600; (g) initial estimate of the object (second test); (h)-(i) reconstruction results – number of iterations; (h) 2, (i) 215.

Figure 16.1: Example of Fienup's IO algorithm

Use of this relaxation step for the problem of reconstructing a band-limited function from its phase results in an order of magnitude improvement over that of the standard error reduction algorithm according to Oppenheim et al.

The IO algorithm is based on the following relationships. Consider a small change $\Delta g(x)$ of the input. Passing through the algorithm provides a change $\Delta G'(u)$, or output $\Delta g'(x) = \mathcal{F}^{-1}(\Delta G'(u))$. Consider the case of kinoform generation where $g'(u) = Ke^{i\phi(u)}$ where K is a constant. Then for $|\Delta G| << |G|$

$$\Delta G'(u) \sim \Delta G^t(u)K/|G(u)| \qquad (16.19)$$

where $\Delta G^t(u) = |G(u)|sin\beta(u)e^{i(\phi(u)+\pi/2)}$. Here $\beta(u)$ is the angle between $\Delta G(u)$ and G(u). Viewing the phase angles $\beta(u)$ and magnitudes $|G(u)|$ as random variables with $|G(u)|$ identically distributed and independent of $\beta(u)$ and taking $\beta(u)$ to be uniformly distributed, the expected value of $\Delta G^t(u)$ is

$$E(\Delta G^t(u)) = \frac{1}{2}\Delta G(u) \qquad (16.20)$$

and so

$$E(\Delta g'(x)) = E(\mathcal{F}(\Delta G')) = \qquad (16.21)$$

$$\mathcal{F}(E(\Delta G')) = \mathcal{F}(E(\Delta G^t))E(K/|G|)) =$$

$$\mathcal{F}(\frac{1}{2}\Delta G)E(K/|G|) = \frac{1}{2}\Delta g(x)E(K/|G|).$$

That is, the expected change in output is β times the change of input where $\beta = \frac{1}{2}E(K/|G|)$.

The variance of change of output similarly can be checked to be

$$E(|\Delta g'(x)| - |E(\Delta g'(x)|^2) \sim$$

$$\frac{1}{4}(2E(K^2/|G|^2) - (E(K/|G|))^2)\frac{1}{A}\int_{-\infty}^{\infty}|\Delta g(x')|^2dx' \qquad (16.22)$$

where A is the area of the support of the object. Thus the variance of the change of the output at x is proportional to the integrated squared change of the entire input.

In general as we have noted before, there is no guarantee that the IO algorithm or its variations will converge. And for certain types of objects these iterative algorithms stagnate on images that are not fully reconstructed. Three known modes of stagnation are: (1) simultaneous twin images, (2) stripes superimposed on the image, and (3) unintentional truncation by the support constraint. Methods of overcoming these problems have been discussed by Fienup.

The twin image problem arises since both $f(x,y)$ and $f(-x,-y)$ have the same Fourier modulus. When the support of $f(x,y)$ is symmetric, the algorithm is shown by Fienup to output a partially reconstructed image having features of both $f(x, y)$ and $f(-x, -y)$. Fienup reports an improvement if the user employs a temporary support constraint with reduced area for a few iterations.

The stripes problem arises due to phase error symmetry in regions of the Fourier plane. Fienup has discussed two methods of solving this form of stagnation, a voting method and a patching method. The reader is directed to Fienup's papers for details.

The truncated support problem arises since $f(x - x_0, y - y_0)$ has the same Fourier modulus as $f(x, y)$. The image may be partially reconstructed but is not in perfect registration with the support constraint. The support constraint causes an error, which results in stagnation. Methods to overcome this mode of stagnation include enlarging the support or dynamically translating either the support constraint or the image. A proposed method for determining the amount of translation is discussed by Fienup in his papers.

Chapter 17

Topics and Applications of Signal Recovery

17.1 Maximum Entropy Estimation

Consider the Fourier decomposition of the power spectrum, which is real and positive,

$$F(w) = \sum f(n)e^{-inaw} \tag{17.1}$$

where f(n) = af(na). We assume F(w) = 0 for $|w| > \pi/a$. Assume that a segment g(t) of f(t) is known, viz., f(n) = g(n) = ag(na) for $|n| \le M = [T/a]$. Note that $f(-n) = f(n)^*$. Then F(w) is selected to maximized the entropy function $H = \int_{-\pi/a}^{\pi/a} log F(w)dw$ subject to the 2M+1 constraints g(n). By the Fejer-Riesz theorem it follows that

$$F(w) = p/|1 + \sum_{K=1}^{M} y(k)e^{-ikaw}| \tag{17.2}$$

and y(1),...y(M), p are given by the equations

$$\sum_{k=1}^{M} g(n-k)g(k) = -g(n), n = 1, ...M \tag{17.3}$$

$$p = g(0) + \sum_{k=1}^{M} g(-k)y(k).$$

Lim and Malik proposed a generalization of this algorithm. Let $x(n_1, n_2)$ be a signal whose power spectrum $P_x(w_1, w_2)$ is desired. Let $R_x(n_1, n_2)$ be the autocorrelation of $x(n_1, n_2)$. Then the maximum entropy power

221

spectrum estimation (MEPSE) problem is: given $R_x(n_1, n_2)$ for (n_1, n_2) in A (a set of points) determine $\hat{P}_x(w_1, w_2)$ such that the entropy

$$H = \int_{-\pi}^{\pi} \int log \hat{P}_x(w_1, w_2) dw_1 dw_2 \qquad (17.4)$$

is maximized and $R_x(n_1, n_2) = \mathcal{F}^{-1}(\hat{P}_x(w_1, w_2)$ for n_1, n_2 in A.

THEOREM 17.1. The MEPSE problem is equivalent to finding the function $\hat{P}_x(w_1, w_2)$ which is of the form

$$\hat{P}_x(w_1, w_2) = 1/[\sum_{n_1, n_2 \in A} \lambda(n_1, n_2) e^{-iw_1 n_1 - iw_2 n_2}] \qquad (17.5)$$

with $R_x(n_1, n_2) = \mathcal{F}^{-1}(\hat{P}_x(w_1, w_2))$.

This result suggests the iterative algorithm for $\hat{P}_x$:
(1) estimate $\lambda(n_1, n_2)$
(2) set $R_y(n_1, n_2) = \mathcal{F}^{-1}(1/\mathcal{F}(\lambda(n_1, n_2)))$
(3) correct $R_y(n_1, n_2)$ by $R_x(n_1, n_2)$ for n_1, n_2 in A
(4) set $\lambda(n_1, n_2) = \mathcal{F}^{-1}(1/\mathcal{F}(R_y(n_1, n_2)))$
(5) set $\lambda(n_1, n_2) = 0$ for n_1, n_2 not in A
(6) return to(2).
The major problem with this simplistic approach is that zero crossings can occur in $\mathcal{F}(\lambda)$ or $\mathcal{F}(R_y)$. The more robust approach to correct for this is presented in the paper by Lim and Malik.

17.2   Stellar Speckle Interferometry

The method of Labeyrie in stellar speckle interferometry consists of processing many short-exposure images which can be used to build a diffraction limited estimate of the Fourier modulus of an astronomical object. This method is applicable to sky-bright objects, i.e., objects which are real and nonnegative.

Labeyrie's method takes a number of short-exposure images of an astronomical object:

$$d_m(x) = f(x) \star s_m(x) \qquad (17.6)$$

where f(x) is the spatial or angular brightness distribution of the object and $s_m(x)$ is the point-spread function due to the combined effects of atmosphere and the telescope for the mth exposure. It is assumed that the exposure time is short enough to "freeze" the atmosphere. In addition only a narrow spectral band is used. The Fourier transform of each short exposure is computed:

$$D_m(u) = \int_{-\infty}^{\infty} d_m(x) e^{2\pi i u x} dx. \qquad (17.7)$$

The summed square Fourier modulus or the summed power spectrum is computed:

$$\sum_{m=1}^{M} |D_m(u)|^2 = |F(u)|^2 \sum_{m=1}^{M} |S_m(u)|^2. \qquad (17.8)$$

The factor $\sum |S_m(u)|^2$ is the square of the MTF of the speckle interferometry and can be computed by performing the same process on an isolated unresolved star which is imaged through the same atmospheric conditions. By dividing, one obtains the desired Fourier modulus squared of the object.

Fienup and Feldkamp have applied Fienup's algorithm to reconstruct the binary SAO 94163 system, which is a trivial application of the iterative method. It is clearly more useful for more complicated objects. In addition they proposed a method for detection transfer function compensation, noise bias subtraction and speckle MTF compensation. For compensation of the speckle MTF when reference data is not available, an improvement over the Worden subtraction method is proposed which consists of clipping the Fourier modulus spike at the very low spatial frequencies.

17.3 Radio Astronomy

In radio astronomy one samples the magnitude (and perhaps the phase) of the complex visibility function V(u,v). V is a function of the vector spacing $s = (s_x, s_y)$ between the two observing antennas expressed in the units $u = s_x/\lambda, v = s_y/\lambda$. The complex visibility is related to the radio brightness b(l,m) which by the van Cittert-Zernike theorem is

$$V(u,v) = \int \int \hat{b}(l,m)e^{-2\pi i(lu+mv)}dldm \qquad (17.9)$$

where $\hat{b}$ is the normalized radio brightness

$$\hat{b}(l,m) = b(l,m)/\int \int b(l,m)dldm. \qquad (17.10)$$

Here l,m are the direction cosines of the source element with respect to the x,y axes of the plane transverse to the direction of the source.

17.4 The CLEAN Method

The CLEAN method in radio astronomy is the approach to solve the convolution equation $d = b \star t$ where d is the "dirty" map (or image), b is the "dirty" beam (or point spread function) and t if the brightness distribution

(or object). The convolution equation in the discrete case can be written
as the matrix form

$$d = Bt \qquad (17.11)$$

where

$$B =$$

$$1 b_1 b_2 ... b_{N-1}$$

$$b_1 1 b_1 ... b_{N-2}$$

$$...$$

$$b_{N-1} ...... 1.$$

The "dirty" beam is assumed to be symmetric so B is a Toeplitz matrix.

The iterative technique used in CLEAN is that of Southwell-Temple:
let t' be the current estimate to t and let $r = d - Bt'$ denote the residual.
The method finds the largest absolute value of residuals, say it is $r(m)$ at
position m. Then the correction $\delta t' = (0, ...0, gr(m), 0, ..0)$ is defined and
we set $t'' = t' + \delta t'$. Temple showed that this method converges starting
from any initial approximation if B is symmetric and positive definite.

In terms of the visibility function $V_o(u)$ the dirty map is the Fourier
transform

$$d(x_n) = \sum_{k=1}^{R} w(u_k) V_o(u_k) e^{-2\pi i x_n u_k} \quad n = 1, ... N. \qquad (17.12)$$

If we let $\Delta x_{nm} = x_n - x_m$, then the dirty beam is given by the Fourier
transform of the weights:

$$b(\Delta x_{nm}) = \sum_{k=1}^{R} w(u_k) e^{2\pi i \Delta x_{nm} u_k}. \qquad (17.13)$$

Since $w(-u_k) = w(u_k)$ we see that the beam is symmetric, $b(-\Delta x_{nm}) =
b(\Delta x_{nm})$. By definition $\sum_{k=1}^{R} w(u_k) = 1$, so $b(0) = 1$. Setting $\Delta x_{nm} = l\Delta x$
and $\Delta x = 1/(N\Delta u)$ and letting $b(n) = b(n\Delta x)$, the equation for B has the
form 17.11. The equation for components for d becomes

$$d(n) = \sum_{k=-N/2}^{N/2-1} w(k) V_{ok} e^{-2\pi i k n/N}. \qquad (17.14)$$

If the number R of non-zero weights is less than N, then B is only
positive semidefinite. Forsythe and Warsaw have shown that the CLEAN
algorithm or Temple's method converges to t+z where z is an element of
the null space of B, i.e., $Bz = 0$.

17.5 The Knox Method

Several algorithms have been developed which treat the problem of estimating the magnitude and phase of the optical transfer function associated to a blurred image. Let f denote the object and let g denote the blurred image, then

$$g(x,y) = \int\int h(x-\xi, y-\eta)f(\xi,\eta)d\xi d\eta + n(x,y). \qquad (17.15)$$

Taking Fourier transforms we have

$$G(u,v) = H(u,v)F(u,v) + N(u,v). \qquad (17.16)$$

We call $H(u,v)$ the optical transfer function and $h(x,y)$ is the point spread function (PSF).

The restoration technique of Stockham, Cole and Cannon divides the degraded image into subimages which may overlap. If one assumes the extent of the PSF is small compared to the extent of the subimage and ignoring the noise term, then for the ith subimage

$$g_i \sim \int\int_i h(x-\xi, y-\eta)f_i(\xi,\eta)d\xi d\eta, \qquad (17.17)$$

or $G_i \sim HF_i$. In terms of the magnitude and phases we have

$$|G_i|e^{i\theta_{\alpha_i}} \sim |H|e^{i\theta_H}|F_i|e^{i\theta_{F_i}}. \qquad (17.18)$$

The estimator for $|H|$ is obtained by forming the logarithm of

$$|G_i| \sim |H||F_i| \qquad (17.19)$$

and averaging over N subimages to obtain

$$\frac{1}{N}\sum_{i=1}^{N} ln|G_i| \sim ln|H| + \frac{1}{N}\sum_{I=1}^{N} ln|F_i|. \qquad (17.20)$$

To estimate $|F_i|$ it is proposed that one can use an undegraded image of the same prototype class as the degraded image to obtain

$$\frac{1}{N}\sum ln|P_i| = \frac{1}{N}\sum ln|F_i|. \qquad (17.21)$$

Thus, we have

$$ln|H| \sim \frac{1}{N}\sum(ln|G_i| - ln|P_i|). \qquad (17.22)$$

In Coles' approach one assumes the OTF has zero phase (i.e., $\theta_H = 0$).

A method to estimate θ_H was proposed by Knox. He was concerned with obtaining clear photographs of astronomical objects, which have been blurred by atmospheric turbulence. The technique proposed by Knox was to use many short exposure photographs of the same object together with short exposure photos of a nearby star. Let the ith short exposure photo be denoted g_i. Again, we have $G_i = H_i F$ where h_i is the short exposure PSF. Consider the autocorrelation

$$G_i(u,v)G_i^*(u+\Delta v, u+\Delta v) =$$

$$H_i(u,v)H_i(u+\Delta u, v+\Delta v)F(u,v)F^*(u+\Delta u, v+\Delta v); \qquad (17.23)$$

then averaging over multiple images, we have

$$phase(\overline{autoG_i}) = phase(\overline{autoH_i}) + phase(autoF). \qquad (17.24)$$

The term $phase(\overline{autoH_i})$ can be assumed to be negligible for atmospheric turbulence; thus, we have

$$\theta_F(u,v) - \theta_F(u+\Delta u, v+\Delta v) \sim phase(\overline{autoG_i}). \qquad (17.25)$$

Since $\theta_F(0,0)$ can be arbitrarily selected, say equal to 0, we see that the phases can be calculated recursively from the phase difference estimates so described.

A combination of the Knox and Cannon method was proposed by Morton and Andrews. We refer the interested reader to Morton's thesis. Gonsalves attacked the OTF problem by the Gerchberg-Saxton approach where one assumes knowledge of both $|h|$ and $|H|$. Frieden and Currie considered the problem of calculating the OTF phase from the OTF magnitude in the case the OTF has only real zeros.

17.6   Filter Design

As early as 1931 van Cittert proposed an iteration for deconvolving a sequence y(n,m) which has been convolved with a distortion function b(n,m). This problem arose in the context of gamma ray spectrum analysis. In the frequency domain this iteration is given by

$$Y_i(u,v) = X(u,v) - (1 - B(u,v))Y_{i-1}(u,v) \qquad (17.26)$$

where X(u,v) = Y(u,v)B(u,v) is the degraded spectrum.

Quatieri and Dudgeon considered the application of van Cittert's approach to construct 2-D digital filters. Viz., let a 2-D rational frequency response be

$$H(u,v) = A(u,v)/B(u,v) \qquad (17.27)$$

where $A(u, v) = \sum_n \sum_m a(n, m)e^{-iun - ivm}$ and similarly for B. We assume a(n,m) and b(n,m) have finite extent and we assume b(0,0) = 1. If X(u,v) and Y(u,v) are the Fourier transforms of the filter's input and output, they propose the following iterative approach:

$$Y_i(u, v) = A(u, v)X(u, v) + C(u, v)Y_{i-1}(u, v) \qquad (17.28)$$

$$Y_{-1}(u, v) = 0.$$

Here C(u,v) = 1-B(u,v). If $|C(u, v)| < 1$, then $Y_i(u, v)$ converges to Y(u,v).

EXERCISE 17.2. Let

$$G(y_{i-1}(n, m)) = a(n, m) \star x(n, m) + c(n, m)y_{i-1}(n, m)$$

and let

$$Tg(n, m) = \begin{cases} g(n, m) & \text{n,m in S} \\ 0 & \text{otherwise} \end{cases} \qquad (17.29)$$

where S is an NxN square region. Show that if $|C| < 1$, then $Q = TG$ is nonexpansive.

Optimum Fourier transform division filters with magnitude constraints were treated by Gallagher and Liu following their joint work on kinoform construction. We refer the reader to their paper for details.

17.7 Microwave Holographic Metrology

Reflector surface errors can considerably affect the beam width, boresight location, gain and side lobe structure of an antenna's radiation pattern. The performance of large reflector antennas can be improved by identifying the location and amount of their surface distortions and correcting them. To improve the resolution of the surface map data Rahmat-Samii has proposed the following application of Papoulis' algorithm.

First, recall that the components of the far-field radiation are described by

$$T(\theta, \phi) = \int_S J(r')e^{ikr' \cdot \hat{r}} J_s dx' dy' \qquad (17.30)$$

where J_s is the Jacobian $(1 + (\partial f / \partial x')^2 + (\partial f / \partial y')^2)^{1/2}$, J is the induced surface current ($J = 2\hat{n} \times H^i$ where H^i is the incident magnetic field generated by the feed), z = f(x,y) describes the reflector surface. Here $r', \hat{r} = z'\cos\theta + ux' + vy'$ where $u = \cos\theta\sin\phi, v = \sin\theta\cos\phi$. Letting $\tilde{J}(x', y') = J(r')J_s$ we have

$$T(u, v) = \int_S \tilde{J}(x', y')e^{ikz'}e^{ik(ux' + vy')} dx' dy'. \qquad (17.31)$$

Expanding this in θ and keeping only the 0th order term we have

$$T(u,v) = \int \tilde{J}(x',y')e^{ikz'}e^{ik(ux'+vy')}dx'dy'.$$

Once T is determined, then the far field radiated pattern is given by

$$E = -ik\eta e^{-ikr}(T_\theta \hat{\theta} + T_\phi \hat{\phi})/(2\pi r) \qquad (17.32)$$

where $k = 2\pi/\lambda, \eta = 120\pi$.

The θ, ϕ components of T are related to the Cartesian coordinates by

$$T_\theta = cos\theta cos\phi T_x + cos\theta sin\phi T_y - sin\phi T_z \qquad (17.33)$$

$$T_\phi = -sin\phi T_x + cos\phi T_y.$$

And the copolar and cross-polarization components of the far-field pattern are related to the θ, ϕ components by

$$T_{copol} = sin\phi T_\theta + cos\phi T_\phi \qquad (17.34)$$

$$T_{crosspol} = cos\phi T_\theta - sin\phi T_\phi.$$

For small θ we see that

$$T_{copol} = T_y \qquad (17.35)$$

$$T_{crosspol} = T_x.$$

Assuming the dominant polarization is in the y direction, we have the scalar expression

$$T = \int_S \tilde{J}(x',y')e^{ikz'}e^{ik(ux'+vy')}dx'dy'. \qquad (17.36)$$

Consider a parabolic reflector antenna with center of the feed located at

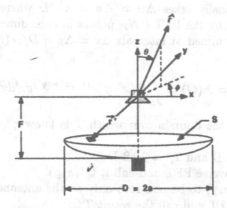

Figure 17.1: Geometry of a parabolic reflector antenna

the focal point. Let $\varepsilon(x,y)$ describe the surface irregularities in the normal direction. One can show that

$$-r' + z' = -2F + 2\varepsilon\cos\xi \tag{17.37}$$

in the coordinates of Figure 17.1 for a parabolic reflector of diameter $D = 2a$ and focal length F. And for a parabolic reflector $\cos\xi = (1 + \frac{x^2+y^2}{4F^2})^{1/2}$. Since the phase center of the feed is located at the focal point, the phase term $e^{-ikr'}$ appears in

$$\tilde{J}(x',y')e^{ikz'} = |\tilde{J}(x',y')|e^{-ikr'}e^{ikz'}. \tag{17.38}$$

If the total distortion phase error is defined as $\delta = 4\pi(\varepsilon/\lambda)\cos\xi$, we see that 17.36 can be written as

$$T(u,v) = e^{-i2kF}\int_S |\tilde{J}(x',y')|e^{i\delta}e^{i(ux'+vy')}dx'dy'. \tag{17.39}$$

Thus, if the amplitude and phase of the reflector pattern can be measured, then

$$|\tilde{J}(x,y)|e^{i\delta} = e^{i2kF}\mathcal{F}^{-1}(T(u,v)) \tag{17.40}$$

and the surface distortion is given by

$$\varepsilon(x,y)/\lambda = \frac{1}{4\pi}(1 + \frac{x^2+y^2}{4F^2})^{1/2}phase(e^{i2kF}\mathcal{F}^{-1}[T(u,v)]). \tag{17.41}$$

The microwave holography method involves trying to improve the estimate of $\varepsilon(x,y)/\lambda$ based on measurements of $T(u,v)$ taken over a limited range. For a reflector of diameter D the sampling interval is λ/D. To overcome aliasing one nominally takes $\Delta u = \Delta v = \kappa\lambda/D$ where $.5 < \kappa < 1$. If 17.41 is evaluated by the FFT (N_F points in each dimension), the values of $\varepsilon(x,y)$ are determined at intervals $\Delta x = \Delta y = D/(\kappa(N_F - 1))$.

Let

$$T(u,v) = \mathcal{F}(Q) = \int_S Q(x',y')e^{ik(ux'+vy')}dx'dy' \qquad (17.42)$$

and let D denote the domain over which T is known; then the initial approximation is

(1) $T_e^o = T$ on D and $T_e^o = 0$ off D
(2) take the inverse FFT and call it $Q^o(x,y)$
(3) allow $Q^o(x,y)$ to be non-zero only on the antenna domain
(4) take the FFT and call the result $T^{o'}$
(5) set $T_e^1 = T$ on D and $T_e^1 = T^{o'}$ off D
(6) return to (2).

Rahmat-Samii has applied this technique to the 64 meter radio astronomy antenna DSS-43 using the 34 meter antenna DSS-42 as a reference. The 64 meter antenna was offset in a grid path centered on the radio source 3C273. The measurements were taken in a 11x11 sampling grid.

17.8   Misell Algorithm

Another iterative algorithm for determining the amplitude and phase from the image intensity in an optical system was proposed by Misell. This method is based on taking two images recorded at different lens defocus values.

In the transmission electron microscope (either conventional or scanning) in the isoplanatic approximation (with magnification M = 1) and under coherent illumination, the elastic electron wave transmitted by the specimen, $\psi_o(x)$, is modified by the lens aberration function $G_1(x)$ for the objective lens to given the image wave function:

$$\psi_1(x) = \int \psi_o(x')G_1(x-x')dx'. \qquad (17.43)$$

One measures the image intensity $j_1(x) = |\psi_1(x)|^2$. The problem is to solve for ψ_o.

In the weak phase and weak amplitude approximation, the object is given by

$$\psi_o(x) = 1 + i\eta(x) - \epsilon(x) \qquad (17.44)$$

where the phase η and absorption ϵ terms are assumed to be much less than unity. The method of Erickson and Klug allows both η and ϵ to be determined by recording two micrographs for different defocus values say

$\Delta f_1, \Delta f_2$. This technique applies only to bright-field microscopy which allows the unscattered wave to interfere with the elastic scattered wave.

In Misell's approach two micrographs are recorded at different defocus values $\Delta f_1, \Delta f_2$. Recall that the image resolution function $G_1(x)$ includes terms for spherical aberration C_s, defocusing Δf and axial astigmatism C_A. Viz., the Fourier transform of G_1 is

$$F(G_1(x)) = T(v) = B(v)e^{-ik_o W_1(v)} \qquad (17.45)$$

where

$$W_1(v) = C_s v^4 \lambda^4/4 + \Delta f_1 v^2 \lambda^2/2 - C_A(v_x^2 - v_y^2)\lambda^2/2 \qquad (17.46)$$

for electrons of wavelength λ, spatial frequency v, $v\lambda = \theta$ = the angle of scattering and wave number $k = 2\pi/\lambda$. Taking a second micrograph at defocus Δf_2 gives the wave equation

$$\psi_2(x) = \int \psi_o(x) G_2(x - x')dx' \qquad (17.47)$$

where $F(g_2(x)) = B(v)e^{-ikW_2(v)}$. However, if we let $\Delta f = \Delta f_2 - \Delta f_1$ then $F(G_2(x)) = F(G_1(x))e^{-ik\Delta f v^2 \lambda^2/2}$. Thus, ψ_1 and ψ_2 are related by the convolution integral equation

$$\psi_2(x) = \int \psi_1(x') G(x - x')dx' \qquad (17.48)$$

where $G(x) = F^{-1}(e^{-ik\Delta f v^2 \lambda^2/2})$. Misell's algorithm is then:

(1) select random phase $\eta_1(x)$ and using the measured $|\psi_1(x)|$ form $\psi_1(x) = |\psi_1(x)|e^{i\eta_1(x)}$

(2) convolve as in 17.48 to form $\psi_2'(x) = |\psi_2'|e^{i\eta_2(x)}$

(3) replace the measured $|\psi_2(x)|$ to form $\psi_2(x) = |\psi_2(x)|e^{i\eta_2(x)}$

(4) form $\psi_1'(x) = \psi_2(x) \star G'(x)$ where $g'(x) = F^{-1}(e^{ik\Delta f v^2 \lambda^2/2})$, say $\psi_1'(x) = |\psi_1'(x)|e^{i\eta_1(x)}$

(5) replace $|\psi_1'|$ by measured $|\psi_1|$ to form $\psi_1(x) = |\psi_1|e^{i\eta_1(x)}$

(6) return to step (2).

Consider now the following one dimensional dispersion relation result:

THEOREM (Hoenders) 17.2. Let the complex function g(y) be defined by $-1 \leq y \leq 1$ and suppose $g'(y)$ exists everywhere in this interval and is of bounded variation. Let h(z) be given by

$$h(z) = \int_{-1}^{1} e^{izy} g(y)dy \qquad (17.49)$$

(so h(z) is an entire function band-limited to [-1,1].) Thus, if the number of zeros a(n) of h(z) in the upper half plane is finite, then the phase and modulus of h(z) are related by

$$\frac{1}{\pi} \int ln(h(x')x')dx'/(x - x') = -argh(x) - arg(xe^{ix}) \prod_n \frac{x - a(n)^*}{z - a(n)} h(z)$$

$$(17.50)$$

Using this result we have

THEOREM (Hoenders) 17.3. Let g,h be as above and let

$$h_1(z) = \prod_n \frac{z - a(n)^*}{z - a(n)} h(z) \qquad (17.51)$$

where $\{n\}$ denotes a finite number of roots of $h(z)h^*(z^*)$. Then h_1 is band-limited where

$$h_1(z) = \int_{-1}^{1} e^{izy} g(y)dy \qquad (17.52)$$

and g(y) and $g_1(y)$ are related by the Volterra integral equation

$$g_1(y) = g(y) -$$

$$i \sum_{n''} (a(n) - a(n''))^* \prod_{\{n_1, n_1 \neq n''\}} \frac{a(n) - a(n)^*}{a(n'') - a(n)} \int_{-1}^{y} e^{ia(n'')(-y-\tau)} g(\tau)d\tau +$$

$$(17.53)$$

$$i \sum_{n'} (a(n') - a(n')^*) \prod_{n_1, n_1 \neq n''} \prod \frac{a(n') - a(n)^*}{a(n') - a(n)} \int_{y}^{1} e^{ia(n')(-y+\tau)} g(\tau)d\tau.$$

Here n" labels all poles in the lower half plane in 17.49 and n' labels all the poles in the upper half plane in 17.49

From the Titchmarsh-Cartwright formula the zeros of h(z) are distributed asymptotically as $a(n) \sim n\pi + \frac{1}{2}ln(g(-1)/g(1))$. Thus, we can deduce that the number of zeros of h(z) located in the upper half plane is finite if $|g(-1)| < |g(1)|$. Having a priori information about the intensity at the rim of the diaphragm in the Fraunhofer plane and a priori knowledge that the object and image are band-limited allows us to conclude that the intensity in the image plane uniquely determines the image wave function (up to a phase factor).

Consider the Misell approach again. Let

$$h_k(x) = \int_{-1}^{1} e^{ixy + i\Delta z_k y^2/f^2} g(y)dy \qquad (17.54)$$

then it follows from Hoender's results above that

COROLLARY (Hoenders) 17.4. Consider the intensity images taken at two different defocus values. Then these two measurements are sufficient to determine $g(y)$ (up to an overall phase factor).

The proof follows by examining the sets of functions, say $\{g_1(y)\}, \{g_2(y)\}$ generated by the dispersion relations in theorem 17.2 Only those functions for which $g_1(y)/g_2(y) = e^{i(\Delta z_1 - \Delta z_2)y^2/f^2}$ are consistent with the a priori knowledge that both exposures are taken with two different values of the defocus. The application of theorem 17.3 shows that $g(y)$ is unique up to a phase factor.

A generalization of Hoender's result has been given by Kiedron. Consider an unknown complex object $t(y)$. Let $P(y) = t(y)rect(y/2y_o)$. Then the complex amplitude in the image plane is given by

$$f(x) = \int_{y_o}^{y_o} P(y)h(x, y)dy \qquad (17.55)$$

where the kernel or point spread function is

$$h(x, y) = \int_{-u_o}^{u_o} a(u)e^{2\pi i(w(y, u) - (x - y)u)}du. \qquad (17.56)$$

Assume $w = 0$ and $a(u) = 1$, so $h(x, y) = 2u_o sinc(2u_o(x - y))$. The propagation can be expressed as two successive Fourier transforms

$$F(u) = \int_{-y_o}^{y_o} P(y)e^{2\pi i u y}dy \qquad (17.57)$$

(i.e. the complex amplitude in the exit pupil plane) and

$$f(x) = \int_{u_o}^{u_o} F(u)e^{-2\pi i x u}du. \qquad (17.58)$$

The problem is to find all complex band-limited, amplitude functions which produce the same intensity in the image plane

$$|g(x)|^2 = |f(x)|^2. \qquad (17.59)$$

By Walthers' theorem

$$g(x) = f(x) \prod_{finite} \frac{x - z_n^*}{x - z_n} \qquad (17.60)$$

where the linear phase term is eliminated due to the a priori knowledge of the cut-off frequency u_o.

Let $A_n = (z - z_n^*) \prod_{n \neq m} \frac{z_n - z_m^*}{z_n - z_m}$ and $R(u_o, z) = \int_{y_o}^{y_o} P(y) e^{2\pi i u_o y} dy$.

THEOREM (Kiedron) 17.5. The spectrum G(u) of g(x) of 17.60 has the form

$$g(u) = \int_{-y_o}^{y_o} P(y)(1 + \sum_n \frac{A_n}{y - z_n}) e^{2\pi i u y} dy \qquad (17.61)$$

where z_n are common zeros of f(z) and $R(u_o, z)$.

THEOREM (Kiedron) 17.6. If $f(u_o) \neq 0$ or $F(-u_o) \neq 0$, the equation 17.59 has a finite number of solutions.

17.9   Radio Holography

The idea to use measurements of the far-field pattern of an antenna to deduce phase errors present in its aperture goes back to Blum, Bennett, Scott, Ryle and others. The method of Scott and Ryle involves direct phase recording. These techniques thus involve auxillary reference antennas and phase sensitive receivers.

Morris has proposed using Misell's algorithm with defocused pairs of far-field power patterns. Considering a square antenna of width D wavelengths, then for a field distribution E(x,y) the far field pattern over a finite data window of width N/D radians is

$$A(a,b) = \sum_{k=0}^{N-1} \sum_{j=0}^{N-1} E(x_k, y_j) e^{i(x_k a + y_j b) 2\pi/N}. \qquad (17.62)$$

And, if A(a,b) were known, the aperture distribution could be reconstructed with surface resolution of D/N wavelenths.

Morris shows that the root mean square phase errors dQ in the aperture plane for the interferometric approach of Scott and Ryle are given by $dQ \sim KN/R$ where R is the signal to noise ratio at boresight and K describes the taper (e.g., $K = \sqrt{5}$ for a -14 dB taper). For the Misell approach Morris shows that $dQ \sim K(2/\pi)^{1/4} N/\sqrt{R}$. Thus, the Misell algorithm demands an SNR which is approximately the square of that needed when phase can be measured directly.

Morris applied the Misell algorithm to simulate antenna data at varying SNRs and for varying array sizes N (32, 64, and 128). Good results were obtained at high SNRs. The reader should also confer Boucher's work, in this general area.

17.10   Pinhole Projections

To perform 3-D object reconstruction in emission and transmission tomography one may have data collected only over a limited angular regime. Consider the imaging system described by

$$\phi(r) = \int \rho(r')\phi_o(r - r')dr' \qquad (17.63)$$

where ϕ_o is the space-invariant point response function. The Fourier transform of this equation gives

$$R(k) = \Phi(k)/\Phi_o(k) \qquad (17.64)$$

where $\Phi_o(k) \neq 0$. To extend R(k) beyond the measurement cone, Tam, Perez-Mendez and Macdonald have proposed the Gerchberg-Papoulis algorithm:

(1) estimate R(k) corrected to those k where $\Phi_o(k) \neq 0$
(2) take the FFT of (1) to give the estimated object $\rho(r)$
(3) correct $\rho(r)$ to be zero outside the known extent of the object
(4) inverse FFT to obtain R(k)
(5) return to (1).

In the presence of noise these authors used the Twomey smoother

$$R(k) = \Phi(k)/[\Phi_o(k) + \gamma(2\pi k)^4/\Phi_o(k)]. \qquad (17.65)$$

The computer simulations of these authors showed reconstruction was possible with as little as imaging with a solid angle of four steradians. The reader is also direct to the paper of Chiu et al.

17.11   Tomography

The algorithm of Sato et al. is an application of the Gerchberg-Papoulis algorithm to the case of tomographic imaging systems with gaps or missing sectors in the Fourier plane. The approach of Sato is:

(1) begin with an initial estimate of the image f_o defined by the known support
(2) take the 2-D Fourier transform of f_o to give F_o
(3) revise F_o by replacing values of F_o by the known F in the measured regions of the Fourier plane; call it G_o
(4) take the inverse 2-D Fourier transform of G_o; call it g_1
(5) revise g_1 by applying the a priori object data in the image plane – i.e., support and threshold criteria
(6) return to (2).

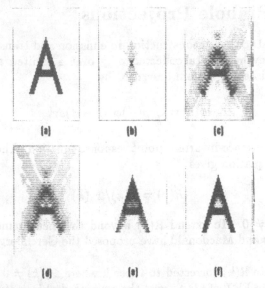

Simulated reconstruction in absence of noise: (a) object; (b) limited-angle projection data; (c) image obtained by inverse transformation of (b); (d) image obtained by inverse transformation of (b) after application of a single-step interpolation operation to the limited data to fill in the missing sector; (e) image after two interations; (f) image after 100 iterations.

Figure 17.2: Example of Sato's algorithm

If we let S denote the image support and D denote the Fourier domain support, then the general equations are:

$$f_{n+1}(x,y) = g_{n+1}(x,y)P_S(x,y) \qquad (17.66)$$

$$f_{n+1} = \begin{cases} t_u(x,y) & \text{if } f_{n+1}(x,y) > t_u(x,y) \\ t_l(x,y) & \text{if } f(x,y) < t_l(x,y) \\ f_{n+1} & \text{otherwise} \end{cases}$$

$$G_n(u,v) = F_n(u,v) + (F(u,v) - F_n(u,v))P_D(u,v).$$

An example of this algorithm is shown in Figure 17.2 taken from Sato's work. An optical implementation of this algorithm has been proposed by Sato et al.

17.12   The Spin-Glass Model

Lawton and Morrison have developed the following spin-glass model to help understand the phase retrieval problem. Consider the 2-D discrete Fourier transform

$$F(p) = \frac{1}{N^2} \sum_{q \in L} f(q)e^{-2\pi i pq/N} \qquad (17.67)$$

where L is the NxN lattice $\{(j,k), j = 0, ...N-1, k = 0, ...N-1\}$. Similarly, let W be a real function and set

$$W(p) = \frac{1}{N^2} \sum_{q} w(q)e^{-2\pi i pq/N}. \qquad (17.68)$$

Here we take W(p) = 1 for p in the support of f, $S(f) \subset L$, and W(p) = 0 otherwise. As a complex function we write

$$F(p) = |F(p)|e^{i\theta(p)}. \qquad (17.69)$$

Consider the sets of real numbers $\{W(p), p \in L\}$ and $\{F(p), p \in L\}$ which are held fixed. Consider the objective function

$$H(\theta) = -\sum_{p \in L} W(p)|f(p)|^2. \qquad (17.70)$$

First, we note that $H(\theta)$ can be expressed in terms of F and w:

THEOREM17.7. $H(\theta) = -\sum_{p \in L} \sum_{q \in L} F(p)F^*(q)w(p-q)$.

If we let $\phi(p)$ denote the phase of w(p) and using the fact that W is real implies w(-p) = w(p), then we have the following Hamiltonian representation of $H(\theta)$:

COROLLARY 17.8.

$$H(\theta) = -\sum_{p \in L} \sum_{q \in L} |F(p)||F(q)||w(p-q)|cos(\theta(p) - \theta(q) + \phi(p-q)). \quad (17.71)$$

In the spin glass system $w(p) = I(p)e^{2\pi i rs/N}$ where I(p) is real valued and symmetric about the origin. In this case we have

$$H(\theta) = -\sum_{p} \sum_{q} |F(q)||F(p)|I(p-q)cos(\theta'(p) - \theta'(q)) \qquad (17.72)$$

where $\theta'(r) = \theta(r) + 2\pi rs/N$.

The following is a very clever result which Lawton and Morrison have incorporated into Fienup's Input-Output algorithm:

THEOREM 17.8. The value $\theta_o(p)$ which minimizes $H(\theta)$ is

$$\theta_o(p) = phase((N^2 W \widehat{-w}(0,0)f)(p)). \qquad (17.73)$$

Lawton and Morrison have used this approach to partition the phase retrieval problem. We leave it to the reader to refer to the Lawton-Morrison paper for details.

17.13   Simulated Annealing

Lawton and Morrison and others have tried simulated annealing to solve the phase retrieval problem. The algorithm is basically:

(1) choose a search strategy for selecting candidates θ'_{k+1} for θ_{k+1} given a current value for θ_k

(2) select $\beta > 0$ and choose θ_1

(3.1) use the search strategy to calculate θ'_{k+1}

(3.2) calculate $\Delta H = H(\theta_{k+1}) - H(\theta_k)$

(3.3) if $\Delta H \le 0$ let $\theta_{k+1} = \theta'_{k+1}$

(3.4) if $\Delta H > 0$ let

$$\theta_{k+1} = \begin{cases} \theta'_{k+1} & \text{with probability } e^{-\beta \Delta H} \\ \theta_k & \text{with probability } 1 - e^{-\beta \Delta H} \end{cases}$$

(4) increase β and apply step (3) until a satisfactory result is obtained. If we are updating one phase variable at a time we have

$$\Delta H = -2|F(p)|Re((e^{-i\theta'_{k+1}} - e^{i\theta_k}))(N^2 W \widehat{-w}(0,0)f(p)). \qquad (17.74)$$

More elaborate schemes have been proposed by Lawton and Morrison.

17.14   Robbins-Monro

The Robbins-Monro stochastic approximation (RMSA) is a recursive technique for estimating the root r of the equation $M(x) = a$, where a is given, $M(x)$ is monotone, real measurable function such that $M(x) > a$ for $x > r$ and $M(x) < a$ for $x < r$. We assume that corresponding to each real x is a random variable Y(x) with a specified distribution function where M(x) is the expected value of Y given x. The recursive procedure starts with $X_1 = x_1$, an arbitrary real number, and defines $X_2, X_3, \ldots$ by

$$X_{n+1} = X_n - a_n(Y_n - a). \qquad (17.75)$$

Here a_n is a sequence of positive constants such that $\sum a_n = \infty$, $\sum a_n^2 < \infty$; e.g., $a_n = A/n$. Under certain conditions on M(x), Robbins and Munro have shown that

$$lim_{n \to \infty} E((X_n - r)^2) = 0. \tag{17.76}$$

Consider the case of a finite segment of a signal f(t)

$$g(t) = P_T(t)f(t) \tag{17.77}$$

which has been corrupted by noise y(t) = g(t)+v(t). We estimate an approximation to y(t) by using a sequence of independent samples at t_k, $k = 1, 2, ...$ Let $\bar{y}(t)$ denote a block sample of y(t) where

$$\bar{y}(t) = \sum_{k=1}^{\infty} a(k)^* \psi_k(t) = \bar{a}^{*T} \bar{\psi}(t) \tag{17.78}$$

where $\psi_k(t) = \phi_k(t)/\sqrt{\lambda_k}$ and $\bar{a}^*$ is an n-vector of unknown coefficients. We want to find the $\bar{a}^*$ which minimizes

$$I(\bar{a}) = \int_{-T}^{T} (\bar{y}(t) - \bar{a}^T \bar{\psi}(t))^2 dt \tag{17.79}$$

where $dt = \prod_{k=1}^{m} dt_k$. To evaluate $\bar{a}^*$ numerically using RMSA, we form

$$\bar{a}_{k+1} = \bar{a}_k - \frac{A}{k} \bar{Y}(\bar{a}_k, t_k) \tag{17.80}$$

where A is now a diagonal gain matrix. The function $\bar{Y}(\bar{a}, t)$ is chosen by

$$E(\bar{Y}(\bar{a}, t)|\bar{a}) = \partial I / \partial \bar{a} = - \int_{-T}^{T} \bar{\psi}\bar{y} + [\int_{-T}^{T} \bar{\psi}\bar{\psi}^T]\bar{a}. \tag{17.81}$$

The function $\bar{Y}(\bar{a}, t)$ which satisfies the above condition is

$$\bar{Y}(\bar{a}, t) = \bar{\beta}(t) + K\bar{a} \tag{17.82}$$

where

$$\bar{\beta}(t) = \begin{cases} \bar{\psi}(t) & t \leq |T| \\ 0 & \text{otherwise} \end{cases}$$

and $K = \int_{-T}^{T} \bar{\psi}\bar{\psi}^T(t)dt$.

Thus, the RMSA algorithm can be written in the form

$$\bar{a}_{k+1} = \bar{a}_k + \frac{A}{k}(\bar{\beta}(t_k) - K\bar{a}_k). \tag{17.83}$$

Under the conditions of Kersten-Kurz this algorithm converges in the mean square sense and with probability one.

17.15   Minimum Variance Algorithm

Consider the Gerchberg-Papoulis algorithm

$$g_{n+1}(t) = (g_n(t) \star k(t))p_o(t) + f(t)p_T(t) \qquad (17.84)$$

$$g_1(t) = f(t)p_T(t)$$

$$p_o(t) = 1 - p_T(t)$$

$$k(t) = sin(\sigma t)/\pi t.$$

However, assume now at each stage of the iteration we do not know $g(t)$ but rather we have

$$y_n(t) = g(t) + v_n(t) \qquad (17.85)$$

where $v_n(t)$ is an additive noise term which is independent and identically distributed for each step n.

At each step, assume a block of samples is taken of all terms in 17.85. At any stage then 17.85 has the form

$$\bar{g}_{n+1} = S_n \bar{\alpha} + \bar{W}_n \qquad (17.86)$$

where $\bar{g}, \bar{\alpha}$ and $\bar{W}$ are m-vectors corresponding to m samples where $\bar{\alpha}$ is a vector of parameters representing the amplitude of the time varying signal S_n, which is represented by a diagonal nxm matrix whose m diagonal elements are block samples of S_n;

$$S_n = diag(S_{11}, S_{22}, ...S_{mm}). \qquad (17.87)$$

Viz., if $\bar{X}_n = \{I_{mxm} \sum_{j=0}^{i} g_n(j)K(i-j)\}\bar{p}_o + \bar{g}$ then $S_{11} = X_{1n}, ...S_{nm} = X_{mn}$. Similarly $\bar{W}_n$ is given by

$$\bar{W}_n = \{[I_{mxn}] \sum_{j=0}^{i} V_n(j)K(i-j)\}p_o + \bar{v}_n. \qquad (17.88)$$

Note that the dimension m grows with each iteration step n.

The stochastic approximation minimum variance least squares approach to estimate $\bar{\alpha}$ in

$$\bar{g}_{n+1} = S_n \bar{\alpha} + \bar{W}_n \qquad (17.89)$$

is given by

$$\bar{\alpha}_{k+1} = \bar{\alpha}_k - \frac{1}{d}A_k(\hat{\alpha}_1, ...\hat{\alpha}_k)(\bar{Y}(\bar{\alpha}_k, \bar{\alpha})) \qquad (17.90)$$

where initially $\bar{\alpha}_k$ is an m-vector. $A_k(.)$ is a diagonal mxm adaptive gain matrix. $\bar{\alpha}_1$ is taken to be the amplitude samples of $g(t)$ for $|t| < T$ and zero for $|t| > T$. Y is obtained by correlating the known signal at step $k = n$ with $\bar{g}_{n+1}$, viz.,

$$\bar{Y}(.) = S_k^T \bar{g}_{k+1} - S_k^T S_k \bar{\alpha}_k. \tag{17.91}$$

The regression function is given by

$$EY = \bar{M}(\bar{\alpha}_k, \bar{\alpha}) = S_k^T S_k (\bar{\alpha} - \bar{\alpha}_k) \tag{17.92}$$

which is linear and has a unique zero root at $\bar{\alpha} = \bar{\alpha}_k$.

In Kadar's thesis he shows that the Kersten-Kurz theorem applies and $k^{1/2}(\bar{\alpha}_k - \bar{\alpha})$ is asymptotically normal with mean zero and specified covariance matrix.

Kadar also proposed to use a batch processing and a nonparametric rank statistic of the form

$$W_k^q(.) = \frac{1}{q^2} \sum_{i=1}^{q} \sum_{j=1}^{q} sign(S_{[i+q(k-1)]}^T g - [S^T S\bar{\alpha}]_{[i+q(k+1)]}) \tag{17.93}$$

where $W_k^q(.)$ is a symmetric version of the Mann-Whitney-Wilcoxon non-parpametric statistic, to robustize the signal reconstruction algorithm. In this case the robust vector SAMVLS algorithm has the form

$$\bar{\alpha}_{k+1} = \bar{\alpha}_k - \frac{1}{k} A_k(.) W^q Y(.). \tag{17.94}$$

For further details we refer the reader to Kadar's thesis.

17.16 Newton-Direction Algorithms

Let U be a nonempty open subset of a Banach space B. We say that function f on U is differentiable on a set S in U if f has f Frechet derivative $f'(s)$ at each point s in S. We say f is a $C'-$ diffeomorphism if f is a homeomorphism of U onto B and f' and $(f^{-1})'$ exist and are continuous on U and B respectively. Consider the problem of solving for f(x) = y. The damped Newton method is as follows: let s(v) = f(v)-y for v in U; choose real numbers, such that: $0 < a < 1/2 < b < 1$; for k = 0,1,... if $f(x_k) = y$ set $x_{k+1} = x_k$; if $f(x_k) \neq y$ determine ϕ_k in B such that $f'(x_k)\phi_k = y - f(x_k)$, choose $\gamma_k > 0$ such that $x_k + \gamma_k \phi_k$ belongs to U and

$$a \le (s^2(x_k) - s^2(x_k + \gamma_k \phi_k))/2\gamma_k s^2(x_k) \le b \tag{17.95}$$

and take $\gamma_k = 1$ when possible. Set $x_{k+1} = y_k + \gamma_k \phi_k$.

If $\gamma_k = 1$ for all k, then this is just Newton's method.

Let $\psi : [0, \infty) \rightarrow [0, \infty)$ be a continuous, strictly increasing function for which $\psi(0) = 0, \psi(a) \rightarrow \infty$ as $a \rightarrow \infty$ and $a^{-1}\psi(a) \ge c$ for a in (0,d) for some positive constants c and d. E.g., $\psi(x) = x$ satisfies these conditions.

THEOREM (Sandberg) 17.9 Let f map a real Hilbert space into itself such that $(f(u) - f(v), u - v) \geq \|u - v\|\psi(\|u - v\|)$ for all u,v in H. Assume f' exists and is uniformly continuous on closed bounded subsets U of H. Then f is a $C'-$ diffeomorphism of H onto H and for each y in B and each x in U, the damped Newton method provides a sequence which converges to the unique solution of f(x) = y.

Consider the case of Landau-Miranker's results where $H = L^2(R)$. Let $B(\Omega)$ denote the closed subspace of H of all elements for which the Fourier transform vanishes outside $\Omega = [-\omega_o, \omega_o]$. Let $Q : B(\Omega) \to L^2$ where $(Qv)(t) = q(v(t))$; so $Q : R \to R$ is continuously differentiable, q(0) = 0 and $c \leq q'(a) \leq \lambda$ for positive constant c,λ.

We first note that $(Q(u) - Q(v), u - v) \geq c\|u - v\|^2$ for all u,v in $B(\Omega)$. And, if P is the projection of L^2 onto $B(\Omega)$, since P is self-adjoint we have

$$(f(u) - f(v), u - v) \geq c\|u - v\| \tag{17.96}$$

where f = PQ for u,v in $B(\Omega)$. Since $\|P\| = 1$ we have

$$\|f(u) - f(v)\| \leq \lambda\|u - v\| \text{ for u,v in } B(\Omega). \tag{17.97}$$

For each v in $B(\Omega)$ let $R_v : B(\Omega) \to L^2$ be defined by $(R_v u)(t) = q'(v(t))u(t)$ for almost all t and any u in $B(\Omega)$. If we set $\delta(h)$ in $B(\Omega)$ to be

$$\delta(h) = (PQ)(v + h) = (PQ)(v) - R_v h \tag{17.98}$$

for v and h in $B(\Omega)$, one can verify that $f'(v)$ exists and $f'(0) = PR_v$. Furthermore, one can check that f' is uniformly continuous on bounded subsets of $B(\Omega)$. Thus, Sandberg's theorem applies and there is a super-linearly convergent algorithm that recovers band-limited signals that are nonlinearly distorted by Q and subsequently band-limited by P.

17.17   Sandberg's Algorithm

Let K be a subspace of a Hilbert space H and let P denote the projection of H onto K

THEOREM (Sandberg) 17.10. Let Q be a mapping of K into H such that for all f,g in H

$$Re(Qf - Qg, f - g) \geq k_1\|f - g\|^2 \tag{17.99}$$

$$\|PQf - PQg\|^2 \leq k_2\|f - g\|^2$$

where k_1, k_2 are positive constants. Then for each h in H the equation h = PQf possesses a unique solution $(PQ)^{-1}h$ in K given by $(PQ)^{-1}h = lim f_n$

where

$$f_{n+1} = \frac{k_1}{k_2}(h - PQf_n) + f_n \qquad (17.100)$$

where f_o is an arbitrary element of K. Furthermore, for all h_1, h_2 in K

$$\|(PQ)^{-1}h_1 - (PQ)^{-1}h_2\| \leq \frac{1}{k_1}\|h_1 - h_2\|. \qquad (17.101)$$

Proof. Set A = PQ and note that $RE(Af - Ag, f - g) = Re(Qf - Qg, Pf - Pg) = Re(Qf - Qg, f - g) \geq k_1\|f - g\|^2$ for all f,g in K since P is self-adjoint.

The equation h = Af is equivalent to $f = \tilde{A}f$ where

$$\tilde{A}f = ch + f - cAf \qquad (17.102)$$

for any nonzero c. Since $\|\tilde{A}f - \tilde{A}g\|^2 = \|f - g - cAf + cAg\|^2 = \|f - g\|^2 - 2cRe(Af - Ag, f - g) + c^2\|Af - Ag\|^2 \leq (1 - 2ck_1 + c^2k_2)\|f - g\|^2$. Thus, $\|\tilde{A}f - \tilde{A}g\| \leq (1 - k_1^2/k_2^2)\|f - g\|^2$ for $0 \leq (1 - k_1^2/k_2^2) < 1$, i.e., $\tilde{A}$ is a contraction operator for $c = k_1/k_2$. For all f,g in K we have

$$\|Af - Ag\|\|f - g\| \geq |(Af - Ag, f - g)| \geq k_1\|f - g\|^2. \qquad (17.103)$$

So we have $\|Af - Ag\| \geq k_1\|f - g\|$. In particular for $f = A^{-1}h_1$ and $y = A^{-1}h_2$ we have

$$\|h_1 - h_2\| \geq k_1\|A^{-1}h_1 - A^{-1}h_2\|. \qquad (17.104)$$

THEOREM (Beurling) 17.11. For f,g let Q be a mapping of K into H such that (Qf-Qg,f-g) vanishes only if f=g. Then if h = PQz has a solution z in K, it is unique.

Proof. Assume $PQz_1 = PQz_2$. Since P is self-adjoint we have

$$(Qz_1 - Qz_2, z_1 - z_2) = (Qz_1 - Qz_2, Pz_1 - Pz_2) = (PQz_1 - PQz_2, z_1 - z_2) = 0. \qquad (17.105)$$

Hence, $z_1 = z_2$.

17.18   FM Demodulation

Consider a signal x(t) which frequency modulates a carrier to give

$$s(t) = A\sin(w_c t + 2\pi \int_{-\infty}^{t} x(a)da + \theta_o)$$

where $w_c = 2\pi f_c$. The total phase is $\phi(t) = s_c t + 2\pi \int_{-\infty}^{t} x(a)da + \theta_o$ and the instantaneous frequency is $f(t) = \frac{1}{2\pi}d\phi(t)/dt = f_c + x(t)$. We assume x(t) is square integrable and band-limited with bandwidth F_B less than the carrier frequency f_c. We assume $f_c > 2F_B$, and we assume the peak deviation is less than the carrier frequency f_c, i.e., $|x(t)| < f_c$.

The zero crossings of s(t) occur at times t_n where $\phi(t) = n\pi$. Setting the phase deviation $g(t) = \int_{-\infty}^{t} x(a)da$ we see that if the carrier frequency is known, measurement of the zero crossing t_n provides sample values $g(t_n) = n\pi - w_c t_n$ of the phase deviation.

Let $\{t_n\}$ denote the zero crossing times of the modulated signal s(t). Set $Qx = \sum_{-\infty}^{\infty} \frac{u(t-t_n)-u(t-t_{n+1})}{t_{n+1}-t_n} \int_{t_n}^{t_{n+1}} x(a)da$. Here u(t) is the unit step function. One can show that

$$\int_{t_{n+1}}^{t_n} x(a)da = 1/2 - f_c(t_{n+1} - t_n). \qquad (17.106)$$

So Qx is obtained by subtracting the carrier frequency from the reciprocal of twice the derivative of each zero crossing interval.

Wiley, Schwarzlander and Weiner have shown that Sandberg's theorem applies to this case. So we have an iterative formula for FM demodulation.

EXERCISE (Wiley et al.) 17.12. Show that the spacing between zero crossings satisfies $t_{n+1} - t_n > \pi/(2w_c)$, i.e., the spacing between zeros is strictly greater than some minimum.

EXERCISE (Wiley et al.) 17.13. Show that

$$|t_n - n\pi/w_c| < g_{max}/w \qquad (17.107)$$

where $|g(t)| < g_{max}$.

EXERCISE (Duffin-Schaeffer, Boas) 17.14. A sequence of numbers t_n is said to have uniform density d if there are constants L, δ such that $|t_n - n/d| \le L, n = 0, \pm1, \pm2, \ldots$ and $|t_n - t_m| \ge \delta > 0, m \ne n$. So the zero crossings above have uniform density $2f_c$. Show that in this case there are constants A,B such that

$$A \le \sum_n |x(t_n)|^2 / \int_{-\infty}^{\infty} |x(t)^2 dt \le B. \qquad (17.108)$$

17.19 Toeplitz Equations

Define α by

$$\alpha = \frac{1}{E} \int_{-T/2}^{T/2} v(t)|y(t)|^2 dt \qquad (17.109)$$

where v(t) is a real function and y(t) is the trigonometric polynomial

$$y(t) = \sum_{-M}^{M} y(n)e^{inw_o t} \tag{17.110}$$

where $w_o - 2\pi/T$. We leave it to the reader to show that the following algorithm converges to the maximum of α:

(1) $y_N(t) = \sum_{-M}^{M} y_N(n)e^{inw_o t}$

(2) multiply by $v(t)$ to get

$$w_N(t) = y_N(t)v(t) = \sum w(n)e^{inw_o t}$$

where $w_N(t) = \sum_m^M v(n-k)y_N(k)$.

(3) set $y_{N+1}(n) = w_N(n)/\sqrt{E_N}, |n| < M$, where $E_N = \sum_{-M}^{M} |w_N(n)|^2$.

(4) return to (1).

One can check that $y_\infty(t)$ maximizes 17.109 with $\alpha_{max} = \sqrt{E_\infty}$.

EXERCISE (Papoulis) 17.15. Show that α_{max} is the maximum eigenvalue λ of the Toeplitz equations

$$\sum_{k=-M}^{M} v(n-k)y(k) = \lambda y(n) \tag{17.111}$$

for $|n| \leq M$ where v(n) is the nth Fourier coefficient of v(t) for $|t| \leq T/2$.

17.20 Moses-Prosser Algorithm

Consider passing the Fourier transform through an aperture $a < x < \infty$, so

$$F_a(u) = \frac{1}{\sqrt{2\pi}} \int_a^\infty e^{-iux} f(x)dx. \tag{17.112}$$

Set $I_a(u) = |F_a(u)|^2$. Similarly set

$$G_a(u) = \frac{1}{\sqrt{2\pi}} \int_{-\infty}^a e^{-iux} f(x)dx \tag{17.113}$$

and set $J_a(u) = |G_a(u)|^2$. Let $f(x) = |f(x)|e^{i\phi(x)}$.

THEOREM (Moses-Prosser) 17.16. We have

$$e^{i(\phi(a)-\phi(b))} =$$

$$[|f(a)||f(b)|]^{-1} \int_{-\infty}^\infty H(a-b)\partial J_a(u)/\partial a - H(b-a)I_a(u))e^{i(a-b)u}du$$

where H is the Heaviside function

$$H(x) = \begin{cases} 1 & x > 0 \\ 0 & x < 0. \end{cases} \qquad (17.114)$$

Thus, if this setup is available and $F(a) \neq 0$, then one can find $e^{i\phi(b)}$ for all b. This result can be restated as

THEOREM 17.17.

$$f^*(x+s)f(x) = \int_{-\infty}^{\infty} (H(-s)\partial J_x(u)/\partial x - H(s)\partial I_x(u)/\partial x)e^{isu}du. \quad (17.115)$$

COROLLARY (Wiener-Khinchine) 17.18.

$$\int_{-\infty}^{\infty} f^*(x+s)f(x)dx = \int_{\infty}^{\infty} e^{isu}J_\infty(u)du. \qquad (17.116)$$

If we use the Mellin transform

$$F_a(x) = \frac{1}{\sqrt{2\pi}} \int_0^a x^{-is-\frac{1}{2}}f(x)dx \qquad (17.117)$$

with $K_a(x) = |F_a(x)|^2$ and

$$G_a(s) = \frac{1}{\sqrt{2\pi}} \int_a^\infty x^{-is-\frac{1}{2}}f(x)dx \qquad (17.118)$$

and $L_a(s) = |G_a(s)|^2$ we leave it to the reader to derive the analogue of theorem 17.17. In particular, the corollary becomes

$$\int_0^\infty f^*(e^{\lambda x})f(x)dx = e^{-\lambda/2} \int_{-\infty}^\infty e^{-i\lambda s}K_\infty(s)ds. \qquad (17.119)$$

17.21   The Baghlay Algorithm

Consider the all-pass filter synthesis problem:

$$I_\Omega(\phi) = \|f - \mathcal{F}|W e^{i\phi}\|_{L^2(S)} \qquad (17.120)$$

where f(x,y) belongs to $L^2(S)$ and W(u,v) belongs to $L^2(\Omega)$ and they are finite positive functions. The problem is to find the minimum $min_\phi I_\Omega(\phi)$. Consider the 1-D problem. Assume $\phi(u)$ is a convex function and $\phi''(u_o) > 0$; so we have

$$\frac{1}{2\pi} \int_{-u}^{u} W(u)e^{i(\phi(u)-ux)}dx \sim W(u_o)(2\pi\phi''(u_o))^{-1/2}e^{i[\phi(u_o)-u_ox+\pi/4]}$$

$$(17.121)$$

where $\phi'(u_o) = x$. We leave it to the reader to show that the minimization of $I(\phi)$ is equivalent to maximization of $(f, |\widehat{We^{i\phi}}|)$. Under the above conditions we have

$$(f, |\widehat{We^{i\phi}}|) \sim \int_{-X}^{X} f(x) W(u_o)(2\pi\phi''(u_o))^{-1/2}. \qquad (17.122)$$

Thus, at the maximum we have

$$f^2(\phi')d\phi' = (2\pi)^{-1} W^2(u) du \qquad (17.123)$$

and $\phi(u) = x$, which gives the required phase ϕ.

One can extend this result to 2-D to show that ϕ must satisfy the non-linear elliptical equation of the Monge-Ampere type

$$f(u,v)(\phi''_{u,u}\phi''_{v,v} - (\phi''_{u,v})^2) = (2\pi)^{-2} W^2(u,v). \qquad (17.124)$$

17.22 Gonsalves' Algorithm

Consider the case that $|f|$ and $|F|$ are known. We assume $F(u) = e^{i\theta(u)}$ where $\theta(u)$ is the phase aberration over the aperture. Let $\phi(x) = |f(x)|^2$ and let $\Phi(u)$ be the Fourier transform of ϕ. So $\Phi(u)$ is the autocorrelation of the pupil function

$$\Phi(u) = \int_{-\infty}^{\infty} ds F^*(x) F(s+u). \qquad (17.125)$$

Consider a trial function $F_p(u) = |F(u)|e^{i\theta_p(u)}$ where $\theta_p(u) = p_1 u + p_2 u^2 + p_3 u^3 + p_4 u^4$. One forms the function $|f_p(x)|^2$ and hence, $\Phi_p(u)$. The approach is to apply Fletcher-Powell algorithm to adjust $p = (p_1, p_2, p_3, p_4)$ to minimize $M = \int_{-\infty}^{\infty} du |\Phi(u) - \Phi_p(u)|$. For details we refer the reader to Gonsalves' paper.

17.23 Neural Nets and POCS

Consider a set of L neurons or nodes each of which is assigned a (binary) values or state s_k, k=1,...L. The jth neuron is connected to the kth neuron with an interconnect value or transmittance t(j,k). One nominally assumes symmetric interconnects, i.e., t(j,k) = t(k,j). If the sum of the inputs to a neuron exceeds a specified threshold, the neuron fires (i.e., sets its binary value). If the sum is less than the threshold, the neuron turns off (i.e., resets its binary value). This process continues until a stable state is reached.

The synchronous Hopfield model is given as follows. Let f_n, n=1,...N, denote a set of library elements, each of length L. The ith element of f_n, $f_n(i)$ is either 1 or -1. We form the LxL matrix T with elements

$$T(i,j) = \sum_{n=1}^{N} f_n(i)f_n(j) \qquad (17.126)$$

where i,j = 1,...L. If we set $F = [f_1, f_2, ...f_N]$ to be the LxN library matrix, then T is just the outer product $T = FF^T$. The neural net processor is described by the iteration

$$s_{M+1} = C(Ts_m) \qquad (17.127)$$

where $s_o = s$ is an initialization or input set s composed of ±1's. Hence, s_M is a vector of the L neural states at time m, $i_m = Ts_m$ and C is the node operator which determines the next set of states from i_m. The standard Hopfield neural net processor assumes no autointerconnections, i.e., $t(i,i) = 0$. This can be described by equation 17.127 with T replaced by $T_H = -NI_N + FF^T$ where I_N is the NxN identity matrix and C is taken to be sgn. The object is to find a fixed point of 17.127.

A model proposed by Personnaz et al. and elaborated by R. J. Marks and coworkers is the projection neural net where T is selected to be a projection matrix. Let $S(L)$ denote the linear subspace spanned by the library elements. Then the projection onto to the subspace $S(L)$ is given by $T_p = F(F^TF)^{-1}F^T$. The projection neural net is described by

$$s_{M+1} = sgn(T_p s_M). \qquad (17.128)$$

Marks and coworkers have extended this concept to the memory extrapolation neural network. Here one assumes that the states of neurons 1 to $p < L$ are known for some vector f in the library. The node operator is then

$$Ci = C(i(1),...i(p)|i(p+1),...i(L))^T = (f(1),...f(p)|i(p+1),...i(L))^T \qquad (17.129)$$

where $f(k)$ are the elements of f. The first p neurons are referred to as clamped neurons whereas the remainder are floating neurons. Using the partitioning notation we write $i = (i_p|i_q)$ where i_p is the p-tuple of the first elements of i and i_q is a vector of the remaining $q = L - p$ elements. So $Ci = (f_p|i_q)^T$.

The generalized model is that there is the L dimensional Hilbert space. The T matrix orthogonally projects any vector onto the N dimensional subspace $S(L)$ formed by the closure of the library vectors. The clamping operator C orthogonally projects onto the $q = L - p$ dimensional linear variety formed by the set of all tuples whose first p elements are equal to f_p. By the von Neumann alternating projection theorem, 17.127 will converge to a point common to both linear varieties. Clearly the library vector f is common to both spaces. And if F_p is full rank (i.e., the first p rows of F form a full rank matrix), then f is the only point of intersection.

In summary, in contrast to the Hopfield neural net the extrapolation neural net allows for library vectors with continuous elements. And in contrast to Hopfield, initially known neural states are imposed on the net each iteration; i.e., the known states act as the net stimulus and the remaining nodes catalog the response. And given a portion of one of the library vectors, this net extrapolates the remainder. Sufficient conditions for convergence are stated above.

Marks and coworkers have suggested also the use of relaxation techniques and thresholding techniques to improve the convergence rates of these iterative algorithms. We refer the reader to the references for details.

17.24   Synthetic Aperture Radar

The basic principle underlying Computer-Aided Tomography is the projection slice theorem. Viz., if $G(X,Y) = \mathcal{F}g(x,y)$ and if the projection of g at the angle θ is

$$p_\theta(u) = \int_{-\infty}^{\infty} g(ux\cos\theta - v\sin\theta, u\sin\theta + v\cos\theta)dv \qquad (17.130)$$

with Fourier transform

$$P_\theta(U) = \int_{\infty}^{\infty} p_\theta(u)e^{-iuU}\,du, \qquad (17.131)$$

then the projection slice theorem states that

$$P_\theta(U) = G(U\cos\theta, U\sin\theta). \qquad (17.132)$$

In other words, the Fourier transform of the projection at the angle θ is a slice of the 2-D transfrom G(X,Y) taken at angle θ with respect to the x axis.

Let $g(x,y)$ denote the reflectivity of a patch of ground illuminated by a synthetic aperture radar or SAR. At the angle θ to a ground patch the radar signal transmitted is a linear FM chirp pulse

$$s(t) = e^{i(\omega_o t + \alpha t^2)} \quad |t| \leq T/2$$

where ω_o is th RF frequency and 2α is the FM rate. The return from the patch centered at (x_o, y_o) at a distance R from the radar is

$$r_o(t) = A|g(x_o,y_o)|\cos(\omega_o(t - 2R/c) + \alpha(t - 2R/c)^2 + Arg(g))dxdy$$

$$= ARe(g(x_o, y_o)s(t - 2R/c))dxdy$$

where c is the speed of light. Points equidistant from the radar can be taken to be a line of scatterers which contribute

$$r_\theta(t) = ARe(\int_{-L}^{L} p_\theta(u)s(t - 2(R + u)/c))du).$$

The returned signal is thus given by the function

$$c_\theta(t) = \frac{A}{2} \int_{-L}^{L} p_\theta(u)e^{i4u^2/c^2} e^{-i2c(\omega_o t + 2\alpha(t - \tau_o))u/c} du$$

where $\tau_o = 2R/c$ is the round-trip delay. Assuming the quadratic phase term can be removed, the spot-light mode in SAR is given by the function

$$\bar{c}_\theta(t) = \frac{A}{2} \int_{-L}^{L} p_\theta(u)e^{-\frac{i2}{c}(\omega_o + 2\alpha(t - \tau_o))u} du = \frac{A}{2} P_\theta(\frac{2}{c}(\omega_o + 2\alpha(t - \tau_o))).$$

So by the projection-slice theorem $\bar{c}_\theta(t)$ is a slice at angle θ of the 2-D transform G of the unknown reflectivity g.

However, $\bar{c}_\theta(t)$ is available only for

$$-T/2 + 2(R + L)/c \le t \le T/2 + 2(R - L)/c.$$

Processing returns from angles $|\theta| \le \Phi$ provides samples of G(X,Y) on the polar grid in an annulus AA with inner and outer radii X_i, X_o where $X_i = \frac{2}{c}(\omega_o - \alpha T + 4\alpha L/c)$ and $X_o = \frac{2}{c}(\omega_o + \alpha T - 4\alpha L/c)$.

The approach to reconstruct the complex reflectivity function g is then:

(1) interpolate the measured Fourier data over a Cartesian grid with AA

(2) restore the missing Fourier data by means of POCS and fill in an NxN array of Fourier samples

(3) compute the inverse FFT to obtain the desired function g.

Possible projections to be used in step (2) are discussed in Sezan-Stark (1987).

17.25   NMR Tomography

The measured signal $s_\phi(t)$ in nuclear magnetic resonance (NMR) tomography is related to the magnetic distribution by

$$s_{\phi_k}(t) = C \int \int M(x, y)e^{-i\gamma|G|(x\cos\phi_k + y\sin\phi_k)t} dxdy. \qquad (17.133)$$

Here C is a constant and γ is the gyromagnetic ratio. As in the last section $s_{\phi_k}(n\Delta t)$ for n = 1,...N,k=1,...K forms samples of the Fourier transform of the nuclear magnetization M(x,y) sampled at $(\rho_n \cos\phi_k, \rho_n \sin\phi_k), \rho_n = \gamma|G|n\Delta t$; and if full-view spectrum is not available, POCS can be used to return the missing spectrum.

17.26 Multiple Threshold Crossings

In her thesis Zakhor considered the representation of signals with threshold crossings as a trade-off between bandwidth and dynamic range where representation of signals via their samples at Nyquist sample interval can be viewed as requiring minimal bandwidth and infinite dynamic range. Curtis and Oppenheim showed that for a band-limited periodic signal with a (2N+1)x(2N+1) region of support in the Fourier domain, $16N^2 + 1$ or more samples of the zero crossings are sufficient to reconstruct the signal to within a scale factor. However, the locations of the zero crossings need to be specified to 16 digits. Thus, we can view the representation of signals with threshold crossings as a trade – if available bandwidth is sufficient to preserve the threshold crossings accurately, then the dynamic range requirements are reduced.

Zakhor's basic results follow from bivariate polynomial interpolation theory.

THEOREM 17.18. Let $l_o, ... l_n$ be distinct lines with line l_i defined by $z = \alpha_i w, \alpha_i \neq 0$, and consider arbitrary distinct points on l_i given by $\{(w_j^{(i)}, z_j^{(i)}), j = 0, ...2i\}$. If none of the interpolation points is (0,0) then for any data set $\{t_j^{(i)} | 0 \leq j \leq 2i, 0 \leq i \leq n\}$ there is a unique bivariate polynomial of the form

$$p(w, z) = \sum_{i=0}^{n} \sum_{j=0}^{n} a(i, j) w^i z^j \qquad (17.134)$$

such that $p(w_j^{(i)}, z_j^{(i)}) = t_j^{(i)}, 0 \leq j \leq 2i, 0 \leq i \leq n$.

COROLLARY 17.19. Consider a band-limited, continuous periodic signal

$$f(x, y) = \sum_{n_1=-N}^{N} \sum_{n_2=-N}^{N} F(n_1, n_2) e^{i2\pi x n_1 + i2\psi y n_2}.$$

Let $l_o, ... l_{2N}$ be distinct lines in the x-y plane with l_i given by $y = x + \beta_i$. Let $\{x_j^{(i)}, y_j^{(i)}, 0 \leq j \leq 2i\}$ be arbitrary distinct samples on l_i. Then for any data set $\{t_j^{(i)} | 0 \leq j \leq 2i, 0 \leq i \leq 2N\}$ we can reconstruct f(x,y) uniquely.

The result in theorem 17.18 can be extended to sampling signals with lines $c_o, ... c_p$ of positive integral slope $z = \alpha_i w^m, \alpha_i \neq 0$ where $m \leq n$ and p is such that

$$\sum_{i=0}^{p} [(m + 1)n - 2mi + 1] \geq (n + 1)^2.$$

The data set is now $\{t_j^{(i)} | j = 0, ...(m + 1)n - 2mi, i = 0, ...p\}$.

EXERCISE 17.20. Show that the one-projection theorem of Mersereau-Oppenheim for a band-limited 2-D signal of order M follows from these remarks– viz., that M samples on one slice of the 2-D transform is sufficient for unique reconstruction of the signal provided the angle of the projection is the critical angle $\theta = tan^{-1}(M)$.

The iterative approach to reconstruction of 1-D signals associated with sampling lines from their intersections with level crossings is as follows: let $g(z)$ be a continuous band-limited periodic signal with period one and $2N+1$ Fourier harmonics from all of its level crossings with thresholds $t_1, ... t_p$:

(1) deduce the range of intensity for $M > 2N + 1$ equally spaced points so $t_l(n) \leq g(n/M) \leq t_u(n), n = 0, ... M - 1$

(2) let $g^{(l)}(n/M)$ denote the value of g in the lth iteration at point n/M with

$$g^{(0)}(n/M) = (t_l(n) - t_u(n))/2, n = 0, ... M - 1$$

(3) $G^{(l)}(k) = DFT(g^{(l)}(n/M)), 0 \leq n, k < M$

(4) $\tilde{g}^{(l+1)}(k) = G^{(l)}(k), 0 \leq k \leq N, M - N \leq k < M$

(5) $\tilde{g}^{(l+1)}(n/M) = DFT^{-1}(\tilde{G}^{(l+1)}(k)), 0 \leq k, n < M$

(6) $g^{(l+1)}(n/M) = \begin{cases} \tilde{g}^{(l+1)}(n/M) & t_l \leq \tilde{g}^{(l+1)}(n/M) \leq t_u(n) \\ t_l(n) & \text{for } \tilde{g} < t_l(n) \\ t_u(n) & \text{for } \tilde{g} > t_u(n) \end{cases}$

(7) go to (3).

If we let H be the space of M point real sequences with

$$(x, y) = \sum_{n=0}^{M} x(n)y(n),$$

then C_1 is the closed convex set of all M point real band-limited sequences whose DFT is 0 in the range $n < k < M - N$. C_2 is the convex set of all M point real sequences y(n) which satisfy $t_l(n) \leq y(n) \leq t_u(n), 0 \leq n < M$. If P_1 and P_2 are the projections on C_1 and C_2, then for $T = P_1 P_2$ the iterative sequence $\{T^n x\}$ converges strongly to a point in $C_O = C_1 \cap C_2$ since C_1 is finite dimensional.

We leave it to the reader to extend this approach to 2-D signals. Further details can be found in Zakhor (1988).

17.27   Block-Iterative Methods

The relationships between Bregman's algorithm, Cimmino's algorithm, the generalized iterative scaling method of Darroch and Ratcliff, and the MART or multiplicative algebraic reconstruction technique have been studied by

Censor and co-workers. Consider the entropy functional

$$ent(x) = -\sum_{j+1}^{n} x_j ln(x_j)$$

for x on R where $0ln0 = 0$. The problem is to find x in R_+^n which maximizes ent(x) subject to $x \in Q$ where $Q = \{x \in R^n | Ax = b\}$ or $Q = \{x \in R^n | Ax \le b\}$.

We recall that in the approach of Herman, Censor et al. there is a row-action operator $R_i : R^n \rightarrow R^n$ for which $x^{k+1} = R_i x^k$ at each step. For example, the Kaczmarz algorithm takes the form

$$R_i(x) = x + (\lambda b_i + < a^i, x > a^i / \|a^i\|^2) \qquad (17.135)$$

which is the relaxed orthogonal projection of x onto the ith hyperplane

$$H_i = \{x \in R^n | < a^i, x >= b_i\}$$

of the system of equalities

$$< a^i, x >= b_i, i = 1, ...m.$$

For inequalities the row-action operator is

$$R_i(x) = x + c(x)a^i$$

where $c(x) = min(0, \lambda b_j - < a^j, x > / \|a^j\|^2)$, which is the relaxed orthogonal projection of x onto the jth half-space

$$Q_j = \{x \in R^I | < a_j, x >\le b_j\}$$

which gives the method of Agmon-Motzkin-Schoenberg (AMS) for solving a system of linear inequalities.

The Cimmino method for linear inequalities uses the iterative step

$$x^{k+1} = x^k + \lambda_k \sum_{j=1}^{J} w_j c_j(x_k)a^j$$

where $0 < w_j < 1$ are weights, j=1,...J with $\sum_{j=1}^{J} w_j = 1$. So in contrast to the AMS method, the method of Cimmino is a simultaneous method which uses the information in all inequalities of the system at each step, whereas the row-action method uses only one inequality at a time. The order in which the inequalities are picked out is determined by a control sequence $\{j(k), k = 0, ...\infty\}$ where $1 \le j(k) \le J$. A control sequence $\{i(k)\}$ is called almost cyclic if for some fixed integer r, $\{1, ...m\} \subset \{i(k), ...i(k+r)\}$ for all k.

In more detail suppose we have a system of equations $Az = Y$ or

$$< a_i, z >= y_i, 1 \le i \le LM.$$

We think of the LM equations as being subdivided into M blocks each of size L. Set $m = n(mod\ M) + 1$. Then the generalized version of ART is

$$z^{(n+1)} = z^{(n)} + \lambda \sum_{l=1}^{L} (y_i - < a_i, z^{(n)} > a_i)/\|a_i\|^2$$

with $i = (m_n - 1)L + 1$. If the block size L is 1, then this reduces to the method of Kaczmarz. If the number of blocks M is 1, then this is the method of Cimmino, which is very similar to the SIRT method of Gilbert. Let $A_m = (a_{(m-1)L+1}^T, ...a_{mL}^T)^T$ and $Y_m = (Y_{(m-1)L+1}, ...Y_{mL})^T$ and let

$$z^{(n+1)} = z^{(n)} + A_{m_n}^T \sum (Y_{m_n} - A_{m_n} z^{(n)}).$$

If $\Sigma^{(m)}$ satisfies

$$\|A^\dagger (I_l - A_m A_m^T \Sigma^{(m)} A_m)\| < 1$$

we have

THEOREM (Eggermont-Herman-Lent) 17.21. Under these conditions if $Az = Y$ is consistent, then for any initial choice the sequence $\{z^{(n)}\}$ converges to a solution of $Az = Y$. If $z^{(0)}$ is in the range of A^T, then

$$lim_{n \to \infty} z^{(n)} = A^\dagger Y.$$

Bregman's method starts with an objective function such as

$$f_1(x) = \frac{1}{2}\|x\|^2$$

or

$$f_2(x) = -ent(x).$$

Then the Bregman projection parameter $B_i(x)$ is determined by

$$\Delta f(y) = \Delta f(x) + B_i(x)a^i$$

$$< y, a^i >= b_i$$

and the dual equations

$$\Delta f(R_i(x)) = \Delta f(x) + c(x)a^i$$

$$D_i(z) = a - c(x)e^i$$

where $c(x) = min(z_i, B_i(x))$. Taking $f = f_1$ gives the Kaczmarz method or ART. Taking $f = f_2$ gives the row-action operator

$$(R_i(x))_j = x_j e^{B_i a_j^i}, j = 1, ...n$$

$$< R_i(x), a_i >= b_i.$$

MART uses the row-action method

$$(R_i(x))_j = x_j (b_i / < a^i, x >)^{a_j^i}, j = 1, ...n.$$

When A is a zero-one matrix clearly MART coincides with Bregman's method.

For the discussion below let $b > 0$, and $1 \geq a_j \geq 0$ for $1 \leq j \leq n$. Let a^i denote the transpose of the ith row of A. Consider the iteration scheme

$$x_j^{k+1} = x_j^k e^{c_k a_j^{i(k)}}, j = 1, .., n$$

with the constraint

$$< a^{i(k)}, x^{k+1} >= b_{i(k)}.$$

For any given x^k such that $x_j^k > 0$, for $j = 1, ...n$ there is an unique choice of x^{k+1}, c_k for which these equalities hold. In MART c_k is given by

$$c_k = sgn(b_{i(k)}) log b_{i(k)} / < a^{i(k)}, x^k > .$$

One can show that a MART iterative step is a secant approximation to a Bregman iterative step and MART converges to the maximum entropy solution. Viz., the solution to the linearly constrained optimization problem

$$min f(x) \text{ such that } Ax = b \text{ and } x \geq 0$$

where A is a real mxn matrix, $b \in R^m, x_j \geq 0, j = 1, ...n, x = (x_j) \in R^n$. The Lagrangian of this problem is

$$L(x, u) = f(x) + < u, Ax - b >$$

$u \in R^m$ is the dual vector of Lagrange multipliers. Minimization implies $\Delta_x L(x, u) = 0$ which gives

$$\Delta f(x) = -A^T u.$$

Let $H_i = \{x \in R^n | < a_i, x >= b_i, i = 1, ...m\}$ and $H \cap_{i=1}^n H_i$. We assume now:

(a1) $H \cap R_+^n \neq 0$ (feasibility)
(a2) $a_j^i \leq 1$ (normalization)

(a3) $0 \geq a^i_j$ for $j = 1,...n$ and $a^i \neq 0$ for all $i = 1,...n$ and $b_i > 0$ for $i = 1,...m$.

The algorithm is:

(1) take u^o in R^m and x^o in R^n so that $1 + \ln x^o_j = -(A^T u^o)_j$, $j = 1,...n$

(2) $x^{k+1}_j = \frac{1}{e} e^{-(A^T u^{k+1})_j}$, $j = 1,,...n$

$$a^{k+1}_i = u^k_i \text{ for i not in } I_{t(k)}$$

$$a^{k+1}_i = u^k_i - w^{t(k)}_i d^k_i \text{ for i in } I_{t(k)}$$

$$d^k_i = \ln b_i / < a^i, x^k >$$

$$0 < w^t_i < 1, t = 1,...M$$

$$\sum_{i \in I_t} w^t_i = 1.$$

Here the equations $Ax = b$ are assigned to blocks by partitioning $I = \{1,...m\}$ into M blocks

$$0 < m_o < m_1 < ...m_M = m$$

$$I_t = \{m_{t-1} + 1, m_{t-1} + 2, ...m_t\}.$$

The control sequence $\{t(k)\}$ is assumed almost cyclic.

THEOREM (Censor-Segman) 17.22. Under the assumptions (a1)-(a3) the sequence $\{x^k\}$ generated by the algorithm converges to the unique solution for $f(x) = -ent(x)$.

In case $M = m$ where every equation of $Ax = b$ constitutes an individual block the algorithm is precisely MART. For the relationship to Darroch-Ratcliff we refer the reader to Censor-Segman (1987).

For a proof that -ent(x) is a Bregman function, the construction of MART for inequalities and the proof of convergence of MART under inequality constraints, the reader is directed to Censor et al. (1987).

The similarity of the MART algorithm and the expectation maximization (EM) method of Dempster, Laird and Rubin is noted in Censor-Herman (1987). The EM method has been used by Lagendijk et al. (1988) for solving the problem of simultaneous image identification and restoration.

Appendix A

The Geometry of Projections on Convex Sets

Given a category $\mathcal{A}$ of objects $\{A, B, ..\}$ and maps $\alpha : A \rightarrow B$ then α is called an $\mathcal{A}$-isomorphism if and only if there exists a morphism $\beta : B \rightarrow A$ in $\mathcal{A}$ such that $\alpha\beta = id_B$ and $\beta\alpha = id_A$. A morphism α in $\mathcal{A}$ is called a retraction if and only if there exists a morphism β in $\mathcal{A}$ such that $\alpha\beta = id_B$. As an example let H be a normed vector space, let F be a subspace of H and let $\epsilon : F \rightarrow H$ be the embedding map. Let P be the projection of H onto F, $P : H \rightarrow F$. Then $P\epsilon = id_F$.

Let H be now a real Hilbert space and let C be a closed convex set in H. For each x in H there is a unique closest point in C called its projection which we denote by $P_C x$. Thus,

$$\|x - P_C x\| \leq \|x - y\| \tag{A.1}$$

for all y in C is the characteristic property of $P_C x$. As we have noted in the text (A.1) is equivalent to the property

$$< x - P_C x, P_c x - y >\, \geq 0 \tag{A.2}$$

for all y in C. And any projection on a closed convex set (POCS) can be shown to satisfy the inequality

$$< (I - P)x - (I - P)x', Px - Px' >\, \geq 0 \tag{A.3}$$

for all x,x' in H. Thus, any POCS P satisfies

$$\|Px - Px'\|^2 \leq\, < x - x', Px - Px' > \tag{A.4}$$

257

or by Schwarz inequality

$$\|Px - Px'\| \le \|x - x'\| \tag{A.5}$$

i.e., the POCS P is nonexpansive. Similarly, one checks that I-P is nonexpansive. It also follows that both P and I-P are monotone operators. We let the reader check that:

THEOREM A1. P is a POCS if and only if 2P-I is nonexpansive.

COROLLARY A2. The difference of any two POCSs is nonexpansive.

In summary we have

THEOREM A3. Any mapping P from H into H which satisfies:
(a) $R(P) \subset D(P)$
(b) $P^2 = P$
(c) $P^{-1} - I$ is monotone
is the restriction of a POCS.

If P satisfies A3, then one checks that the set of fixed points of P is the closed convex set

$$C = \{y| < x - Px, Px - y > \ge 0 \text{ for all x in C}\}. \tag{A.6}$$

For if y is in C, then the inequality defining C is satisfied by x = y - i.e., $< y - Py, Py - y > \ge 0$ or Py = y. The reader can check the converse and that C is a closed convex set.

THEOREM A4. The inverse image of a point by a POCS is either empty or a closed convex cone with vertex at the point.
   Proof. The only points for which $P_C^{-1}y \ne 0$ are the points y in C. Let y be such a point. Then $P_C^{-1}y$ is the set of x such that $P_C x = y$, i.e., the x's such that $(I - P_C)x + y = x$. But this is the set of fixed points of the nonexpansive map $(I - P_C)x + y$ which is closed and convex. Since P_C is idempotent, y belongs to $P_C^{-1}y$. To conclude we must show that all points $x(t) = P_C x + t(x - P_C x)$ for $t \ge 0$ project onto $P_C x$. But

$$< x(t) - P_C x, P_c x - y > = t < x - P_C x, P_C x - y > \ge 0 \tag{A.7}$$

for all y in H and $t \ge 0$. So $P_C x(t) = P_C x$ by A2.

Thus, a POCS is a retraction r of C onto the fixed-point set which for each x in C maps each point of the ray $\{r(x) + t(x - r(x))|t \ge 0\} \cap C$ onto the same point r(x). In general, if C is a nonempty closed convex subset of a Banach space E, and if F is a nonempty closed subset of C, then F is said to be a nonexpansive retract of C if there exists a retraction of C onto F which is a nonexpansive mapping.

From the above discussion we see that if E is a Hilbert space and $T : C \to C$ is a nonexpansive mapping with fixed point set F(T), then F(T) is a nonexpansive retract of C. More generally Bruck has shown that if E is a reflexive, strictly convex Banach space, and $T : C \to C$ is nonexpansive, then F(T) is a nonexpansive retract of C.

The notion of a projection on a convex set extends to any strictly convex, reflexive space; however, the projection operator is not, in general, nonexpansive. For further details see Bruck (1973). Bruck cites the following interesting example of a nonexpansive projection of C onto F even if F is not convex; viz., let $E = R^2, \|(x, y)\| = max(|x|, |y|)$ and define $f : E \to E$ by $f(x, y) = (x, |x|)$. Then f is a nonexpansive projection of E onto $F = \{(x, y) | y = |x|\}$ but F is not convex.

Returning to maps satisfying A3 and leaving all points of C fixed, we leave it to the reader to verify that P_C is the only one having its range contained in C. So if P satifies A3 and leaves C fixed pointwise, then

$$< x - Px, Px - P_C x >\geq 0 \tag{A.8}$$

for all x in H.

DEFINITION A5. For any cone C with vertex at v, the cone

$$C^\perp = \{x^\perp | < x^\perp - v, x - v >\leq 0 \text{ for all x in C}\} \tag{A.9}$$

is called the dual or polar cone of C. $C^\perp$ is a closed convex cone with vertex at v and the only point in common between C and $C^\perp$ is their vertex v. And if C is a closed cone, then $C^\perp$ is the set of points whose projection on C is v.

Let coC denote the convex hull of C.

THEOREM A6. For any cone C $C^{\perp\perp} = \widehat{coC}$.

So if C is a closed convex cone, then C and $C^\perp$ are dual cones.

We leave it to the reader to check that for a closed convex set C, $I - P_C$ is a projection if and only if C is a cone with a vertex at the origin.

DEFINITION A7. The support cone of a closed convex set C at point x is the smallest closed convex cone with vertex at the origin containing C-x. We denote it by $T_C(x)$.

The support cone is the closure of the set of all projecting rays

$$T_C(x) = \{u | u = t(x' - x) \text{ for x' in C and } t \geq 0\}^-. \tag{A.10}$$

We can view $T_C(x)$ as the tangent cone to C at x for if x belongs to C, then u belongs to $T_C(x)$ if and only if for each $t > 0$ there is an x(t) in C such that

$$x + tu = x(t) + o(t) \tag{A.11}$$

as $t \to 0$.

We leave it to the reader to check this and to check that $T_C(x)$ is convex.

THEOREM A8. If x belongs to C, then

$$P_C^{-1}x = T_C^\perp(x) + x. \qquad (A.12)$$

DEFINITION A9. For any point x of a closed convex set C, the set

$$N_C(x) = P_C^{-1}x - x = T_C^\perp(x) \qquad (A.13)$$

is called the vertex of C at x or the normal space to C at x; any u in $N_C(x)$ is said to be normal to C at x.

We see that u belongs to $N_C(x)$ if and only if $x = P_C(x+u)$. The figure below may help identify the pertinent features.

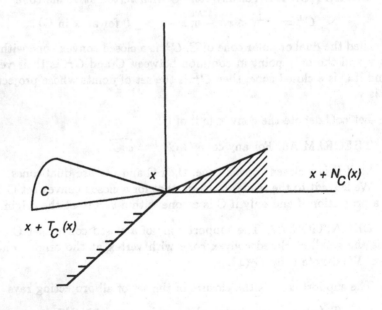

EXERCISE A10. Let B be the unit ball of a Hilbert space X and let x belong to B. Show that $T_C(x) = X$ if x belongs to the interior of B and $T_C(x) = \{x\}^-$ if $\|x\| = 1$.

EXERCISE A11. If A is a continuous linear operator $A : X \to Y$, and if $C = A^{-1}y$ is an affine subspace, then if $Ax = y$ show that $T_C(x) = ker(A)$.

EXERCISE A12. Show that $\{x + N_C(x)\}$ for x in C is a collection of closed convex cones whose union is H.

DEFINITION A13. A point x of a closed convex set C is said to be an inner point if $x + T_C(x)$ coincides with the closed linear manifold spanned by C. The set of inner points of C is called the core, C^o. And the other points form the shell of C.

THEOREM A14. For any closed convex set C, C^o is a convex set; and if C is separable, then C^o is dense in C.

COROLLARY A15. Any separable closed convex set has a nonempty core.

THEOREM A16. If a closed convex set C has an interior point, then all inner points are interior points and $C^o = int(C)$.

EXAMPLE A17. Let C be the positive cone $C = \{x| < x, \phi_\alpha >\geq 0, \alpha \in I\}$ relative to an orthonormal base $\{\phi_\alpha\}_{\alpha \in I}$ in H. Then $C^o = \{x| < x, \phi_\alpha >> 0, \alpha \in I\}$; and C^o will be empty if H is not separable.

DEFINITION A18. The face of a closed convex set C perpendicular to a vector u is the set

$$F_C(u) = \{y|y \in C < y, u >= sup_{x \in C} < x, u >\}. \qquad (A.14)$$

The reader can check that $F_C(u) = \{y|y = PC(y + u)\}$ which is equivalent to $F_C(u) = (I - P_C)^{-1}u - u$. Thus, vertices and faces are dual notions in the sense that if C_1 and C_2 are dual closed cones with vertex at x, then faces and vertices of one are the vertices and faces of the other.

THEOREM A19. In a separable space among all faces of C containing a pont x there is a minimal one, i.e., a face contained in all others; viz., $F_C(u)$ for any u in $(N_C(x))^o$; and any non-empty face of a closed convex set is the face of any x^o in $(F_C(u))^o$.

DEFINITION A20. The recession cone of a closed convex set C is the set

$$C^\infty = \{u|x + tu \in C \text{ for all } x \text{ in } C \text{ and } t \geq 0\}. \qquad (A.15)$$

So C^∞ is a closed convex cone with vertex at the origin and it is the largest such cone that can be placed inside C.

THEOREM A21. A vector u belongs to $(C^\infty)^\perp$ if and only if

$$lim_{t \to \infty} P_C(\frac{x_o + tu}{t}) = 0 \qquad (A.16)$$

for all x_o in H.

THEOREM A22. For any closed convex set C

$$R(\widehat{I - P_C}) = (C^\infty)^\perp. \qquad (A.17)$$

DEFINITION A23. For two convex sets C_1 and C_2 the translation set of C_1 into C_2 is the set $T(C_1, C_2) = \{u | C_1 + u \subset C_2\}$.

THEOREM A24. If C is a closed convex set, then $T(C, C) = C^\infty$.

THEOREM A25. For any closed convex set C

$$P_C(x + h) = x + h + o(h) \qquad (A.18)$$

for x in C, h in $T_C(x)$ where $o(h)/\|h\| \to 0$ as $h \to 0$ over any locally compact cone of increments in $T_C(x)$.

DEFINITION A26. A real-valued function $\alpha : H \to R$ is said to be compact differentiable at x interior to its domain if there is a vector u such that

$$\alpha(x + h) = \alpha(x) + <h, u> + o(x; h) \qquad (A.19)$$

where $o(x; h)/\|h\| \to 0$ as $h \to 0$ over any compact set of h's. The vector u is called the compact gradient and is denoted $\nabla\alpha(x)$.

THEOREM A27. A Lipshitz mapping $P : H \to H$ is a POCS if and only if it satisfies the differential equation

$$(I - P)x = \nabla\frac{1}{2}\|(I - P)x\|^2 \qquad (A.20)$$

for all x in H.

For a proof see Zarantonello (1971).

Since $P_C x = x - (I - P_C)x = \nabla\frac{1}{2}[\|x\|^2 - \|(I - P_C)x\|^2]$ we have:

COROLLARY A28. Projections and their complements are gradient mappings.

And since both P_C and $I - P_C$ are monotone we have

COROLLARY A29. Both $\|x - P_C x\|^2$ and $\|x\|^2 - \|x - P_C x\|^2$ are convex functions.

Appendix B

Reference Summary

Chapter I. There are several reviews of phase retrieval: Saxton (1978), Fienup (1981) and Bates and Mnyama (1986). For 1.1 see Gerchberg-Saxton (1972), Ville (1956), and Cadzow (1979,1981). For 1.2 see Gallagher-Liu (1973). See also Akahori (1973) and Powers-Goodman (1975). For 1.4 see Lee (1981), Taylor and Whinnery (1951), Clarke (1968), and Gallagher-Sweeney (1979). For 1.6 see Boucher (1980), Gonsalves (1976) and Maeda-Murata (1981). For 1.7 see Wolf (1962). For 1.8 see Bricogne (1974,1984). For 1.9 see Rogers (1980) and Readhead-Wilkinson (1978). An excellent review is contained in Thompson, Moran and Swenson (1986). For 1.12 see Bertero-De Mol (1981), Harris (1964) and di Francia (1955,1969).

Chapter II. Any standard reference on abstract algebra will supplement section 2.0. For 2.1 see Requicha (1980).

Chapter III. The basic references are Boas (1954), Levin (1964), Paley-Wiener (1934) and Titchmarsh (1926). For 3.1 see Khurgin-Yakovlev (1977). For 3.2 see Polya (1918) and Cartwright (1930). For 3.3 see Fiddy et al. (1981, 1982). For 3.4 see Hormander (1963), Fejer(1916). For 3.5 Huffman (1962). For 3.6 see Hayes (1982). For 3.7 see Bruck-Sodin (1979). For 3.8 and 3.9 see Hofstetter (1964). For 3.10 see Boas-Kac (1945). For 3.11 see Mehta et al. (1966). For 3.12 and 3.13 see Wolf (1962), Nussenzveig (1967), Baltes et al. (1970), and Kano-Wolf (1962). For 3.14 see Burge et al. (1976). For 3.15 see Akutowicz (1956,1957). For 3.16 see Walther (1963), Crimmins-Fienup (1981). For 3.17 see Barakat-Newsam (1985).

Chapter IV. The classic reference is Patterson (1939). The basic references for homometric distributions are Rosenblatt-Seymour (1982) and Rosenblatt (1984,1987).

Chapter V. The basic review is Voelcker (1966). For 5.2 see Bedrosian (1962). For 5.3 see Logan (1978) and Barnard (1964). For 5.4 see Zakai (1965) and Masry (1977). For 5.5 see Requicha (1980). For 5.6 see Voelcker-

Requicha (1973) and Logan (1984). For 5.7 see Kay-Sudhaker (1986).

Chapter VI. For 6.0 see Shitz-Zeevi (1985), Hayes-Lim-Oppenheim (1980) and Hayes (1982). For 6.1 see Li-Kurkjian (1983). For 6.2 and 6.3 see Hayes-Lim-Oppenheim (1980). For 6.4 see Shitz-Zeevi (1985) and Oppenheim-Lim-Curtis (1983).

Chapter VII. For 7.0 the basic references are Landau-Miranker (1961), Zames(1959), and Logan (1978). For 7.1 and 7.2 see Landau (1965,1967). For 7.3 see Slepian-Pollack (1961,1962). Another verison of the Landau-Pollak result is the dimensionality theorem which states that if f(t) is identically zero outside $[-T/2,T/2]$, $\int |F(u)|^2 du = 1$ and the energy of f(t) outside $[-W,W]$ is at most η_W^2, i.e. $\int_{-W}^{W} |F(u)|^2 du \geq 1 - \eta_W^2$, then there exists a set of orthonormal functions $\psi_i(t)$ such that $\int_{-\infty}^{\infty} [f(t) - \sum_{i=0}^{L-1} f_i \psi_i(t)]^2 dt < \epsilon^2$, $f_i = \int_{-\infty}^{\infty} f(t)\psi_i(t)dt$, $\epsilon^2 = 12\eta_W^2$, and L=[2TW+1]. For further discussions of the dimensionality theorem and bandwidth requirements with block-orthogonal signaling, see Wozencraft and Jacobs, Principles of Communications Engineering (Wiley, New York, 1967). For 7.4 see Widom (1964) and Slepian (1965). For 7.5 see Odlyzko (1987) and des Cloizeaux-Mehta (1972). For 7.6 see Masry (1973).

Chapter VIII. For 8.0 see Groetsch (1977). For 8.1 see Bertero- Pike (1982). For 8.3 see McWhirter-Pike (1978). For 8.4 see Bertero et al. (1982). For 8.5 see Yao (1967) and Nashed- Wahba (1974).

Chapter IX. For 9.0 see Kaczmarz (1937), Tanabe (1971,1974), and Trummer (1984). For nonlinear extension of Kaczmarz see Meyn (1983) and McCormick (1975,1977). For 9.1 see Cimmino (1938) and Ansorge (1984). For 9.2 see Kammerer-Nashed (1972), Fridman (1956), Bialy (1959) and Landweber (1951). For 9.3 see Strand (1974) and Twomey (1965). For 9.4 see Huang et al. (1975) and Ramakrishnam et al. (1979). For 9.5 see van Cittert (1931), Prost-Goute (1977), Silverman-Pearson (1973), and Thomas (1981). For 9.6 see Gerchberg-Saxton (1974) and Papoulis (1975). For 9.7 see DeSantis-Gori (1975). For 9.8 and 9.9 see Papoulis (1975). For 9.10 see Sanz-Huang (1983). For 9.11 and 9.12 see Sanz-Huang (1984). For 9.13 and 9.14 see Xu-Chamzas (1983). For 9.15 see Papoulis-Chamzas (1979,1979). For 9.16 see Papoulis (1975). For 9.17 see Yeh-Chin (1985). For 9.18 to 9.20 see Groetsch (1984). For 9.21 see Abbiss et al. (1983,1983,1983). For 9.22 see Wahba (1977). For 9.23 see Biraud (1969) and Davies (1983).

Chapter X. For 10.1 see Hayes (1982),Sanz-Huang (1985). For 10.2 see Plancherel-Polya (1937), Hayes-McClellan (1982), Manolitsakis (1982), Stefanescu (1985) and Barakat-Newsam (1984). For 10.3 see Lawton (1981). For 10.4 see Lawton-Morrison (1987). For 10.5 see Fiddy-Brames-Dainty (1983) and Nieto-Vesperinas-Dainty (1985,1986).

Chapter XI. For 11.1 see Canterakis (1983). For 11.2 see Berenyi et al. (1985). For 11.3 see Burge et al. (1976) and Nakajima- Asakura (1982). See also Roman-Marathay (1963). For 11.4 see Nakajima-Asakura (1986). For

11.5 see Curtis (1985), Curtis-Shitz-Oppenheim (1987), Sanz-Huang (1983). For 11.6 see Israelevitz-Lim (1987). For 11.7 see Rotem-Zeevi (1986).

Chapter XII. A basic introduction is given in Ortega-Rheinboldt (1970). For 12.0 see Browder (1965,1967), Browder- Petryshyn (1966,1967), Dotson (1970), Opial (1967), Halperin (1962) and Amemiya-Ando (1965). For 12.1 see Tom et al. (1981). For 12.4 see Quatieri-Oppenheim (1981). For 12.6 see Sayegh (1987).

Chapter XIII. For 13.0 see Youla (1981), Gubin et al. (1967). The method of alternating projections is due to von Neumann in the early 1930s for the case $k = 2$, although it was independently discovered by Nakano in 1940 and Wiener in 1955. For $k > 2$ Kaczmarz obtained the result in the case of linear equations. Halperin (1962) proved strong convergence for the general case. Results on error bounds for the method of alternating projections are in Aronszajn (1950), Youla (1978), Deutsch (1983) and Kaylar and Weinert (1988). For 13.1 to 13.3 see Youla (1978). For 13.5 see Goldburg-Marks (1985). For 13.6 see Herman et al. (1973), Herman (1980) and Youla (1981). For 13.7 see Stark et al. (1982). For 13.10 and 13.11 see Lent- Tuy (1981). The relationship of shape-from-shading algorithms and the method of projection onto convex sets is discussed in R. T. Frankot - R. Chellapa, A method for enforcing integrability in shape from shading algorithms, IEEE PAMI 10 (1988) 439-451.

Chapter XIV. For 14.0 and 14.1 see Levi-Stark (1984). For 14.2 see Trussell-Civanlar (1984). For 14.3 see Civanlar-Trussell (1986). For 14.4 see Miller (1973). For 14.5 see Nashed (1974), Kammerer-Nashed (1971) and Jain (1978). For 14.6 see Manry-Aggarwal (1976) and Daniel (1976). For 14.7 see Barakat-Newsam (1985).

Chapter XV. For 15.0 see Fienup et al. (1982,1982), Fienup (1986) and Crimmins (1987).

Chapter XVI. For 16.0 see Gallagher-Liu (1973), Fienup (1979), Fienup (1975) and Dainty-Fienup (1987).

Chapter XVII. For 17.1 see Lim-Malik (1981), Ables (1974), and Lim (1986). For 17.2 see Labeyrie (1970), Fienup-Feldkamp (1980), Feldkamp-Fienup (1980), and Fienup (1978,1979). For 17.3 and 17.4 see Schwarz (1978) and Rogers (1980). For 17.5 see Stockham et al. (1975), Knox (1976), Morton-Andrews (1979), Gonsalves (1976) and the paper of Frieden-Currie (1976). See also Rino (1976) and Trussell-Hunt (1978). For 17.6 see Quatieri-Dudgeon (1982) and Liu-Gallagher (1974). See also Algazi-Suk (1975), Liu-Gallagher (1974,1974), and Requicha-Voelcker (1970). For 17.7 see Rahmat-Samii (1984,1985). For 17.8 see Missell (1973), Hoenders (1975) and Kiedron (1987). For electron microscopy see Saxton (1978), Huiser et al. (1976,1977). For 17.9 see Morris (1985). For 17.10 see Tam et al. (1981). For 17.11 see Sato et al. (1981). For 17.12 and 17.13 see Lawton-Morrison (1987). For 17.14 and 17.15 see Kadar (1977). For 17.16 and 17.17 see

Sandberg (1980). For 17.18 see Wiley et al. (1977) and Duffin-Schaeffer (1952). For 17.19 see Papoulis (1975). For 17.20 see Moses-Prosser (1983). For 17.21 see Baghlay, Saxton (1978) and Fowle (1964). For 17.22 see Gonsalves (1976). For 17.23 see Marks et al. (1987). For 17.24 see Sezan-Stark (1987). For 17.25 see Hinshaw-Lent (1983). For 17.26 see Zakhor (1988). For 17.27 see the recent works of Censor and coworkers.

For Appendix A see Zarantonello (1971) and Bruck (1973).

Appendix C

References

J. B. Abbiss et al., Regularization of an iterative algorithm for the extrapolation of band-limited signals (1983 URSI Symp. on Electromag. Theory).

J. B. Abbiss et al., Regularized iterative and noniterative procedure for object restoration in the presence of noise: an error analysis, J. Opt. Soc. Am. 73 (1983) 1470-1475.

J. B. Abbiss, C. De Mol, and H. S. Dhadwal, Regularized iterative and non-iterative procedures for object restoration from experimental data, Opt. Acta 30 (1983) 107-124.

J. G. Ables, Maximum entropy spectral analysis, Astron. Astrophys. Suppl. 15 (1974) 383-393.

A. Abo-Taleh and M. M. Fahmy, Design of FIR two-dimensional digital filters by successive projections, IEEE CAS 31 (1984) 801-805.

H. Akahori, Comparison of deterministic phase coding with random phase coding in terms of dynamic range, Appl. Opt. 12 (1973) 2336-2343.

S. Agmon, The relaxation method for linear inequalities, Can. J. Math. 6 (1954) 382-392.

E. J. Akutowicz, On the determination of the phase of a Fourier integral, I, Trans. Amer. Math. Soc. 83 (1956) 179-192; II, Proc. Amer. Math. Soc. 8 (1957) 234-238.

V. R. Algazi and M. Suk, On the frequency weighted least-square design of finite duration filters, IEEE CAS-22 (1975) 943-953.

J. P. Allebach, N. C. Gallagher and B. Liu, Aliasing error in digital holography, Appl. Opt. 15 (1976) 2183-2188.

I. Amemiya and T. Ando, Convergence of random products of contractions in Hilbert space, Acta Sci. Math. 26 (1965) 239-244.

R. Ansorge, Connections between the Cimmino-method and the Kaczmarz method for the solution of singular and regular systems of equations, Computing 33 (1984) 367-375.

N. Aronszajn, Theory of reproducing kernels, Trans. Amer. Math. Soc. 68 (1950) 337-404.

H. A. Arsenault and K. Chalasinka-Macukow, The solution to the phase retrieval problem using the sampling theorem, Opt. Comm. 47 (1983) 380-386.

, Fast iterative solution to exact equations for the two-dimensional phase retrieval problem, J. Opt. Soc. Am. A2 (1985) 46-50.

L. E. Atlas, T. Homma, R. J. Marks II, A neural network model for vowel classification (preprint).

R. D. Baghlay, About numerical solutions of a multiextreme problem of all-pass filter synthesis (preprint).

H. P. Baltes, Progess in Inverse Optical Problems, in Inverse Scattering Problems (Springer-Verlag, Berlin, 1980) 1-13.

, Introduction, in Inverse Source Problems in Optics (Springer-Verlag, Berlin, 1978) 1-11.

H. P. Baltes and H. A. Ferwerda, Inverse problems and coherence, IEEE AP-29 (1981) 405-406.

H. P. Baltes, R. Muri, and F. K. Kneubuhl, Spectral densities of cavity resonances and black body radiation standards in the submillimeter wave region, Sympo. Submill. Waves (1970) 667-691.

R. Barakat and G. Newsam, Necessary conditions for multiple solutions to the two-dimensional phase recovery problem, J. Math. Phys. 25 (1984) 3190-3193.

, Algorithms for reconstruction of partially known, bandlimited Fourier transform pairs from noisy data: I. the prototypical linear problem, J. Opt. Soc. Am. A2 (1985) 2027-2039.

, On the existence of multiple solutions to the two-dimensional phase recovery problem (preprint).

, Numerically stable iterative method for inversion of wave-front aberrations from measured point-spread-function data, J. Opt. Soc. Am. 70 (1980) 1255-1263.

R. D. Barnard, On the spectral properties of single-sideband angle-modulated signals, Bell Sys. Tech. J. 43 (1964) 2811-2838.

H. O. Bartlet, A. W. Lohmann and B. Winnitzer, Phase and amplitude recovery from bispectrum, Appl. Opt. 23 (1984) 3121- 3129.

R. H. T. Bates, Contributions to the theory of intensity interferometry, Mon. Not. R. Astron. Soc. 142 (1969) 413-428.

R. H. T. Bates, Holographic approach to radiation pattern measurement I., Int. J. Eng. Sci. 9 (1971) 1193-1208.

R. H. T. Bates et al. Self-consistent deconvolution, Optik 44 (1976) 183-201; 253-272.

R. H. T. Bates, On phase problems I,II, Optik 51 (1978) 161-170; 223-234.

R. H. T. Bates et al., Generalization of shift-and-add imaging, SPIE 556 (1985) 263-269.

R. H. T. Bates and M. J. McDonnell, Image Restoration and Reconstruction (Clarendon, Oxford, 1986).

R. H. T. Bates and W. R. Fright, Reconstructing images from their Fourier intensities, in Advances in Computer Vision and Image Processing 1 (1984) 227-264.

R. H. T. Bates, Uniqueness of solutions to two-dimensional Fourier phase problems for localized and positive images, Comp. Vis. Graph. Imag. Proc. 25 (1984) 205-217.

R. H. T. Bates, Fourier phase problems are uniquely solvable in more than one dimension I, Optik 61 (1982) 247-262; II, Optik 62 (1982) 131-142.

, Astronomical speckle imaging, Phys. Rep. 90 (1982) 203-297.

R. H. T. Bates and D. G. H. Tan, Fourier phase retrieval when the image is complex, SPIE 558 (1985) 54-59.

, Towards reconstructing phases of inverse scattering signals, J. Opt. Soc. Am. 2A (1985) 2013- 2017.

R. H. T. Bates and D. Mnyama, The status of practical Fourier phase retrieval, in Advances in Electronics and Electron Physics (Academic Press, New York, 1986) 1-64.

R. H. T. Bates and W. R. Fright, Two-dimensional phase restoration, in Fourier Techniques and Applications (Plenum, London, 1985) 121-148.

, Composite two-dimensional phase- restoration procedure, J. Opt. Soc. Am. 73 (1983) 358-365.

R. H. T. Bates, W. R. Fright, and W. A. Norton, Phase restoration is successful in the optical as well as the computational laboratory, in Indirect Imaging (Cambridge University Press, Cambridge, 1984) 119-124.

R. H. T. Bates, Practical phase retrieval, SPIE 413 (1983) 208-211.

E. Bedrosian, The analytic signal representation of modulated waveforms, Proc. IRE 50 (1962) 2071-2076.

M. Bendinelli et al., Degrees of freedom and eigenfunctions, for noisy image, J. Opt. Soc. Am. 64 (1974) 1498-1502.

H. Berenyi and M. A. Fiddy, Object reconstruction over the integer field from noisy Fourier intensity data, Opt. Comm. 59 (1986) 342-344.

H. M. Berenyi, C. L. Byrne and M. A. Fiddy, Object estimation from limited Fourier magnitude samples, SPIE 808 (1987).

H. M. Berenyi, H. V. Deighton and M. A. Fiddy, The use of bivariate polynomial factorization algorithms in two-dimensional phase problems, Opt. Acta 32 (1985) 689-701.

H. M. Berenyi and M. A. Fiddy, Application of homometric sets for beam manipulation, J. Opt. Soc. Am. A3 (1986) 373-375.

M. Bertero, Regularization methods for linear inverse problems, in Inverse Problems (Springer-Verlag, Berlin, 1986).

M. Bertero, P. Brianzi and E. R. Pike, On the recovery and resolution of exponential relaxation rates from experimental data II,III, Proc. Roy. Soc. A383 (1982) 15, A398 (1985) 23.

M. Bertero, P. Boccacci and E. R. Pike, Resolution in diffraction limited imaging, a singular value analysis II. the case of incoherent illumination, Opt. Acta 29 (1982) 1599-1611.

M. Bertero, P. Brianzi and E. R. Pike, On the recovery and resolution of exponential relaxation rates from experimental data, Inv. Prob. 1 (1985) 1.

M. Bertero, P. Brianzi, P. Parker and E. R. Pike, ..III.,the effect of sampling and truncation of the data, Opt. Acta 31 (1984) 181-201.

M. Bertero, P. Brianzi and E. R. Pike, Super-resolution in confocal scanning microscopy, Inv. Prob. 3 (1987) 195.

M. Bertero, P. Brianzi, E. R. Pike and L. Rebolia, Linear regularizing algorithms for positive solutions of linear inverse problems (preprint).

M. Bertero, C. De Mol, E. R. Pike and J. G. Walker, ..IV.,the case of uncertain localization or non-uniform illumination of the object, Opt. Acta 31 (1984) 923-946.

M. Bertero, C. De Mol and E. R. Pike, Linear inverse problems with discrete data: I, Inv. Prob. 1 (1985) 301-330; II, (preprint).

M. Bertero and V. Dovi, Regularized and positive-constrained inverse methods in the problem of object restoration, Opt. Acta 28 (1981) 1635-1649.

M. Bertero and C. De Mol, Ill-posedness, regularization and number of degrees of freedom, Atti Fond. "Giorgio Ronchi", 36 (1981) 619-632.

M. Bertero, C. De Mol and G. A. Viano, Restoration of optical objects using regularization, Opt. Lett. 3 (1978) 51-53.

, The stability of inverse problems, in Inverse Scattering Problems in Optics (Springer-Verlag, Berlin, 1980) 161-214.

, On the regularization of linear inverse problems in Fourier optics, in Applied Inverse Problems (Srpinger-Verlag, Berlin, 1978)

M. Bertero and C. De Mol, Stability problems in inverse diffraction, IEEE AP-29 (1981) 368-372.

M. Bertero, C. De Mol, F. Gori and L. Ronchi, Number of degrees of freedom in inverse diffraction, Opt. Acta 30 (1983) 1051- 1065.

M. Bertero and F. A. Grunbaum, Commuting differential operators for the finite Laplace transform, Inv. Prob. 1 (1985) 181.

M. Bertero, F. A. Grunbaum and L. Rebolia, Spectral properties of a differential operator related to the inversion of the finite Laplace transform, Inv. Prob. 2 (1986) 131.

M. Bertero and E. R. Pike, Resolution in diffraction-limited imaging, a singular value analysis I., the case of coherent illumination, Opt. Acta 29 (1982) 727-746.

, Singular value analyses of inversion of Laplace and optical imaging transforms (preprint).

M. Bertero, G. A. Viano and C. De Mol, Resolution beyond the diffraction limit for regularized object restoration, Opt. Acta 27 (1980) 307-320.

H. Bialy, Iterative behandlung linearen funktion algleichungen, Arch. Rat. Mech. Anal. 4 (1959) 166-176.

R. Biovin and A. Biovin, Optimized amplitude filtering for superresolution over a restricted field, Opt. Acta 27 (1980) 587-610.

Y. Biraud, A new approach for increasing the resolving power by data processing, Astron. Astrophys. 1 (1969) 124-127.

R. P. Boas, Some theorems on Fourier transform and conjugate trigonometric integrals, Trans. Amer. Math. Soc. 40 (1936) 287-308.

, Entire Functions (Academic Press, New York, 1954).

R. P. Boas and M. Kac, Inequalities for Fourier transforms of positive functions, Duke Math. J. 12 (1945) 189-206.

F. E. Bond and C. R. Cahn, On sampling the zeros of bandwidth limited signals, IRE IT-4 (1958) 110-113.

F. E. Bond, C. R. Cahn and J. C. Hancock, A relation between zero crossings and Fourier coefficients for band-limited functions, IRE IT-6 (1960) 51-55.

R. H. Boucher, Convergence of algorithms for phase retrieval from two intensity distributions, SPIE 231 (1980) 130-141.

G. R. Boyer, Pupil filters for moderate superresolution, App. Opt. 15 (1976) 3089-3093.

B. J. Brames, Unique phase retrieval with explicit support information, Opt. Lett. 11 (1986) 61-63.

L. M. Bregman, The method of successive projection for finding a common point of convex sets, Sov. Math. Dokl. 6 (1965) 688- 692.

L. M. Bregman, The relaxation method of finding the common point of convex sets and its application to the solution of problems in convex programming, USSR Comp. Math. Math. Phys. 7(3) (1967) 200-217.

G. Bricogne, Geometric sources of redundancy in intensity data and their use for phase determination, Acta Cryst. A30 (1974) 395-400.

, Maximum entropy and the foundations of direct methods, Acta Cryst. A40 (1984) 410-445.

F. E. Browder, Convergence theorems for sequences of nonlinear operators in Banach spaces, Math. Z. 100 (1967) 201-225.

, Convergence of approximants to fixed points of nonexpansive nonlinear mappings in Banach spaces, Arch. Rat. Mech. Anal. 24 (1967) 82-90.

, Nonexpansive nonlinear operators in a Banach space, Proc. Nat. Acad. Sci. USA 54 (1965) 1041-1044.

, Fixed point theorems for nonlinear semicontractive mappings in Banach spaces, Arch. Rat. Mech. Anal. 21 (1966) 259-269.

F. E. Browder and W. V. Petryshyn, Construction of fixed points of nonlinear mappings in Hilbert space, J. Math. Anal. Appl. 20 (1967) 197-228.

, The solution by iteration of nonlinear functional equations in Banach spaces, Bull. Amer. Math. Soc. 72 (1966) 571-575.

R. E. Bruck, Jr., Nonexpansive projections on subsets of Banach spaces, Pac. J. Math. 47 (1973) 341-355.

Y. M. Bruck and L. G. Sodin, On the ambiguity of the image reconstruction problem, Opt. Comm. 30 (1979) 304-308.

, Reconstruction of images scattered in inhomogeneous medium (preprint, Kharkov, 1981).

J. Bucklew and N. Gallagher, Jr., Detour phase error in the Lohmann hologram, Appl. Opt. 18 (1979) 575-580.

J. A. Bucklew and B. E. A. Saleh, Theorem for high-resolution high-contrast image synthesis, J. Opt. Soc. Am. A2 (1985) 1233-1236.

R. E. Burge et al., The phase problem, Proc. Roy. Soc. London, A350 (1976) 191-212.

R. E. Burge and M. A. Fiddy, Object reconstruction in electron microscopy from the real part of the scattered wave, Optik 54 (1979) 21-26.

R. E. Burge et al., The phase problem in scattering phenomena: the zeros of entire functions and their significance, Proc. Roy. Soc. London A360 (1978) 25-45.

C. L. Byrne and R. M. Fitzgerald, Reconstruction from partial information with applications to tomography, SIAM J. Appl. Math. 42 (1982) 933-940.

, Some new approaches to one and two dimensional power spectrum estimation, Proc. ASSP Spec. Est. Workshop II (1983) 202-204.

, Spectral estimators that extend the maximum entropy and maximum likelihood methods, SIAM J. Appl. Math. 44 (1984) 425-442.

C. L. Byrne et al., Image restoration and resolution enhancement, J. Opt. Soc. Amer. 73 (1983) 1481-1487.

, Image restoration incorporating prior knowledge, Opt. Lett. 73 (1983) 1466.

C. L .Byrne and D. M. Wells, Limit of continuous and discrete finite-band Gerchberg iterative spectrum extrapolation, Opt. Lett. 8 (1983) 526-527.

C. L. Byrne and M. A. Fiddy, Estimation of continuous object distributions from limited Fourier magnitude data, J. Opt. Soc. Amer. A4 (1987) 112-117.

J. A. Cadzow, An extrapolation procedure for band-limited signals, IEEE ASSP 27 (1979) 4-12.

, Observations on the extrapolation of band-limited signal problem, IEEE ASSP-29 (1981) 1208-1209.

, Improved spectral estimation from incomplete sampled data observations, Proc. 1978 RADC Spec. Est. Workshop (1978) 109-123.

D. Cahana and H. Stark, Bandlimited image extrapolation with faster convergence, Appl. Opt. 20 (1981) 2780-2786.

S. Cambanis and E. Masry, Zakai's class of bandlimited functions and processes, SIAM J. Appl. Math. 30 (1976) 10-21.

M. Cannon, Blind deconvolution of spatially invariant image blurs with phase, IEEE ASSP-24 (1976) 58-63.

N. Canterakis, Magnitude-only reconstruction of two-dimensional sequences with finite regions of support, IEEE ASSP-31 (1983) 1256- 1262.

M. L. Cartwright, The zeros of certain integral functions, Q. J. Math. 1 (193) 38-59; 2 (1931) 113-129.

W. T. Cathey et al., Image gathering and processing for enhanced resolution, J. Opt. Soc. Am. A1 (1984) 241-250.

Y. Censor and T. Elfving, New methods for linear inequalities, Lin. Alg. App. 42 (1982) 199-211.

Y. Censor, Row-action methods for huge and sparse systems and their applications, SIAM Rev. 23 (1981) 444-466.

Y. Censor, Intervals in linear and nonlinear problems of image reconstruction, in Mathematical Aspects of Computerized Tomography (Springer-Verlag, New York, 1981) 152-159.

Y. Censor, Finite series-expansion reconstruction methods, Proc. IEEE 71 (1983) 409-419.

Y. Censor and G. T. Herman, Row generation methods for feasibility and optimization problems involving sparse matrices and their applications, in Sparse Matrix Proceedings 1978 (SIAM, Philadelphia, 1979) 197-219.

Y. Censor et al., Strong underrelaxation in Kaczmarz's method for inconsistent systems, Numer. Math. 41 (1983) 83-92.

Y. Censor et al., Special purpose algorithms for linearly constrained entropy maximization, in Maximum Entropy and Bayes Spectral Analysis and Estimation Problems (D. Reidel, Dordrecht, 1987) 241-254.

Y. Censor and A. Lent, Optimization of log x-entropy over linear equality constraints, SIAM J. Cont. Opt. 25 (1987) 921-933.

Y. Censor and G. T. Herman, On some optimization techniques in image reconstruction from projections, Appl Num. Math. 3 (1987) 365-391.

Y. Censor and J. Segman, On block-iterative entropy maximization, J. Inf. Opt. Sci. 8 (1987) 275-291.

Y. Censor et al., On iterative methods for linearly constrained entropy maximization (preprint, 1987).

Y. Censor et al., On the use of Cimmino's simultaneous projections method for computing a solution of the inverse problem in radiation treatment planning (preprint, 1988).

Y. Censor et al., On maximization of entropies and a generalization of Bregman's method for convex programming (preprint, 1988).

G. Cesini et al., An iterative method for restoring noisy images, Opt. Acta 25 (1978) 501-508.

J. N. Chapman, The application of iterative techniques to the objects investigation of strong phase objects in microscopy, Philos. Mag. 32 (1975) 541-552.

L. M. Cheng, A. S. Ho, and R. E. Burge, The use of prior knowledge in image reconstruction (preprint).

K. F. Cheung et al., Neural net associative memories based on convex set projections (preprint).

R. T. Chin, C. L. Yeh and W. S. Olson, Restoration of multichannel microwave radiometric images, IEEE PAMI-7 (1985) 475-484.

M. Y. Chiu et al., Three-dimensional radiographic imaging with a restricted view angle, J. Opt. Soc. Amer. 69 (1979) 1323-1333.

G. Cimmino, Calcolo approssiomoto, Ric. Sci. progr. tecn. econ. naz. 1 (1938) 326-333.

M. R. Civanlar and H. J. Trussell, Digital signal restoration using fuzzy sets, IEEE ASSP-34 (1986) 919-936.

J. Clarke, Steering of zeros in the directional pattern of a linear array, IEEE AP-16 (1968) 267-268.

J. des Cloizeaux and M. L. Mehta, Some asymptotic expressions for prolate spheroidal functions, J. Math. Phys. 13 (1972) 1745-1754.

T. R. Crimmins, Phase retrieval for discrete functions with support constraints, J. Opt. Soc. Amer. A4 (1987) 124-134.

T. R. Crimmins and J. R. Fienup, Uniqueness of Phase retrieval for functions with sufficiently disconnected support, J. Opt. Soc. Amer. 73 (1983) 218-221.

, Phase retrieval for functions with disconnected support (1981).

, Ambiguity of phase retrieval for functions with disconnected support, J. Opt. Soc. Am. 71 (1981) 1026 -1029.

S. R. Curtis, Reconstruction of multidimensional signals from zero crossings (thesis, MIT, 1985).

S. R. Curtis, A. V. Oppenheim and J. S. Lim, Signal reconstruction from Fourier transform sign information, IEEE ASSP-33 (1985) 643-657.

S. R. Curtis, S. Shitz and A. V. Oppenheim, Reconstruction of nonperiodic two-dimensional signals from zero crossings, IEEE ASSP-35 (1987) 890-893.

S. R. Curtis and A. V. Oppenheim, Reconstruction of multidimensional signals from zero crossings, J. Opt. Soc. Amer. (1987).

J. C. Dainty, M. A. Fiddy and A. H. Greenway, On the danger of applying statistical reconstruction methods in the case of missing phase information, in Image Formation from Coherence Functions in Astronomy

(Reidel,Dordrecht, 1979) 95-101.

J. C. Dainty and M. A. Fiddy, The essential role of prior knowledge in phase retrieval, Opt. Acta 31 (1984) 325-330.

J. C. Dainty and J. R. Fienup, Phase retrieval and image reconstruction for astronomy, in Image Recovery (Academic Press, New York, 1987) 231-275.

W. J. Dallas, Digital computation of image complex amplitude from image and diffraction intensity: an alternative to holography, Optik 41 (1975) 45-59.

J. W. Daniel, Improving the Manry-Aggarwal method for designing multi-dimensional FIR digital filters (preprint, Un. Texas, 1976).

, Conjugate gradients beat the Manry-Aggarwal method for filters (preprint, Un. Texas, 1977).

A. M. Darling, A. M. Hall and M. A. Fiddy, Stable, noniterative object reconstruction from incomplete data using prior knowledge, J. Opt. Soc. Am. 73 (1983) 1466-1469.

A. M. Darling, H. V. Deighton and M. A. Fiddy, Phase ambiguities in more than one dimension, SPIE 413 (1983) 197-201.

J. N. Darroch and D. Ratcliff, Generalized iterative scaling for log-linear models, Ann. Math. Stat. 43 (1972) 1470-1480.

I. Daubechies, Time-frequency localization operators: a geometric phase space approach, IEEE IT-34 (1988) 605-612.

A. R. Davies, On the maximum likelihood regularization of Fredholm convolution equations of the first kind, in Proc. of Durham Symp. on Numerical Treatment of Integral Equations (Academic Press, New York, 1982) 95-105.

, On a constrained Fourier extrapolation method for numerical deconvolution, in Proceedings of the Oberwolfach Meeting on Improperly Posed Problems and their Numerical Treatment (Birkhauser-Verlag,Boston,1983) 65-80.

A. R. Davies et al., A comparison of statistical regularization and Fourier extrapolation, methods for numerical deconvolution, in Numerical Treatment of Inverse Problems in Differential and Integral Equations (Birkhauser-Verlag,Boston, 1983) 320-324.

M. Defrise and C. De Mol, A note on stopping rules for iterative regularization methods and filtered SVD, in Inverse Problems: An Interdisciplinary Study (Academic Press, New York, 1987) 261.

H. V. Deighton et al., Practical phase retrieval based on theoretical models for multi-dimensional band-limited signals, SPIE 558 (1985) 65-72.

H. V. Deighton, M. S. Scivier and M. A. Fiddy, Solution of the two dimensional phase retrieval problem, Opt. Lett. 10 (1985) 250-251.

A. P. Dempster et al., Maximum likelihood from incomplete data via the EM algorithm, J. Roy. Stat. Soc. 39 (1977) 1-38.

A. R. DePierro and A. N. Iusem, A simultaneous projections method for linear inequalities, Lin. Alg. Appl. 64 (1985) 243-253.

, A relaxed version of Bregman's method for convex programming, J. Opt. Thy. App. 5 (1986) 421-440.

P. DeSantis and F. Gori, On an iterative method for super resolution, Opt. Acta 22 (1975) 691-695.

F. Deutsch, Applications of von Neumann's alternating projections algorithm, in Mathematical Methods in Operations Research (Sofia, 1983) 44-51.

A. J. Devaney and R. Chidlaw, On the uniqueness question in the problem of phase retrieval from intensity measurements, J. Opt. Soc. Am. 68 (1978) 1352-1354.

W. G. Dotson, Jr., On the Mann iterative process, Trans. Amer. Math. Soc. 149 (1970) 65-73.

A. J. J. Drenth, A. M. J. Huiser and H. A. Ferwerda, The problem of phase retrieval in light and electron microscopy of strong objects , Opt. Acta 22 (1975) 615-628.

D. E. Dudgeon, An iterative implementation for 2-D digital filters, IEEE ASSP-28 (1980) 666-680.

R. J. Duffin and A. C. Schaeffer, A class of non-harmonic Fourier series, Trans. Amer. Math. Soc. 72 (1952) 341-366.

P. P. Eggermont et al., Iterative algorithms for large partitioned linear systems with applications to image reconstruction, Lin. Alg. App. 40 (1981) 37-67.

C. Y. Espy and J. S. Lim, Effects of additive noise on signal reconstruction from Fourier transform phase, IEEE ASSP-31 (1983) 894-898.

J. Evans, P. Kersten and L. Kurz, Robustized recursive estimation with applications, Inf. Sci. 11 (1976) 69-92.

L. Fejer, Uber trigonometrische polynome, Z. reine u. angew. Mat. 146 (1916) 53-82.

G. B. Feldkamp and J. R. Fienup, Noise properties of images reconstructed from Fourier modulus, SPIE 231 (1980) 84-93.

H. A. Ferwerda, The phase reconstruction problem, in Inverse Source Problems in Optics (Springer-Verlag, Berlin, 1978) 13-39.

H. A. Ferwerda, Optics in four dimensions-1980, AIP Conf. Proc. 65 (1981) 402-411.

H. A. Ferwerda and B. J. Hoenders, On the reconstruction of a weak phase-amplitude object IV, Optik 39 (1974) 317-326.

, On the theory of the reconstruction of a weak phase-amplitude object from its images, especially in electron microscopy I., Isoplanatic imaging, Opt. Acta 22 (1975) 25-34; II., Non-isoplanatic imaging, Opt. Acta 22 (1975) 35-56.

M. A. Fiddy, Object reconstruction from partial information (preprint, 1984).

, The phase retrieval problem, SPIE 413 (1983) 176-181.

, The role of analyticity in image recovery, in Image Recovery: Theory and Application (Academic Press, New York, 1987).

, Inversion of optical scattered field data, J. Phys. D19 (1986) 301-317.

M. A. Fiddy and A. H. Greenway, Phase retrieval using zero information, Opt. Comm. 29 (1979) 270-272.

M. A. Fiddy and T. J. Hall, On the non-uniqueness of super-resolution techniques applied to sampled data, J. Opt. Soc. Amer. 71 (1981) 1406-1407.

M. A. Fiddy and G. Ross, Analytic Fourier optics, the encoding information by complex zeros, Opt. Acta 26 (1979) 1139-1146.

M. A. Fiddy et al., Encoding of information in inverse optical problems, Opt. Acta 29 (1982) 23-40.

M. A. Fiddy et al., Encoding of information in inverse scattering problems, IEEE AP-29 (1981) 406-408.

M. A. Fiddy, B. J. Brames and J. C. Dainty, Enforcing irreducibility for phase retrieval in two dimensions, Opt. Lett. 8 (1983) 96-98.

J. R. Fienup, Reconstruction of an object from the modulus of its Fourier transform, Opt. Lett. 3 (1978) 27-29.

, Improved synthesis and computational methods for computer-generated holograms (thesis, Stanford Univ., 1975).

, Experimental evidence of uniqueness of phase retrieval from intensity data, in Indirect Imaging (Cambridge University Press, London, 1984) 99-109.

, Reconstruction of a complex valued object from the modulus of its Fourier transform using support constraint, J. Opt. Soc. Am. A4 (1987) 118-123.

, Phase retrieval using boundary conditions, J. Opt. Soc. Am. A3 (1986) 284-288.

, Reconstruction of objects having latent reference points, J. Opt. Soc. Am. 73 (1983) 1421-1426.

, Phase retrieval algorithms: a comparison, Appl. Opt. 21 (1982) 2758-2769.

, Space object imaging through the turbulent atmosphere, Opt. Eng. 18 (1979) 529-534.

, Comparison of phase retrieval algorithms, in Advances in Computer Vision and Image Processing I (1984) 191-225.

, Image reconstruction for stellar interferometry, in Current Trends in Optics (Taylor and Francis, London, 1981) 95-102.

, Reconstruction and synthesis applications of an iterative algorithm, SPIE 373 (1981) 147-160.

J. R. Fienup, T. R. Crimmins and W. Holsztynski, Reconstruction of the support of an object from the support of its autocorrelation, J. Opt. Soc. Am. 72 (1982) 610-624.

, Reconstruction of object having latent reference points, J. Opt. Soc. Am. 73 (1982) 1421-1433.

J. R. Fienup and G. B Feldkamp, Astonomical imaging by processing stellar speckle interferometry data, SPIE 243 (1980) 95-102.

J. R. Fienup and C. C. Wackerman, Phase retrieval stagnation problems and solutions, J. Opt. Soc. Am. A3 (1986) 1897-1907.

J. T. Foley and R. R. Butts, Uniqueness of phase retrieval from intensity measurements, J. Opt. Soc. Am. 71 (1981) 1008-1014.

G. E. Forsythe and W. R. Wasow, Finite Difference Methods for Partial Differential Equations (Wiley, New York, 1960).

E. N. Fowle, The design of FM pulse compression signals, IEEE IT-10 (1964) 61-67.

J. N. Franklin, Well-posed stochastic extensions of ill-posed linear problems, J. Math. Anal. 31 (1970) 682-716.

V. Fridman, A method of successive approximation for Fredholm integral equations of the first kind, Usp. Mat. Nauk 11 (1956) 232-234.

B. R. Frieden, Band-unlimited reconstruction of optical objects and spectra, J. Opt. Soc. Am. 57 (1967) 1013-1019.

B. R. Frieden, On arbitrarily perfect imagery with a finite aperture, Opt. Acta 16 (1969) 795-807.

B. R. Frieden, the extrapolating pupil, image synthesis and some thought applications, Appl. Opt. 9 (1970) 2489-2496.

B. R. Frieden, Image enhancement and resolution, in Picture Processing and Digital Filtering (Springer-Verlag, New York, 1978).

, Evaluation, design and extrapolation methods for optical signals based on use of the prolate functions, Progress in Optics 9 (1971) 311-407.

B. R. Frieden and D. G. Currie, On unfolding the autocorrleation function, J. Opt. Soc. Am. 66 (1976) 1111A.

W. R. Fright and R. H. T. Bates, Fourier phase problems are uniquely solvable in more than one dimension III, Optik 62 (1982) 219-230.

, Composite two-dimensional phase restoration procedure, J. Opt. Soc. Am. 73 (1983) 358-365.

R. A. Gabel and B. Liu, Minimization of reconstruction errors with computer generated binary holograms, Appl Opt. 9 (1970) 1180- 1190.

N. C. Gallagher and B. Liu, Method for computing kinoforms that reduces image reconstruction error, Appl. Opt. 12 (1973) 2328-2335.

N. C. Gallagher and D. W. Sweeney, Infared holographic optical elements, IEEE QE-15 (1979) 1369-1381.

K. L. Garden and R. H. T. Bates, Fourier phase problems are uniquely solvable in more than one dimension, II, Optik 62 (1982) 131-142.

R. W. Gerchberg, Super-resolution through error energy reduction Optica Acta 21 (1974) 709-720.

R. W. Gerchberg and W. O. Saxton, A practical algorithm for the determination of phase from image and diffraction plane pictures, Optik 35 (1972) 237-246.

, Phase determination from image and diffraction plane pictures in the electron microscope, Optik 34 (1971) 275-284.

, Comment on ' a method for the solution of the phase problem in electron microscopy', J. Phys. D6 (1973) L31-L32.

, Wavephase from image and diffraction plane pictures, in Image Processing and Computer-aided Design in Electron Optics (Academic Press, London, 1973) 66-81.

P. F. C. Gilbert, Iterative methods for 3-D reconstruction of an object from projections, J. Theo. Biol. 36 (1972) 105-117.

M. L. Goldberger, H. W. Lewis and K. M. Watson, Use of intensity correlations to determine the phase of a scattering amplitude, Phys. Rev. 132 (1963) 2764-2787.

M. Goldburg and R. J. Marks II, Signal synthesis in the presence of an inconsistent set of constraints, IEEE CAS-32 (1985) 647-663.

R. A. Gonsalves, Phase retrieval from modulus data, J. Opt. Soc. Am. 66 (1976) 961-964.

, Phase retrieval by differential intensity measurements, J. Opt. Soc. Am. A4 (1987) 166.

,Phase retrieval and diversity in adaptive optics, Opt. Eng. 21 (1982) 829-832.

, Fundamentals of wavefront sensing by phase retrieval, SPIE 351 (1982) 56.

R. Gordon et al., Algebraic reconstruction techniques (ART) for three-dimensional electron microscopy and x-ray photography, J. Theo. Biol. 29 (1970) 471-481.

F. Gori, Integral equations for incoherent imagery, J. Opt. Soc. Am. 64 (1974) 1237-1243.

F. Gori and G. Guattari, Degrees of freedom of images from point-like-element pupils, J. Opt. Soc. Am. 64 (1974) 453-458.

F. Gori and G. Guattari, Signal restoration for linear systems with weighted inputs, Inv. Prob. 1 (1985) 67-85.

F. Gori and C. Palma, On the eigenvalues of the sinc2 kernel, J. Phys. A11 (1975) 1709-1719.

A. H. Greenway, The phase problem in astronomy, J. Optics (Paris) 10 (1979) 308-310.

, Proposal for phase recovery form a single intensity distribution, Opt. Lett. 1 (1977) 10-12.

M. A. Grimm and A. W. Lohmann, Super-resolution for one-dimensional objects, J. Opt. Soc. Am. 56 (1966) 1151-1156.

C. W. Groetsch, The Theory of Tikhonov Regularization (Pitnam Books Ltd., London, 1984).

, Generalized Inverses of Linear Operators: Representation and Approximation (Dekker, New York, 1977).

F. A. Grunbaum, A study of Fourier space methods for limited angle image reconstruction, Numer. Func. Anal. Opt. 2 (1980) 31-42.

F. A. Grunbaum, Second-order differnetial operators commuting with convolution integral operators, J. Math. Anal. Appl. 91 (1983) 80-93.

L. G. Gubin, P. T. Polyak amd E. V. Raik, The method of projections for finding the common point of convex sets, USSR Comp. Math. and Math. Phys. 7 (1967) (6),1-24.

T. J. Hall, A. M. Darling and M. A. Fiddy, Image compression and restoration incorporating prior knowledge, Opt. Lett. 7 (1982) 467- 468.

I. Halperin, The product of projection operators, Acta Sci. Math. 23 (1962) 96-99.

B. Halpern, Fixed points of nonexpanding maps, Bull Amer. Math. Soc. 73 (1967) 957-961.

K. Hanson, Bayesian and related methods in image reconstruction from incomplete data, in Image Recovery (Academic Press, Orlando, 1987).

J. L. Harris, Diffraction and resolving power, J. Opt. Soc. Am 54 (1964) 931-936.

M. H. Hayes, The reconstruction of a multidimensional sequence from the phase or magnitude of its Fourier transform, IEEE ASSP-30 (1982) 140-154.

, Signal reconstruction from spectral phase or spectral magnitude, in Advances in Computer Vision and Image Processing 1 (1984) 145-189.

, The unique reconstruction of multidimensional sequences from Fourier transform magnitude or phase, in Image Recovery (Academic Press, New York, 1987) 195-230.

M. H. Hayes, J. S. Lim and A. V. Oppenheim, Signal reconstruction from phase or magnitude, IEEE ASSP-28 (1980) 672-680.

M. H. Hayes and J. H. McClellan, Reducible polynomials in more than one variable, Proc. IEEE 70 (1982) 197-198.

M. H. Hayes and T. F. Quatieri, The importance of boundary conditions in the phase retieval problem, Proc. ICASSP 82 3 (1982) 1545.

, Recursive phase retrieval using boundary conditions, J. Opt. Soc. Am. 73 (1983) 1427-1433.

D. A. Hayner and W. K. Jenkins, The missing cone problem in computer tomography, Advances in Computer Vision and Image Processing 1 (1984) 83-144.

C. W. Helstrom, Image restoration by the method of least squares, J. Opt. Soc. Am. 57 (1967) 297-304.

G. T. Herman, A. Lent and S. W. Rowland, ART: Mathematics and Applications, J. theo. Biol. 42 (1973) 1-32.

G. T. Herman and A. Lent, Iterative reconstruction algorithms, Comput. Biol. Med. 6 (1976) 237-294.

G. T. Herman, Image Reconstruction from Projections (Academic Press, New York, 1980).

G. T. Herman and H. Levkowitz, Initial performance of block-iterative reconstruction algorithms, in NATO ASI (Springer-Verlag, Berlin, 1988) 305-317.

G. T. Herman et al., Multilevel image reconstruction, in Multiresolution Image Processing and Analysis (Springer-Verlag, Berlin, 1984) 121-135.

G. T. Herman, Application of maximum entropy and Bayesian optimization methods to image reconstruction from projections, in Maximum Entropy and Bayesian Methods in Inverse Problems (D. Reidel, Dordrecht, 1985) 319-338.

E. Hille and J. D. Tamarkin, On the characteristic values of linear integral equations, Acta Math. 57 (1931) 1-76.

W. S. Hinshaw and A. H. Lent, An introduction to NMR imagery, Proc. IEEE 71 (1983) 338-350.

B. J. Hoenders, On the solution of the phase retrieval problem, J. Math. Phys. 16 (1975) 1719-1725.

B. J. Hoenders and H. A. Ferwerda, On the reconstruction of a weak phase-amplitude object, Optik 37 (1973) 542-556.

B. J. Hoenders, The uniqueness of inverse problems, in Inverse Source Problems in Optics (Springer-Verlag, Berlin, 1978) 41-82.

D. F. Hoerl and J. P. Allebach, Testing digital filters: numerically derived signals improve frequency measurements (preprint, Univ. Delaware).

E. M. Hofstetter, Construction of time-limited functions with specified autocorrelation functions, IEEE IT-10 (1964) 119-126.

L. Hormander, Linear Partial Differential Operators (Springer- Verlag, Berlin, 1963).

T. S. Huang, J. W. Burnett, and A. G. Deczky, The importance of phase in image procesing filters, IEEE ASSP-23 (1975) 529-542.

T. S. Huang et al., Iterative Image Restoration Appl. Opt. 14 (1975) 1165-1168.

T. T. Huang et al., Image representation by one-bit Fourier phase, IEEE ASSP-36 (1988) 1292-1304.

D. A. Huffman, The generation of impulse equivalent pulse trains, IRE IT-8 (1962) S10-S16.

A. M. J. Huiser and H. A. Ferwerda, The problem of phase retrieval in light and electron microscopy of strong objects, II, Optica Acta 23 (1976)

445-456; III, Optica Acta 47 (1977) 1-8.

A. M. J. Huiser et al., On phase retrieval in electron microscopy from image and diffraction patterns, Optik 45 (1976) 303-316; II, Optik 46 (1976) 407-420.

A. Huiser and P. van Toorn, Ambiguity of the phase-reconstruction problem, Opt. Lett. 5 (1980) 499-501.

B. R. Hunt, Deconvolution of linear systems by constrained regression and its relation to Wiener theory, IEEE AC-16 (1971) 703-705.

Y. Ichioka and N. Nakajima, Iterative image restoration considering visibility, J. Opt. Soc. Am. 71 (1981) 983-988.

V. K. Ivanov, The approximate solution of operator equations of the first kind, USSR Comp. Math. Math. Phys. 6 (no. 6) (1966) 197-205.

D. Izraelevitz and J. S. Lim, A new direct algorithm for image reconstruction from Fourier transform magnitude, IEEE ASSP-35 (1987) 511-519.

A. K. Jain, Spectal estimation and signal extrapolation in one and two dimensions, in Proc. RADC Spec. Est. Workshop (1978) 195-214.

A. K. Jain and S. Ranganath, Extrapolation algorithms for discrete signals with application in spectral estimation, IEEE ASSP-29 (1981) 830-845.

P. A. Jansson, R. H. Hunt and E. K. Plyler, Resolution enhancement of spectra, J. Opt. Soc. Am. 60 (1970) 596-599.

L. Joyce and W. L. Root, Precision bounds in superresolution processing, J. Opt. Soc. Am. A1 (1984) 149-168.

S. Kaczmarz, Angenaberte auflosung von systemen linearen gleichungen, Bull. Acad. Polon. Sci. Lett. A35 (1937) 355-357.

I. Kadar, Robustized estimation and image processing (thesis, Polytechnic Institute of New York, 1977).

W. J. Kammerer and M. Z. Nashed, Iterative methods for best approximate solutions of linear integral equations of the first and second kinds, J. Math. Anal. Appl. 40 (1972) 547-573.

, Steepest descent for singular linear operators with nonclosed range, Appl. Anal. 1 (1971) 143-159.

, On the convergence of the conjugate gradient method for singular linear operator equations, SIAM J. Num. Anal. 4 (1971) 165-181.

Y. Kano and E. Wolf, Temporal coherence of black body radiation, Proc. Phys. Soc. 80 (1962) 1273-1276.

D. Kaplan and R. J. Marks II, Noise sensitivity of interpolation and extrapolation matrices, Appl. Opt. 21 (1982) 4489-4492.

A. K. Katsaggelos, A unified approach to iterative image restoration, SPIE (1987).

S. M. Kay and R. Sudhaker, A zero crossing-based spectrum analyzer, IEEE ASSP-34 (1986) 96-104.

S. Kaylar and H. L. Weinert, Error bounds for the method of alternating projections, Math. Cont. Sig. Sys. 1 (1988) 43-59.

I. Kazuyoshi and O. Yoshihiro, Phase estimation based on maximum likelihood criterion, Appl. Opt. 22 (1983) 3054-3057.

D. Kermisch, Image reconstruction from phase information only, J. Opt. Soc. Am. 60 (1970) 15-17.

, A deterministic analysis of the maximum entropy image restoration method and of some related methods, J. Opt. Soc. Am. 67 (1977) 1154-1159.

P. Kersten and L. Kurz, Robustized vector Robbins-Monro algorithm with applications to M-interval detection, Inform. Sci. 11 (1976) 121-140.

, Bivariate M-interval classifiers in pattern recognition, Inf. Contr. (1977).

Y. I. Khurgin and V. P. Yakovlev, Progress in the Soviet Union on the theory of bandlimited functions, Proc. IEEE 65 (1977) 1005- 1029.

P. Kiedron, Conditions sufficient for a one dimensional unique recovery of phase ..., Opt. App. 10 (1980) 149-154; 253-265; 483- 486.

, On the 2-D solution ambiguity of the phase recovery problem, Optik 59 (1981) 303-309.

, Propagation of phase in optical systems, SPIE 808 (1987).

A. Klug, R. A. Crowther and D. J. DeRosier, The reconstruction of a three-dimensional structure from projections and its application to electron microscopy, Proc. Roy. Soc. London A317 (1970) 319-340.

K. T. Knox, Image retrieval from astronomical speckle patterns, J. Opt. Soc. Am. 66 (1976) 1236-1239.

K. T. Knox and B. J. Thompson, Recovery of images from atmospherically -degraded short-exposure photographs, Astrophys. J. 193 (1974) L45-L48.

D. Kohler and L. Mandel, Source reconstruction from the modulus of the correlation function, J. Opt. Soc. Am. 63 (1973) 126.

O. M. Kosheleva, Sufficiency proof and error estimates for closure phase method (Leningrad preprint).

O. M. Kosheleva and V. Ja. Kreinovic, A new method of reconstructing nonnegative functions from the results of approximative measurements (Leningrad preprint).

R. Kovacs and L. Solyman, Theory of aperture aerials based on the properties of entire functions of exponential type, Acta Phys. Budapest 6 (1956) 161-184.

W. Kryloff, On functions which are regular in a half-plane, Rec. Math. (Mat. Sbornik) NS 6 (1939).

B. V. K. Kumar, Effect of signal bandwidth on the accuracy of signal reconstruction from its phase, IEEE ASSP-32 (1984) 1238-1239.

A. Labeyrie, Attainment of diffraction limited resolution in large telescopes by Fourier analyzing speckle patterns in star images, Astr. and Astrophys. 6 (1970) 85-87.

R. L. Lagendijk, J. Biemond and D. E. Boekee, Regularized iterative image restoration with ringing reduction, IEEE ASSP-36 (1988) 1874-1888.

R. L. Lagendyk et al., Recursive and iterative methods for image identification and restoration (preprint, 1988).

H. J. Landau, The eigenvalue behavior of certain convolution equations, Trans. Amer. Math. Soc. 115 (1965) 242-256.

, Necessary density conditions for sampling and interpolation of certain entire functions, Acta Math. 117 (1967) 37-52.

, On the recovery of a band-limited signal, after instantaneous companding and subsequent band limiting, Bell Sys. Tech. J. 39 (1960) 351-364.

H. J. Landau and W. L. Miranker, The recovery of distorted band-limited signals, J. Math. Anal. Appl. 2 (1961) 97-104.

H. J. Landau and H. Widom, Eigenvalue distribution of time and frequency limiting, J. Math. Anal. Appl. (1980) 469-481.

L. Landweber, An iteration formula for Fredholm integral equations of the first kind, Amer. J. Math. 73 (1951) 615-624.

R. G. Lane, W. R. Fright and R. H. T. Bates, Direct phase retrieval, IEEE ASSP-35 (1987) 520-525.

R. G. Lane and R. H. T. Bates, Automatic multidimensional deconvolution, J. Opt. Soc. Am. A4 (1987) 180-188.

A. Lannes et al., Resolution and robustness in image processing: a new regularization principle, J. Opt. Soc. Am. A4 (1987) 189.

A. Lannes et al., Stabilized reconstruction in signal and image processing I., J. Mod. Op. 34 (1987) 161-226, II. 321-370.

W. Lawton, Uniqueness results for the phase-retrieval problem for radial functions, J. Opt. Soc. Am. 71 (1981) 1519-1522.

W. Lawton and J. Morrison, Factoring trigonometric polynomials regarded as entire functions of exponential type, J. Opt. Soc. Am. A4 (1987) 105.

R. M. Leahy and C. E. Goutis, An optimal technique for constraint-based image restoration and reconstruction, IEEE ASSP-34 (1986) 1629-1642.

W. H. Lee, Method for converting a gaussian laser beam into a uniform beam, Opt. Comm. 36 (1981) 469-471.

A. Lent and H. Tuy, An iterative method for the extrapolation of band-limited functions, J. Math. Anal. Appl. 83 (1981) 554-565.

L. B. Lesem, P. Hirsch and J. A. Jordan, Jr., Computer generation and reconstruction of holograms (preprint).

, The kinoform: a new waveform reconstruction device, IBM Res. Dev. 13 (1969) 150-154.

A. Levi and H. Stark, Restoration from phase and magnitude by generalized projections, in Image Recovery: Theory and Application (Academic Press, New York, 1987) 277-320.

———, Signal restoration from phase by projection onto convex sets, J. Opt. Soc. Am. 73 (1983) 810-822.

———, Image restoration by the method of generalized projections with application to restoration from magnitude, J. Opt. Soc. Am. A1 (1984) 932-943.

B. Ya. Levin, Distribution of Zeos of Entire Functions (Amer. Math. Soc., Providence, 1964).

R. M. Lewitt and R. H. T. Bates, Image reconstruction from projections, I. Optik 50 (1978) 19-33; III, 189-204; IV, 269-278.

R. M. Lewitt, Reconstruction algorithms: transform methods, Proc. IEEE 71 (1983) 390-408.

Y. T. Li and A. L. Kurkjian, Arrival time determination using iterative signal reconstruction from the phase of the cross spectrum, IEEE ASSP-31 (1983) 502-504.

J. S. Lim and N. A. Malik, A new algorithm for two-dimensional maximum entropy power spectrum estimation, IEEE ASSP-29 (1981) 401-413.

J. S. Lim, Multidimensional spectral estimation, in Advances in Computer Vision and Image Processing, 2 (1986) 275-322.

B. Liu and N. C. Gallagher, Convergence of a spectrum shaping algorithm, Appl. Opt. 13 (1974) 2470-2471.

———, Optimum Fourier-transform division filters with magnitude constraint, J. Opt. Soc. Am. 64 (1974) 1227- 1236.

B. F. Logan, Jr., Signals designed for recovery after clipping, ATT Bell Lab. Tech. J. 63 (1984) 261-285; 287-306; 379-399.

———, Theory of analytic modulation systems, Bell Sys. Tech. J. 57 (1978) 491-576.

———, Information in the zero crossings of bandpass signals, Bell Sys. Tech. J. 56 (1977) 487-510.

A. W. Lohmann and D. P. Paris, Super-resolution for nonbirefringent objects, Appl. Opt. 3 (1964) 1037-1043.

W. Lokusz, Optical systems with resolving power exceeding the classical limit, I. J. Opt. Soc. Am. 56 (1966) 1463-1472; II. ibid. 57 (1967) 932-941.

S. F. McCormick, An iterative procedure for the solution of constrained nonlinear equations, Numer. Math. 23 (1975) 371-385.

———, The methods of Kaczmarz and iterative row orthogonalization for solving linear equations and least square problems in Hilbert space, Indiana Univ. Math. J. 26 (1977) 1137-1150.

J. D. McDonnell, R. J. Marks II, and L. E. Atlas, An introduction to neural networks for solving combinatorial search problems (preprint).

J. G. McWhirter and E. R. Pike, On the numerical inversion of the

Laplace transform and similar Fredholm integral equations of the first kind, J. Phys. A11 (1978) 1729-1745.

J. Maeda and K. Murata, Retrieval of wave aberration from point spread function or optical transfer function data, Appl. Opt. 20 (1981) 274-279.

J. N. Mait and W. T. Rhodes, Iterative design of pupil functions for bipolar incoherent spatial filtering, SPIE 292 (1981) 66-70.

H. Maitre, Iterative super resolution, some new fast methods, Opt. Acta 28 (1981) 973-980.

R. J. Mammone, Spectral extrapolation of constrained signals, J. Opt. Soc. Am. 73 (1983) 506-526.

R. Mammone and G. Eichman, Restoration of discrete Fourier spectra using linear programming, J. Opt. Soc. Am. 72 (1982) 987-992.

I. Manolitsakis, Two-dimensional scattered fields: a description in terms of zeros of entire functions, J. Math. Phys. 23 (1982) 2291- 2298.

M. T. Manry and J. K. Aggarwal, The measurement of phase distortion due to filtering in digital pictures, IEEE ASSP-25 (1977) 534-541.

, The design of multi-dimensional FIR digital filters by phase correction, IEEE CAS-23 (1976) 185-199.

, Design and implementation of two-dimensional FIR digital filters with nonrectangular arrays, IEEE ASSP-26 (1978) 314-318.

R. J. Marks, II, Gerchberg's extrapolation algorithm in two dimensions, Appl. Opt. 20 (1981) 1815-1820.

, Coherent optical extrapolation of 2-D band-limited signals: processor theory, Appl. Opt. 19 (1980) 1670-1672.

, Posedness of bandlimited image extension problem in tomography, Opt. Lett. 7 (1982) 376-377.

R. J. Marks II and T. Reightley, On iterative evaluation of extrema of integrals of trigonometric polynomials, IEEE ASSP-33 (1985) 1039- 1040.

R. J. Marks II and D. K. Smith, Gerchberg-type linear deconvolution and extrapolation, SPIE 373 (1981) 161-178.

R. J. Marks II and M. J. Smith, Closed-form object restoration, Opt. Lett. 6 (1981) 522-524.

R. J. Marks II, Class of continuous level associative memory neural nets, Appl. Opt. 26 (1987) 2005-2010.

R. J. Marks II and D. K. Smith, An iterative coherent processor for bandlimited signal extrapolation, SPIE 231 (1980) 106.

R. J. Marks II and L. E. Atlas, Content addressable memories: a relationship between Hopfield's neural net and an iterative matched filter (preprint).

R. J. Marks II, L. E. Atlas and K. F. Cheung, A class of continuous level neural nets, SPIE 813 (1987) 29-30.

D. Marr, S. Ullman and T. Poggio, Bandpass channels, zero crossings and early visual information processing, J. Opt. Soc. Am. 69 (1979)

914-916.

J. T. Marti, On the convergence of the discrete ART algorithm for reconstruction of digital pictures from their projections, Computing 21 (1979) 105-111.

E. Masry, The recovery of distorted band-limited stochastic processes, IEEE IT-19 (1973) 398-403.

——, Distortionless demodulation of narrow-band single-sideband angle modulated signals, IEEE IT-23 (1977) 582-591.

E. Masry and S. Cambanis, Bandlimited processes and certain nonlinear transformations, J. Math. Anal. Appl. 53 (1976) 59-77.

J. E. Mazo and J. Salz, Spectral properties of single-sideband angle modulation, IEEE COM-10 (1968) 52-61.

B. P. Medoff, Image reconstruction from limited data, in Image Recovery (Academic Press, Orlando, 1987).

C. L. Mehta, E. Wolf and A. P. Balachandran, Some theorems on the unimodular-complex degree of optical coherence, J. Math. Phys. 7 (1966) 133-138.

G. A. Merchant and T. W. Parks, Reconstruction of signal from phase: efficient algorithms, segmentation and generalizations, IEEE ASSP-31 (1983) 1135-1147.

R. M. Mersereau, Recovering multi-dimensional signals from their projections, Comp. Graph Im. Proc. 1 (1973) 179-195.

R. M. Mersereau and A. V. Oppenheim, Digital reconstruction of multidimensional signals from their projections, IEEE ASSP 62 (1974) 1319-1338.

K. H. Meyn, Solution of under determined nonlinear equations by stationary iterative methods, Numer. Math. 42 (1983) 161-172.

R. P. Millane et al., Towards direct phase retrieval in macro- molecular crystallography, Biophys. J. 49 (1986) 60-62.

K. Miller, Least squares method for ill-posed problems with a prescribed bound, SIAM J. Math. Anal. 1 (1970) 52-74.

K. S. Miller, Complex linear least squares, SIAM Review 15 (1973) 706-726.

G. N. Minerbo, MENT: a maximum entropy algorithm for reconstructing a cource from projection data, Comp. Graph. Imag. Proc. 10 (1979) 48-68.

D. L. Misell, An examination of an iterative method for the solution of the phase problem in optics and electron optics, J. Phys. D6 (1973) 2200-2225.

D. L. Misell, A method for the solution of the phase problem in electron microscopy, J. Phys. D6 (1973) L6-L9.

, The phase problem in electron microscopy, in Advances in Optical and Electron Microscopy 7 (1978) 185-279.

W. D. Montgomery, Phase retrieval and the polarization identity, Opt. Lett. 2 (1978) 120-121.

, Optical applications of von Neumann's alternating-projection theorem, Opt. Lett. 7 (1982) 1-3.

V. A. Morozov, Generalized Inverses and Applications (Springer-Verlag, Berlin, 1984).

C. E. Morris et al., Fast reconstruction of linearly distorted signals, IEEE ASSP 36 (1988) 1017-1025.

C. E. Morris et al., An iterative deconvolution algorithm with quadratic convergence, J. Opt. Soc. Am. A4 (1987) 200-207.

D. Morris, Phase retrieval in the radio holography of reflector antennas and radio telescopes, IEEE AP-33 (1985) 749-755.

J. B. Morton and H. C. Andrews, A posteriori method of image restoration, J. Opt. Soc. Am. 69 (1979) 280-290.

H. E. Moses and R. T. Prosser, Phases of complex functions from amplitudes of functions and the amplitudes of the Fourier and Mellin transforms, J. Opt. Soc. Am. 73 (1983) 1451-1454.

T. S. Motzkin and I. J. Schoenberg, The relaxation method for linear inequalities, Can. J. Math. 6 (1954) 393-404.

D. C. Munson and J. L. C. Sanz, Phase only image reconstruction from offset Fourier data, Opt. Eng 25 (1986) 655-661.

D. C. Munson et al., A tomographic formulation of spotlight-mode synthetic aperture radar, Proc. IEEE 71 (1983) 917-925.

D. C. Munson and J. Sanz, Image reconstruction from frequency offset Fourier data, Proc. IEEE 76 (1984) 66.

P. S. Naidu and B. Paramasivaiah, Estimation of sinusoids from incomplete time series, IEEE ASSP-32 (1984) 559-562.

N. Nakajima and T. Asakura, Study of the algorithm in the phase retrieval using the logarithmic Hilbert transform, Opt. Comm. 41 (1982) 89-94.

, Two-dimensional phase retrieval using the logarithmic Hilbert transform and the estimation technique of zero information, J. Phys. D19 (1986) 319-331.

, Phase retrieval from the image intensity using an exponential filter, Optik 64 (1983) 37-49.

, A new approach to two-dimensional phase retrieval, Optica Acta 32 (1985) 647-658.

, Study of zero location by means of exponential filter, Optik 60 (1982) 289-305.

, Extraction of the influence of zeros, Optik 63 (1983) 99-108.

P. J. Napier and R. H. T. Bates, Inferring phase information from modulus information in two-dimensional aperture synthesis, Astr. Astrophys. Supp. 15 (1974) 427-430.

M. Z. Nashed and G. Wahba, Generalized inverses in reproducing kernel spaces: an approach to regularization of linear operator equations, SIAM J. Math. Anal. 5 (1974) 974-987.

, Convergence rates of approximate least squares solutions of linear integral and operator equations of the first kind, Math. Comp. 28 (1974) 69-80.

M. Z. Nashed, Operator-theoretic and computational approaches to ill-posed problems with applications to antenna theory, IEEE AP-29 (1981) 220-231.

, Generalized inverses, normal solvability and iteration for singular operator equations, in Nonlinear Functional Analysis and Applications (Academic Press, New York, 1971) 311-359.

, Continuous and semicontinuous analogues of Cimmino and Kaczmarz with applications to the inverse Radon transform, in Mathematical Aspects of Computerized Tomography (Springer-Verlag, Berlin, 1981) 160-178.

K. M. Nashold and B. E. A. Saleh, Image construction through diffraction-limited systems: an iteative approach, J. Opt. Soc. Am. A2 (1985) 635-643.

G. Newsam and R. Barakat, The essential dimension as a well-defined number of degrees of freedom of finite convolution operators appearing in optics (preprint).

M. Nieto-Vesperinas, Dispersion relations in two dimensions: application to the phase problem, Optik 56 (1980) 377-384.

M. Nieto-Vesperinas and J. C. Dainty, A note of Eisenstein's irreducibility criterion for two dimensional sampled objects, Opt. Comm. 52 (1985) 94-98.

, Phase recovery for two dimensional digital objects by polynomial factorization, Opt. Comm. 58 (1986) 83-88.

S. J. Norton, Iterative reconstruction algorithms, convergence as a function of spatial frequency, J. Opt. Soc. Am. 2A (1985) 6-13.

H. M. Nussenzveig, Phase problem in coherence theory, J. Math. Phys. 8 (1967) 561-572.

E. L. O'Neill and A. Walther, The question of phase in image formation, Opt. Acta 10 (1963) 33-40.

A. M. Odlyzko, On the distribution of spacings between zeros of the zeta function, Math. Comp. 28 (1987) 273-308.

Y. Ohta, Nonlinear accretive mappings in Banach space, SIAM J. Math. Anal. 10 (1979) 337-353.

E. Oja and H. Ogawa, Parametric projection filter for image and signal restoration, IEEE ASSP-34 (1986) 1643-1653.

Z. Opial, Weak convergence of the successive approximations for non-expansive mappings in Banach spaces, Bull. Amer. Math. Soc. 73 (1967) 591-597.

A. V. Oppenheim, M. H. Hayes and J. S. Lim, Iterative procedures for signal reconstruction from phase, SPIE 231 (1980) 121-129; Opt. Eng. 21 (1982) 122-127.

A. V. Oppenheim and J. S. Lim, The importance of phase in signals, Proc. IEEE 69 (1981) 529-541.

A. V. Oppenheim, J. S. Lim and S. R. Curtis, Signal synthesis from partial Fourier domain information, J. Opt. Soc. Am. 73 (1983) 1413-1420.

J. M. Ortega and W. C. Rheinboldt, Iterative Solution of Nonlinear Equations in Several Variables (Academic Press, New York, 1970).

R. E. A. C. Paley and N. Wiener, Fourier Transforms in the Complex Domain (Amer. Math. Soc., Providence, 1934).

A. Papoulis, A new algorithm in spectral analysis and band-limited extrapolation, IEEE CAS-22 (1975) 735-742.

A. Papoulis and C. Chamzas, Detection of hidden periodicities by adaptive extrapolation, IEEE ASSP-27 (1979) 492-500.

, Improvement of range resolution by spectral extrapolation, Ultrasonic Imag. 1 (1979) 121-135.

A. L. Patterson, Homometric structures, Nature 143 (1939) 939-940.

, Ambiguities in the x-ray analysis of crystal structure, Phys. Rev. 65 (1944) 195-201.

W. M. Patterson III, Iterative Methods for the Solution of a Linear Operator Equation in Hilbert Space (Springer-Verlag, Berlin, 1974).

L. Pauling and M. D. Shappell, The crystal structure of bixbyite and the C-modification of the sesquioxides, Z. Krist. 90 (1935) 517- 542.

L. Personnaz, I. Guyon and G. Dreyfus, Information storage and retrieval in spin-glass like neural networks, J. Phys. Lett. 46 (1985) L359-L365.

B. L. Phillips, A technique for the numerical solution of certain integral equations of the first kind, J. Ass. Comp. Mach. 9 (1962) 84-97.

M. Plancherel and G. Polya, Fonctions entieres et integrales de Fourier multiples, Comm. math. Hel. 9 (1937) 224-248; 10 (1938) 110-163.

T. Poggio, H. K. Nishihara and K. R. K. Nielson, Zero crossings and spatio temporal interpolation in vision (preprint, MIT, 1982).

G. Polya, Uber die nullstellen gewisser ganzer funktionen, Math. Z. 2 (1918) 352-383.

R. S. Powers and J. W. Goodman, Error rates in computer generated holographic memories, Appl. Opt. 14 (1975) 1690-1701.

R. Prost and R. Goute, Deconvolution when the convolution kernel has no inverse, IEEE ASSP-25 (1977) 542-548.

T. F. Quatieri and D. E. Dudgeon, Implementation of 2-D digital filters by iterative methods, IEEE ASSP-30 (1982) 473-487.

T. F. Quatieri and A. V. Oppenheim, Iterative techniques for minimum phase signal reconstruction from phase or magnitude, IEEE ASSP-29 (1981) 1187-1193.

T. F. Quatieri et al., Convergence of iterative signal reconstruction algorithms, Inter. Conf. ASSP (1981) 35-38.

Y. Rahmat-Samii, Microwave holography of large reflector antennas–simulation algorithms, IEEE ASSP-33 (1985) 1194-1203.

, Surface diagnosis of large reflector antennas using microwave holographic metrology – an iterative approach, Radio Sci. 13 (1984) 1205-1217.

R. S. Ramakrishnan et al., Orthogonalization, Bernstein polynomials and image restoration, Appl. Opt. 18 (1979) 464-468.

A. C. S. Readhead and R. N. Wilkinson, The mapping of compact radio sources from VLBI data, Astrophys. J. 223 (1978) 25-36.

T. C. Redshaw, The numerical inversion and extrapolation of integral transforms (thesis, Univ. Wales, 1982).

A. A. G. Requicha, The zeros of entire functions: theory and engineering applications, Proc. IEEE 68 (1980) 308-328.

A. A. G. Requicha and H. B. Voelcker, Design of nonrecursive filters by specification of frequency domain zeros, IEEE AU-18 (1970) 464- 470.

C. L. Rino, Bandlimited image restoration by linear mean-square estimation, J. Opt. Soc. Am. 59 (1969) 547-550.

J. A. Ritcey et al., A signal space interpretation of neural nets, 1987 IEEE Inter. Symp. on Circuits and Systems 2 (1987) 370-373.

A. E. E. Rogers, Method of using closure phases in radio aperture synthesis, SPIE 231 (1980) 10-17.

P. Roman and A. S. Marathay, Analyticity and phase retrieval, Il Nuovo Cim. 30 (1963) 1452-1464.

W. L. Root, Ill-posedness and precision in object field reconstruction problem, J. Opt. Soc. Am. A4 (1987) 171.

L. Ronkin, Introduction to the Theory of Entire Functions of Several Complex Variable (Amer. Math. Soc., Providence, 1974).

J. Rosenblatt, Phase retrieval, Comm. math. Phys. 95 (1984) 317- 343.

, Determining a distribution from the modulus of its Fourier transform, Complex Var. Appl. (1987).

J. Rosenblatt and P. Seymour, The structure of homometric sets, SIAM J. Alg. Disc. Math. 3 (1982) 343-350.

G. Ross, Iterative methods in information processing for object restoration, Optica Acta 29 (1982) 1523-1542.

G. Ross et al., The propagation and encoding of information in the scattered field by complex zeros, Opt. Acta 26 (1979) 229-238.

G. Ross, M. A. Fiddy and H. Moezzi, The solution to the inverse scattering problem, based on fast zero location from two measurements, Opt. Acta 27 (1980) 1433-1444.

G. Ross, M. A. Fiddy and M. Nieto-Vesperinas, The inverse scattering problem in structural determinations, in Inverse Scattering Problems (Springer-Verlag, Berlin, 1980) 15-71.

D. Rotem and Y. Y. Zeevi, Image reconstruction from zero crossings, IEEE ASSP-34 (1986) 1269-1277.

C. K. Rushforth, Signal restoration, functional analysis and Fredholm integral equations of the first kind, in Image Recovery (Academic Press, New York, 1987) 1-27.

C. K. Rushforth et al., Least-squares reconstruction of objects with missing high frequency components, J. Opt. Soc. Am. 72 (1982) 204-211.

C. K. Rushforth and R. L. Frost, Comparison of some algorithms for reconstructing space-limited images, J. Opt. Soc. Am. 70 (1980) 1539-1544.

C. K. Rushforth and R. W. Harris, Restoration, resolution and noise, J. Opt. Soc. Am. 58 (1968) 539-545.

M. S. Sabri and W. Steenaart, Comments on "an extrapolation procedure for bandlimited signals", IEEE ASSP-28 (1980) 254.

, An approach to band-limited signal extrapolation – the extrapolation matrix, IEEE CAS-25 (1978) 74-78.

B. E. A. Saleh, Image synthesis: discovery instead of recovery, in Image Recovery (Academic Press, New York, 1987) 463-498.

I. W. Sandberg, On Newton-direction algorithms and diffeomorphisms, Bell Sys. Tech. J. 60 (1981) 339-346.

, On the properties of some systems that distort signals, I, Bell Sys. Tech. J. 42 (1963) 2033-2045.

, Diffeomorphisms and Newton-direction algorithms, Bell Sys. Tech. J. 59 (1980) 1721-1733.

J. L. C. Sanz, Mathematical considerations for the problem of Fourier transform phase retrieval from magnitude, SIAM J. Appl. Math. 45 (1985) 651-664.

, On the reconstruction of band-limited multidim- ensional signals from algebraic sampling contour, Proc. IEEE 73 (1985) 1334-1336.

J. L. C. Sanz and T. S. Huang, Unique reconstruction of a band- limited multi-dimensional signal from its phase or magnitude, J. Opt. Soc. Am. 73 (1983) 1446-1450.

, A unified Hilbert space approach to iterative linear signal restoration, J. Opt. Soc. Am. 73 (1983) 1455-1465.

, A unified approach to noniterative linear signal restoration, IEEE ASSP-32 (1984) 403-409.

, Support-limited signal and image extrapolation, in Advances in Computer Vision and Image Processing 1 (1984) 1-82.

, On the Papoulis-Gerchberg algorithm, IEEE CAS (1983).

, On iterative procedures for super- resolution (preprint).

, Discrete and continuous bandlimited signal extrapolation, IEEE ASSP-31 (1983) 1276-1285.

, Iterative time-limited signal extrapolation, IEEE ASSP-31 (1983) 643-649.

, Continuation techniques for a certain class of analytic functions (preprint).

, Some aspects of band-limited signal extrapolation models, discrete approximations and noise, IEEE ASSP-31 (1983) 1492-1501.

, Polynomial system of equations and its application to the study of the effect of noise on multidimensional Fourier transform phase retrieval from magnitude, IEEE ASSP-33 (1985) 997-1004.

J. L. C. Sanz, T. S. Huang and F. Cukierman, Stability of unique Fourier transform phase reconstruction, J. Opt. Soc. Am. 73 (1983) 1442-1445.

J. L. C. Sanz, T. S. Huang and T. Wu, A note on iterative Fourier transform phase reconstruction from magnitude, IEEE ASSP-32 (1984) 1251-1254.

T. K. Sarkar, D. D. Weiner and V. K. Jain, Some mathematical considerations in dealing with the inverse problem, IEEE AP-29 (1981) 373-379.

T. Sato et al., Tomographic image reconstruction from limited projections using iterative revisions in image and transform spaces, Appl. Opt. 20 (1981) 395-399.

T. Sato et al., Tomographic image reconstructions from limited projections using coherent optical feedback, Appl. Opt. 20 (1981) 3073-3076.

R. J. Sault, Two procedures for phase estimation from visibility magnitudes, Aust. J. Phys. 37 (1984) 209-229.

W. O. Saxton, Digital processing of electron images - a survey of motivations and methods, Elec. Micr. 1 (1980) 486-493.

, Computer Techniques for Image Processing in Electron Microscopy (Academic Press, New York, 1978).

, Recovery of specimen information for strongly scattering objects, in Computer Processing of Electron Microscope Images (Springer-Verlag, Berlin, 1980) 35-87.

S. I. Sayegh et al., An algorithm to find two dimensional signals with specified zero crossings, IEEE ASSP-35 (1987) 107-111.

S. I. Sayegh et al., Image design, IEEE ASSP 33 (1985) 460-465.

S. I. Sayegh and B. E. A. Saleh, Image design: generation of a prescribed image at the output of band-limited optical systems, IEEE ASSP 33 (1985) 467-470.

R. Schaefer, R. Mersereau and M. Richards, Constrained iterative restoration algorithms, Proc. IEEE 69 (1981) 432-450.

H. Schaeffer, Uber die methode sukzessiver approximationens, J. Deutsch. Math. Ver. 59 (1957) 131-140.

A. C. Schell, Enhancing the angular resolution of incoherent sources, Radio Elec. Eng. 29 (1965) 21-26.

H.-J. Schlebusch and W. Splettstosser, On a conjecture of J. L. C. Sanz. and T. S. Huang, IEEE ASSP-33 (1985) 1628-1630.

C. van Schooneveld, Resolution enhancement: the 'maximum entropy method' and the 'high resolution method', in Image Formation from Coherence Functions in Astronomy (Reidel, Dordrecht, 1979) 197-218.

U. J. Schwarz, Mathematical statistical description of the iterative beam removing technique (method CLEAN), Astron. Astrophys. 65 (1978) 345-356.

M. S. Scivier and M. A. Fiddy, Phase ambiguities and the zeros of mul multidimensional band-limited functions, J. Opt. Soc. Am. A2 (1985) 693-697.

, Ambiguities in magnitude only reconstruction of band-limited signals, Opt. Lett. 10 (1985) 369- 371.

M. S. Scivier, T. J. Hall and M. A. Fiddy, Phase unwrapping using the complex zeros of a bandlimited function and the presence of ambiguities in two dimensions, Opt. Acta 31 (1984) 619-623.

M. I. Sezan and H. Stark, Image restoration by convex projections in the presence of noise, Appl. Opt. 22 (1983) 2781-2789.

, Image restoration by the method of convex projections: part 2, IEEE MI-1 (1982) 95-101.

, Applications of convex projection theory to image recovery in tomography and related areas, in Image Recovery: Theory and Applications (Academic Press, New York, 1987) 415-462.

M. I. Sezan et al., Regularized signal restoration using theory of convex projections, Proc. Int. Conf. ASSP 3 (1987) 1565-1568.

M. I. Sezan and A. M. Tekalp, Iterative image restoration with ringing suppression using method of POCS (preprint, 1988).

M. I. Sezan and H. Stark, Tomographic image reconstruction from incomplete view data by convex projections and direct Fourier inversion, IEEE TMI-3 (1984) 91-98.

J. A. Sheehan, Some properties of the finite convolution, IRE IT-8 (1962) S283-S284.

L. A. Shepp and Y. Vardi, Maximum likelihood reconstruction in positron emission tomography, IEEE MI 1 (1982) 113-122.

S. Shitz and Y. Y. Zeevi, On the duality of time and frequency domain signal reconstruction from partial information, IEEE ASSP-33 (1985) 1486-1498.

H. F. Silverman and A. E. Pearson, On deconvolution using the discrete Fourier transform, IEEE AU-21 (1973) 112-118.

P. Y. Simard and G. E. Mailloux, A projection operator for the restoration of divergence-free vector fields, IEEE PAMI-10 (1988) 248-256.

D. Slepian, On bandwidth, Proc. IEEE 64 (1976) 292-300.

, Prolate spheroidal wave functions, Fourier analysis and uncertainty - V: the discrete case, Bell Sys. Tech. J. 57 (1978) 1371-1430.

, Some asymptotic expansions for prolate spheroidal wave functions, J. Math. Phys. 44 (1965) 99-140.

D. Slepian and H. O. Pollack, Prolate spheroidal wave functions, Fourier analysis and uncertainty - I, Bell System Tech. J. 40 (1961) 43-63; II ibid. 40 (1961) 65-84; III, ibid. 41 (1962) 1295-1336.

D. Slepian and E. Sonnenblick, Eigenvalues associated with prolate spheroidal wave functions of zero order, Bell Sys. Tech. J. 44 (1965) 1745-1759.

D. K. Smith and R. J. Marks, II, Closed form bandlimited image extrapolation, Appl. Opt. 20 (1981) 2476-2483.

M. Soumekh, Image reconstruction techniques in tomographic imaging systems, IEEE ASSP-34 (1986) 952-962.

R. V. Southwell, Stress calculations in framework by the method of "systematic relaxation of constraints", Proc. Roy. Soc. A151 (1935) 56-95.

W. H. Southwell, Wavefront analyzer using a maximum likelihood algorithm, J. Opt. Soc. Am. 67 (1977) 396-399.

H. Stark, D. Cahana and H. Webb, Restoration of arbitrary finite energy objects from limited spatial and spectral information, J. Opt. Soc. Am. 71 (1981) 635-642.

H. Stark, D. Cahana and G. J.Habetler, Is it possible to restore an optical object from its low pass spectrum and its truncated image?, Opt. Lett. 6 (1981) 259-260.

H. Stark, S. Cruze and G. Habetler, Restoration of optical objects subject to nonnegative spatial or spectral constraints, J. Opt. Soc. Am. 72 (1982) 993-1000.

I. S. Stefanescu, On the phase retrieval problem in two dimensions, J. Math. Phys. 26 (1985) 2141-2160.

K. Stewart and T. S. Duranni, Constrained signal reconstruction - a unified approach, in Signal Processing III (Elvesier, Amsterdam, 1986) 1423-1426.

T. G. Stockham, T. M. Cannon and R. B. Ingebretsen, Blind deconvolution through digital signal processing, Proc. IEEE 63 (1975) 678-692.

O. N. Strand, Theory and methods related to the singular-function expansion and Landweber's iteration for integral equations of the first kind, SIAM J. Num. Anal. 11 (1974) 798-825.

O. N. Strand and E. R. Westwater, Minimum-RMS estimation of the numerical solution of a Fredholm integral equation of the first kind, SIAM J. Num. Anal. 5 (1968).

B. J. Sullivan and B. Liu, On the use of singular value decomposition and decimation in discrete-time band-limited signal extrapolation, IEEE ASSP-32 (1984) 1201-1212.

K. C. Tam and V. Perez-Mendez, Tomographic imaging with limited-angle input, J. Opt. Soc. Am. 71 (1981) 582-592.

K. C. Tam et al., 3-D object reconstruction in emission and transmission tomography with limited angular input, IEEE NS-26 (1979) 2797-2805.

K. C. Tam et al., Limited angle 3-D reconstructions from continuous and pinhole projections, IEEE NS-27 (1980) 445-458.

K. Tanabe, Projection method for solving a singular system of linear equations, Num. math. 17 (1971) 203-214.

, Characterization of linear stationary iterative processes for solving a singular system of linear equations, Num. math. 22 (1974) 349-359.

L. S. Taylor, The phase retrieval problem, IEEE AP-29 (1981) 406-416.

, Object plane relations in the phase retrieval problem, J. Opt. Soc. Am. 70 (1980) 1554-1556.

T. T. Taylor and J. R. Whinnery, Applications of potential theory to the design of linear arrays, J. Appl. Phys. 22 (1951) 19-29.

G. Temple, The general theory of relaxation methods applied to linear systems, Proc. Roy. Soc. A169 (1938) 476-500.

G. Thomas, A modified version of van-Cittert's iterative deconvolution procedure, IEEE ASSP-29 (1981) 938-939.

A. R. Thompson, J. M. Moran and G. W. Swenson, Jr., Interferometry and Synthesis in Radio Astronomy (Wiley, New York, 1986).

A. N. Tikhonov, Solution of nonlinear integral equations of the first kind, Soviet Math. Dokl 5 (1964) 835.

A. N. Tikhonov and V. Y. Arsenin, Solution of Ill-Posed Problems (Winston-Wiley, New York, 1977).

E. C. Titchmarsh, The zeros of certain integral functions, Proc. London Math. Soc. 25 (1926) 283-302.

V. T. Tom, T. F. Quatieri, M. H. Hayes and J. H. McClellan, Convergence of iterative nonexpansive signal reconstruction, IEEE ASSP-29 (1981) 1052-1058.

G. Toraldo di Francia, Resolving power and information, J. Opt. Soc. Am. 45 (1955) 497-501.

, Degrees of freedom of an image, J. Opt. Soc. Am. 59 (1969) 799-804.

M. R. Trummer, A note on the ART of relaxation, Computing 33 (1984) 349-352.

M. R. Trummer, Reconstructing pictures from projections: on convergence of the ART algorithm with relaxation, Computing 26 (1981) 189-195.

M. R. Trummer, SMART-an algorithm for reconstructing pictures from projections, J. App. Math. Phys. 34 (1983) 746-753.

H. J. Trussell, Convergence criteria for iterative restoration methods, IEEE ASSP-31 (1983) 129-136.

, A priori knowledge in algebraic reconstruction methods, in Advances in Computer Vision and Image Processing 1 (1984) 265-316

H. J. Trussell and B. R. Hunt, Sectional methods for image restoration, IEEE ASSP-26 (1978) 157-164.

H. J. Trussell and M. R. Civanlar, The feasible solution in signal restoration, IEEE ASSP-32 (1984) 201-212.

, The Landweber iteration and projection onto convex sets, IEEE ASSP-33 (1985) 1632-1634.

, Signal deconvolution by projection onto convex sets (preprint, 1984).

V. F. Turchin, V. P. Kozlov, and M. S. Malkevich, The use of mathematical statistics methods in the solution of incorrectly posed problems, Sov. Phys. Usp. 13 (1971) 681-840.

S. Twomey, The application of numerical filtering to the solution of integral equations encountered in indirect sensing measurements, J. Franklin Inst. 279 (1965) 95-109.

P. H. van Cittert, Zum einfluss der splaltbreite auf die intensitat swerteilung in spektrallinien II, Z. Phys. 69 (1931) 298-308.

P. L. van Hove, M. H. Hayes, J. S. Lim and A. V. Oppenheim, Signal reconstruction form signed Fourier transform magnitude, IEEE ASSP-31 (1983) 1286-1293.

P. van Schiske, Ein und mehrdeutigkeit der phasenbestimmung, Optik 40 (1974) 261-275.

P. van Toorn and H. A. Ferwerda, The problem of phase retrieval in light and electron microscopy of strong objects, IV: checking of algorithms by means of simulated objects, Opt. Acta 23 (1976) 469-481.

V. J. Velasco, On the behavior of iterative algorithms based on convex projections in the presence of noise (preprint).

N. J. Vershad, Resolution, optical-channel capacity and information theory, J. Opt. Soc. Am. 59 (1969) 157-163.

G. A. Viano, On the extrapolation of optical image data, J. Math. Phys. 17 (1976) 1160-1165.

J. A. Ville, Sur le prolongement des signaux a spectre born, Cables et Trans. 1 (1956) 44-52.

J. A. Ville and J. Bousitat, Note sur un signal de duree finie et d'energie filtree maximum, Cables et Trans. 11 (1957) 102-127.

H. B. Voelcker, Toward a unified theory of modulation, Part I, Proc. IEEE 54 (1966) 340-353; Part II, ibid. 54 (1966) 735-755.

, Demodulation of single-sideband signals via envelope detection, IEEE COM-14 (1966) 22-30.

H. B. Voelcker and A. P. G. Requicha, Clipping and signal determination: two algorithms requiring validation, IEEE COM-21 (1973) 738-744.

H. B. Voelcker, Zero-crossing properties of angle modulated signals, IEEE COM-20 (1972) 307-315.

J. G. Walker, Optical imaging with resolution exceeding the Rayleigh criterion, Opt. Acta 30 (1983) 1197-1202.

J. G. Walker, The phase retrieval problem: a solution based on zero location by exponential apodization, Opt. Acta 28 (1981) 735-738.

J. G. Walker, Computer simulation of a method for object reconstruction from tellar speckle interferometry, App. Opt. 21 (1982) 3132-3137.

A. Walther, The question of phase retrieval in optics, Opt. Acta 10 (1963) 41-49.

G. Wahba, Practical approximate solutions to linear operator equations when the data are noisy, SIAM J. Num. Anal. 14 (1977) 651-667.

, in Mathematical Aspects of Computerized Tomography (Springer-Verlag, New York, 1981).

S. J. Wernecke and L. R. d'Addario, Maximum entropy image reconstruction, IEEE C-26 (1977) 351-364.

H. Widom, Exteme eigenvalues of N-dimensional convolution operators, Trans. Amer. Math. Soc. 106 (1963) 391-414.

, Asymptotic behavior of eigenvalues of certain integral equations: II, Arch. Rat. Mech. Anal. 17 (1964) 215-229.

N. Wiener, On the factorization of matrices, Comm. math. Hel. 29 (1955) 97-111.

R. Wiley, On an iterative technique for recovery of band limited signals, Proc. IEEE 66 (1978) 522-523.

, Concerning the recovery of a bandlimited signal or its spectrum from a finite segment, IEEE COM-27 (1979) 251-252.

R. G. Wiley, H. Schwarzlander and D. Weiner, Demodulation procedure for very wide-band FM, IEEE COM-25 (1977) 318-327.

R. G. Wiley, Recovery of bandlimited signals from unequally spaced samples, IEEE COM-26 (1978) 135-137.

E. Wolf, Is a complete determination of the energy spectrum of light possible from measurements of the degree of coherence?, Proc. Phys. Soc. London 80 (1962) 1269-1272.

J. Wollnack and P. Besslich, Signal reconstruction from generalized transform phase (preprint, Un. Bremen, 1987).

R. J. Wombell and M. A. Fiddy, Accuracy of extrapolated spectral values as a function of prior knowledge and regularization, J. Opt. Soc. Am. (1987).

M. C. Won, D. Mnyama and R. H. T. Bates, Improving initial phase estimates for phase retrieval algorithms, Opt. Acta 32 (1985) 377-396.

J. W. Wood, T. J. Hall and M. A. Fiddy, A comparison study of some computational methods for locating the zeros of entire functions, Opt. Acta 30 (1983) 511-527.

W. Xu and C. Chamzas, On the extrapolation of band-limited functions with energy constraints, IEEE ASSP-31 (1983) 1222-1234.

K. Yao, Applications of reproducing kernel Hilbert spaces – band-limited signal models, Inf. Cont. 11 (1967) 429-444.

B. Yeǧnanarayama et al. Significance of group delay functions in signal reconstruction from spectral magnitude or phase, IEEE ASSP-32 (1984) 610-622.

C. L. Yeh and R. T. Chin, Error analysis of a class of constrained iterative restoration algorithms, IEEE ASSP-33 (1985) 1593-1598.

D. C. Youla, Generalized image restoration by the method of alternating orthogonal projections, IEEE CAS-25 (1978) 694-702.

, Image restoration by the method of projections onto convex sets - Part I (preprint, 1981).

, Mathematical theory of image restoration by the method of convex projections, in Image Recovery (Academic Press, New York, 1987) 29-77.

D. C. Youla and H. Webb, Image restoration by the method of projections onto convex sets, IEEE TMI-1 (1982) 81-94.

D. C. Youla and V. Velasco, Extensions of a result on the synthesis of signals in the presence of inconsistent constraints, IEEE CAS-33 (1986) 465-468.

M. Zakai, Band-limited functions and the sampling theorem, Inf. Cont. 8 (1965) 143-158.

A. Zakhor, Reconstruction of multidimensional signals from multiple level threshold crossings (preprint, MIT, 1988).

A. Zakhor and D. Izraelevitz, A note on the sampling of zero-crossings of two dimensional signals, Proc. IEEE 74 (1986) 1285- 1287.

G. D. Zames, Conservation of bandwidth in nonlinear operations, Q. Prog. Report MIT (1959) 55.

E. H. Zarantonello, Projections on convex sets in Hilbert space and spectral theory, in Contributions to Nonlinear Functional Analysis (Academic, New York, 1971).

Index

A

Adaptive extrapolation, 123

Adjoint operator, 87,116

Admissable sequence, 172

Agmon-Motzkin-Schoenberg
algorithm, 253

Akutowicz theorem, 37

Algebraic reconstruction
technique (ART), 103,191

Almost cyclic control
sequence, 253

Amemiya-Ando
theorem, 169,185

Amplitude modulation, 55

Analytic set, 137

Analytic set, irreducible, 131

Analytic signal, 55

Antenna design, 6,102

Arrival time determination, 69

Asymptotically regular, 168

Autocorrelation, 31,211

B

Baghlay's algorithm, 246

Band-limited
extrapolation, 1,114

Band-limited functions, 20,77

Barakat-Newsam
theorem, 46,143,209,210

Beam shaping, 4

Berenyi-Deighton-Fiddy
algorithm, 154

Berlekamp's procedure, 154

Beurling's theorem, 243

Bezout's theorem, 159

Bialy's theorem, 107

Biraud's algorithm, 132,193

Blackbody radiation, 36

Blaschke product, 35

Blind deconvolution, 225

Boas-Kac theorem, 33,133

Bregman's method, 254

Browder's theorem,
167,168,169,172

Browder-Petryshyn
theorem, 166,170

Bruck-Sodin, 29

C

Cadzow's iteration, 121

Canterakis algorithm, 151

Cartwright's theorem, 25

Censor-Segman theorem, 256

Characteristic function, 73

Cimmino method, 104,107

CLEAN algorithm, 223

Coherence tensor, 36

Coherent imaging, 91,129

Collinear trigonometric
polynomials, 146

Compact operator, 89

Compact gradient, 262

Compander, 73

Cone, 192

Conjugate gradient
algorithm, 204

Contraction mapping
theorem, 111

Contraction operator, 111

Convex projections, 181,257

300

Convex set, 180

Core, 261

Crimmins' theorem, 214

Crimmins-Fienup theorem, 44

Curtis' theorem, 161

D

Davies' theorem, 133

Deconvolution, 110,200

Degree of coherence, 35

Degrees of freedom, 8

Dispersion relations, 35

Dotson's theorem, 166

Dual cone, 259

E

Eggermont-Herman-Lent
  theorem, 254

Eisenstein's criterion, 149

Entire function, 19

Entire function, irreducible, 137

Error-reduction
  algorithm, 2,178

Euclidean domain, 14

Exponential type,
  functions of, 18

Exponential single
  sideband modulation, 56

Extrapolation problem, 111,118

F

Face, 261

Fejer-Riesz theorem, 26

Fienup's input-output
  algorithm, 216

Fienup's output-output
  algorithm, 217

Filter design, 207,226

FM demodulation, 243

Free zero, 61

Friedman's theorem, 105

Full carrier lower
  sideband signal, 62

Fundamental theorem
  of algebra, 13

Fuzzy set, 202

G

Gallagher-Liu algorithm, 3

Gaussian unitary ensemble, 82

Generalized inverse, 88

Generalized projection, 199

Gerchberg-Papoulis
  algorithm, 2,112,114,240

Gerchberg-Saxton
  algorithm, 2,215

Goldberg-Marks theorem, 190

Gonsalves' algorithm, 247

Gori's algorithm, 113

Groetsch theorem, 127

Groetsch-Engl theorem, 128

H

Hadamard factorization
  theorem, 18

Hadamard product, 18

Halperin, 168,184

Hayes theorem, 29,69,138

Hayes-McClellan-Lawton
  theorem, 141

Hayes-Lim-Oppenheim
  theorem, 68,71

Hermite-Biehler theorem, 60

Hilbert phase, 35

Hilbert transform, 55

Hoenders' theorem, 231,232,233

Hofstetter's theorem, 31

Holographic metrology, 227

Holography, 36,149

Homometric crystal, 5

Homometric distribution, 47

Homometric distribution
 function, 50

Homometric set, 49

Huffman's algorithm, 27

Huiser-van Toorn theorem, 144

I

Ill-posed, 9

Incoherent illumination, 98

Indicator function, 20,136

Integral domain, 11

Israelevitz-Lim algorithm, 163

J

Jain-Ranganath
 theorem, 122,205

Jump function, 63

K

Kaczmarz method, 103,179

Kadar's algorithm, 240

Kammerer-Nashed, 106,204

Karhunen-Loeve expansion, 126

Kiedron's theorem, 234

Kinoform, 3

Knox algorithm, 225

Kryloff theorem, 37

L

Labeyrie's method, 222

Landau's theorem, 77

Landau-Miranker theorem, 74

Landweber iteration,
 107,116,180

Landweber regularization, 128

Lawton's theorem, 142,144

Lawton-Morrison
 theorem, 147,148

Least squares solution, 88

Lent-Tuy algorithm, 196

Levi-Stark theorem, 199

Levin's theorem, 46

Linear variety, 185

Lipshitz map, 84

Locator set, 212

Logan's theorem, 61,62,63,75

Logarithmic Hilbert
 transform, 155

M

Manry-Aggarwal algorithm, 208

MART algorithm, 255

Mask, 213

Masry's theorem, 59,84,85,86

Maximum entropy power
 spectrum estimation, 221

Maximum modulus, 17

Merseau-Oppenheim
 theorem, 252

Microlithography, 175

Minimal phase, 175

Misell's algorithm, 230

Moment discretization, 130

Moses-Prosser algorithm, 245

Mutual coherence function, 33

N

Nakajima-Asakura
algorithm, 157

Nashed's theorem, 204,205

Neural net, 247

Newton-direction algorithm, 241

Nonexpansive map, 165

Nonnegative operator, 107

Nuclear magnetic resonance
tomography, 250

O

Opial's theorem, 168,170

Oppenheim-Lim-Curtis
theorem, 72

Optical transfer function, 98

Order, 18

Osgood product, 140

Osgood's theorem, 137,140

P

Paley-Wiener condition, 35

Paley-Wiener theorem, 20,37,41

Paley-Wiener-Schwartz
theorem, 26,50

Papoulis-Chamzas
algorithm, 124

Patterson function, 5

Petryshyn's theorem, 105

Picard's condition, 90

Picard's theorem, 165

Plancherel-Polya theorem, 139

Point spread function, 91

Polya's theorem, 21,24

Polynomial,11

Polynomial, degree of, 13

Polynomial, irreducible, 13

Principal ideal domain, 14

Projection slice theorem, 249

Prolate spheroidal function, 79

Q

Quasi-nonexpansive, 166

R

Radial function, 144

Radio astronomy, 6,223

Radio brightness function, 6

Rayleigh resolution, 92,98

Reasonable wanderer, 169

Reference point, 213

Regularization, 9,127,188

Reproducing kernel
Hilbert space, 98

Resolution, 8

Retraction, 257

Robbins-Monro
stochastic approximation, 238

Rosenblatt theorem, 49,50,52,53

Rosenblatt-Seymour
theorem, 48,49

Rosenblatt-Titchmarsh
theorem, 53

Rotem-Zeevi algorithm, 164

Rouché's theorem, 13

S

Sampling expansion, 99

Sandberg's theorem, 242

Sanz-Huang algorithm, 117

Sanz-Huang theorem,
138,139,149,161

Sard's theorem, 141

Sato's algorithm, 235

Shannon number, 8,92

Shape from shading, 265

Shitz-Zeevi theorem, 67,71,72

Simulated annealing, 238

Single sideband modulation, 56

Singular value system, 90

Space factor, 4

Spectral width, 15

Spectrum analyzer, 64

Spin glass model, 237

Steepest descent, 203

Stefanescu's theorem, 142

Strand's theorem, 108

Strictly convex, 183

Strictly nonexpansive, 165

Superdirective ratio, 102

Support cone, 259

Synthetic aperture radar, 249

T

Tangent cone, 259

Tikonov regularization, 128,188

Titchmarsh theorem,
 21,22,23,24,34,41,42,45,155

Toeplitz equation, 244

Tomography, 6,235

Trigonometric polynomial, 15

Trivial associate, 137

Twomey-Tikohnov solution, 109

Type, 18

U

Uniformly convex, 183

Unique factorization domain, 14

V

van Cittert deconvolution, 109

van Cittert-Zernike
 theorem, 223

Vertex, 260

Viano's theorem, 188

Visibility, 34

Voelcker-Requicha
 algorithm, 60,164

von Neumann's
 theorem, 169,184

W

Walther's theorem, 44,52,70

Wave front sensing, 5

Weak convergence, 182

Weierstrass preparation
 theorem, 140

Weierstrass factorization
 theorem, 17

Well-posed, 89

Wiener filter, 207

Wiener-Khinchine theorem, 246

X

X-ray crystallography, 5,48

Xu-Chamzas algorithm, 120

Y

Yao's theorem, 100

Youla's theorem, 186,187,189

Youla-Velasco theorem, 191

Z

Zakhor's algorithm, 252

Zeta function, 36,82